PROTEIN BIOCHEMISTRY, SYNTHESIS, STRUCTURE AND CELLULAR FUNCTIONS

RECENT ADVANCES IN PROTEOMICS RESEARCH

PROTEIN BIOCHEMISTRY, SYNTHESIS, STRUCTURE AND CELLULAR FUNCTIONS

Additional books in this series can be found on Nova's website under the Series tab.

Additional e-books in this series can be found on Nova's website under the e-book tab.

PROTEIN BIOCHEMISTRY, SYNTHESIS, STRUCTURE AND CELLULAR FUNCTIONS

RECENT ADVANCES IN PROTEOMICS RESEARCH

MING D. LI, M.D.

EDITOR

New York

For permission to use material from this book please contact us:
Telephone 631-231-7269; Fax 631-231-8175
Web Site: http://www.novapublishers.com

Library of Congress Cataloging-in-Publication Data

ISBN: 978-1-62948-218-7

Library of Congress Control Number: 2013949351

Published by Nova Science Publishers, Inc. † New York

Contents

Preface

It is evident that proteomics is in an extraordinary growth phase. This is primarily due to a great extent to the fact that the major undertaking of sequencing the human and other important genomes has largely completed and rapidly development of next-generation sequencing technology, which has opened the door for proteomics by providing a sequence-based framework for mining the human, animal and plant genomes.

The primary objective of this book is to provide readers with an updated knowledge of Recent Advances in Proteomics Research across different fields. To reach this goal, I have invited a group of leading researchers who have been not only actively engaged in proteomics related research but also applied such these technologies to research subjects of their interest to contribute this book. The research topics covered in the book include but not limited to proteomic analysis for microbiome, specific cell type from heterogeneous tissues, body fluids, mitochondria, and environmental toxicology, to name a few. Importantly, almost all proteomics techniques currently used in Proteomics Research have been introduced and reviewed in at least one chapter of the book, which include traditional 1D or 2D gel-based proteomics, gel-free shotgun proteomics, label-based or label–free quantitative proteomics, and array-based proteomics. In addition, although it is not the major focus of this book, proteomics-related bioinformatics tools such as database searching, data mining, data analysis, and gene pathway analysis are also briefly covered in some chapters. Additionally, this book also contains chapters that focused on some technical issues related to applying proteomics technologies to specific issues, such as sample preparation for removal of contaminant and depletion of highly abundant proteins. Together, this book provides a broad view of various proteomics technologies commonly used in proteomics research and their successful applications to address different questions across various species. It is our hope that such a book can provide readers with a comprehensive understanding of proteomics technologies and their application to address various scientifically questions.

I am most grateful to the distinguished group of leading experts who have come together to produce this book. These experts, united in their mission to deliver a scholarly and comprehensive book, come from plants, animals and human fields. I am glad for all that they taught me through their contributions, for the knowledge they shall distill to all who read this book.

Ming D. Li
Charlottesville, VA June 2013

In: Recent Advances in Proteomics Research
Editor: Ming D. Li
ISBN: 978-1-62948-218-7

Chapter 1

Protein Level Analysis from Specific Cell Types in Heterogeneous Tissue

Andrew O. Koob*
University of Wisconsin – River Falls, WI, US

Abstract

The ability to measure protein levels from specific cell types in heterogeneous tissue has proved elusive. Homogenization has been the preferred preparation method to study proteomics in tissue. However, homogenization of a sample does not allow for the analysis of individual cell types specifically, and little can be learned about the cell type origin of disease states with current proteomic methodology. Recent advances have changed the prospect of these endeavors moving forward. Laser Capture Microdissection (LCM) allows for the excision of individual cells from tissue. Previously, this method has proved useful in understanding the genomics and gene expression within different cell types because small quantities of DNA and RNA can be analyzed with PCR and microarray methodologies and few cells need to be excised from tissue samples. Labeling cells for specific antibodies, and then excising them individually and analyzing an aggregate for proteins of interest can now help researchers understand differential protein expression levels within different cell types. Recently, Western blotting has been achieved with cells excised individually and aggregated for analysis. This method has proved specific for the cell type but requires a large sample for sensitivity, with many cells dissected. Acquiring fewer cells would be less time consuming and lessen tissue degradation. It has been shown that small amounts of protein can be detected and quantified through DNA oligonucleotides associated with antibodies through immuno-PCR techniques and biobarcoding. Immuno-PCR has been shown in cell culture, and more recently with LCM. The understanding of protein levels in different cell types from heterogeneous tissue has important and beneficial implications in the understanding tissue development and the origin of disorders and disease. Here, recent advances will be discussed that allow for protein to be reliably measured from small quantities in order to

* 410 S. 3rd St., AGS 409, Biology Department, University of Wisconsin – River Falls, River Falls WI 54022. Email: andrew.koob@uwrf.edu

understand differences occurring between cell types in heterogeneous tissue samples. A review of the specificity of these techniques and their future viability will also be considered in terms of their ability to study specific cell types, with special emphasis on human tissue analysis.

Keywords: Single Cell Proteomics, Laser Microdissection, Immuno-LCM, Laser Capture Microdissection, IPCR, Bio-barcoding, Aptamers, Gold Nanoparticles, Mass Spectrometry

Introduction

The development of radiolabeling of antibodies in 1959 allowed for the identification of new proteins, and the knowledge of the location of protein expression at the organelle level [1]. Using immune-technology to tag proteins as antigens, reliable quantification of protein expression for tissue also became possible, especially after the safer enzyme-linked immunosorbant assay (ELISA) in 1971 [2]. A fluorescent based detection in 1979 made this technique more sensitive [3] and the early achievements and subsequent technological variations led to a significant advancement in the understanding of measurement of protein expression levels. Western blotting assays in the 1980s provided an even more reliable method for protein level measurements, due to the separation of the proteins based on their molecular weight within a sample homogenate [4], and advancements in flow cytometry allowed for simple protein detection and cell sorting based on fluorescent probes [5]. These strategies have been applied to cell culture, native proteins, and known biomarkers for disease. Additionally, recent advances in mass spectrometry allow for the detection and characterization of unknown proteins without recognition elements, most recently in minute samples [6, 7].

However, because all of these assays were performed with tissue homogenates or required large samples, protein measurement at the specific cellular level was not possible. Essentially, although there is a better understanding of proteins at the tissue and molecular level, the in situ cellular compartment has been bypassed for protein quantification and measurement.

Understanding the origin of tissue pathology or development requires an ability to measure protein differences and characteristics of specific cell types in heterogeneous tissue [8-10]. Cellular activity and protein expression is reliant on intrinsic and extrinsic factors in the heterogeneous environment. Strategies to isolate cells from tissue for analysis through tissue culture have been shown to diminish this cellular integrity, eliminating the actual purpose of the techniques [11, 12]. First described over 15 years ago, Laser Capture Microdissection (LCM) allows researchers to excise cells individually from a tissue section for analysis of cells as preserved from their native environment [13, 14]. Populations of cells can be individually excised from tissue and an aggregate of cells of the same type, referred to as cell enrichment, can then be analyzed for protein quantification and measurement. This can give researchers an accurate idea of differential protein expression of a specific cell type in situ. The understanding of specific cell types at the proteomic level between disease states and normal controls would be especially beneficial in the neurosciences, where interspersed cells with hundreds of variable differences is common throughout the parenchyma [15].

However, with the advancements to date, there is still no cost-effective sensitive bench top analysis strategy for protein measurement from small specific cell populations aggregated with LCM. The most commonly used techniques, Western Blot and ELISA, for reliably measuring protein levels still require relatively large samples for accuracy.

Conversely, LCM has been a progressive technique in understanding genomics and gene expression at the specific cell level [16]. With the advent of polymerase chain reaction technologies (PCR) in 1985, researchers have been able to detect and measure minute levels of DNA through amplification techniques [17, 18]. Now this material can be more accurately quantified through real time or qPCR [19, 20] after specific cell enrichment from tissue with LCM [21, 22]. Researchers initially demonstrated the ability to analyze gene expression from a single cell. This was achieved with a cell isolated from colon adenocarcinoma [23]. More recently, this technology was used to analyze the gene expression profiles of two different neuron cell types in the dorsal root ganglion [24]. Now, through microarray analysis with LCM and qPCR, researchers can study then entire human transcriptome at any time and cell type [25-30].

LCM is now standard equipment in microscopy centers in research institutions and pathology departments for genomic and gene expression analysis from specific cell types in heterogeneous tissue. Harnessing this ability to study differential protein expression specific cell types enriched with LCM would be immensely valuable in biomedicine to understand changes between normal and disease states, thereby locating new reliable biomarkers that will illuminate new therapeutics. Without the capability of researching protein level differences in specific cell types in disease states, knowing which cells to attack and treat in many diseases is currently not fully understood.

In order to take advantage of the amplification ability of DNA and RNA through PCR to study proteins, antibodies linked to oligonucleotides have been used as a method to attempt sensitive protein quantification [31]. The method, called immune-PCR (IPCR) has recently been used to detect amyloid-$\beta_{1\text{-}40}$ in as little as 3 neurons from the CA3 region of the human hippocampus [32].

Western blotting has been shown after LCM as well to quantify protein levels. Because of the number of cells required to acquire accurate measurements, this is typically achieved by excising an area of ~10,000 congruous epithelial cells or tumor mass from frozen sections around 20 microns thick [33-39]. Western blotting was also used in normal brain tissue, but also as a method to excise the cells of epithelium of the brain [40, 41].

Recently, to determine the specificity of LCM for proteomics research, cells were excised individually from the human brain for sensitive Western blot analysis, and expression was reliably visualized with 100-200 enriched neuron populations from the substantia nigra for tyrosine hydroxylase, an abundantly expressed enzyme in the dopamine production pathway [42]. Because of the complexity of the neuropil, another cell type, astrocytes, was also analyzed. Astrocytes have a much smaller soma, and rapid immunohistochemistry (immuno-LCM) was used to label astrocyte nuclei for microdissection in order to combat normal tissue degradation due to the large numbers of cells that needed to be excised (1700 cells) equivalent in area to 200 neurons. A Western blot probed with an antibody for an astrocyte specific protein confirmed the specificity of LCM, and reliability of the enrichment technique [42]. This was also confirmed by Western blot comparison of excised cytochrome oxidase positive and negative muscle cells for proteins of the electron transport chain [42]. However, because of the time intensive procedure of dissecting 1700 cells, the ability to analyze protein levels from fewer cell populations would be beneficial, and require a more sensitive assay.

It now appears that biochemical analysis of the proteome through quantification and measurement of protein levels from specific cell types enriched with laser capture microdissection can be routinely studied moving forward. Some possible cost-effective bench top technological advances that could reliably measure proteins from cells excised individually from human tissue include immuno-PCR, biobarcoding, aptamer detection probes, and utilization of nanoparticles. Here, a comparison between these different cost effective bench top possibilities for specific cell analysis will be considered, taking into account sensitivity and reliable measurements of proteins.

Laser Capture Microdissection and the Proteome

Isenberg et al., first attempted to use early UV laser technology in the mid-1970s to dissect cells at the subcellular level [43]. They showed that proteins avoided damage when they laser microdissected renal medullary collecting duct cells grown in tissue culture after separating cytoplasmic actomyosin fibrils [43]. However, the instrumentation was large and unpractical for the purpose of mass use or to dissect cells from tissue [44]. In the mid-1990s, the National Institutes of Health developed Laser Capture Microdissection by adding ethylene vinyl acetate (EVA) polymer over the tissue section and then applying infrared laser to the EVA polymer [13, 14, 21, 45]. Cells adhere to the EVA polymer after it is warmed by the laser and then they can be removed individually for analysis later after cooling. This infrared technology (IR-LCM) utilized laser beam diameters at 30 μm, 15 μm, and as small as 7.5 μm (44), to allow for the dissection of a single cell.

To avoid unwanted tissue also adhering to the polymer during IR-LCM methodology, ultraviolet laser technology has more recently been developed for more precise microdissection [46-50]. These systems were coined Laser Microbeam Microdissection and had UV laser microbeams with diameters as little as 0.5 μm that took advantage of catapulting systems from glass slides covered with a polyethylene naphthalate membrane [51] (UV-LCM). UV lasers are beneficial because they have a high photon density and are 'cold lasers', disintegrating the membrane in their beam path, but generating minimal heat which eliminates the risk to excessive damage of proteins of interest [52, 53]. The wanted cell or area is microdissected out of the unwanted tissue and then catapulted under pressure into an overhanging collecting cap, eliminating the need for it to adhere to an EVA polymer [50]. Another advantageous development with UV-LCM in laser microdissection was 'catapulting' under the influence of gravity to a collecting cap for a 0.5 ml Eppendorf tube below the slide [21]. This technique excises out the wanted area on a slide consisting entirely of the membrane polymer in a steel frame, and it falls to a cap immediately below the slide into buffer, thereby maximizing the quality of the sample retrieved [47]. Other polymers now used for membranes for UV-LCM are polyphenylene sulfide and polyethylene terephthalate.

Universally, even though the UV-LCM systems use a catapulting system, these systems have, in many cases, been referred to their original name as Laser Capture Microdissection, or in some cases, simply Laser Microdissection. In these systems, frozen sections of human tissue can be added to 'membrane slides' and the experimenter can then draw a circle around the cell that is marked for dissection with an associated computer system while visualizing the tissue under high magnification. The automated laser beam then makes a precise 0.5 μm path along the line drawn, excising out the tissue area inside as it falls into the buffer in the cap (Figure 1). The automated system can dissect several cells out of tissue individually in quick succession, and obtain an enriched sample population of specific cell types from heterogeneous tissue for subsequent analysis.

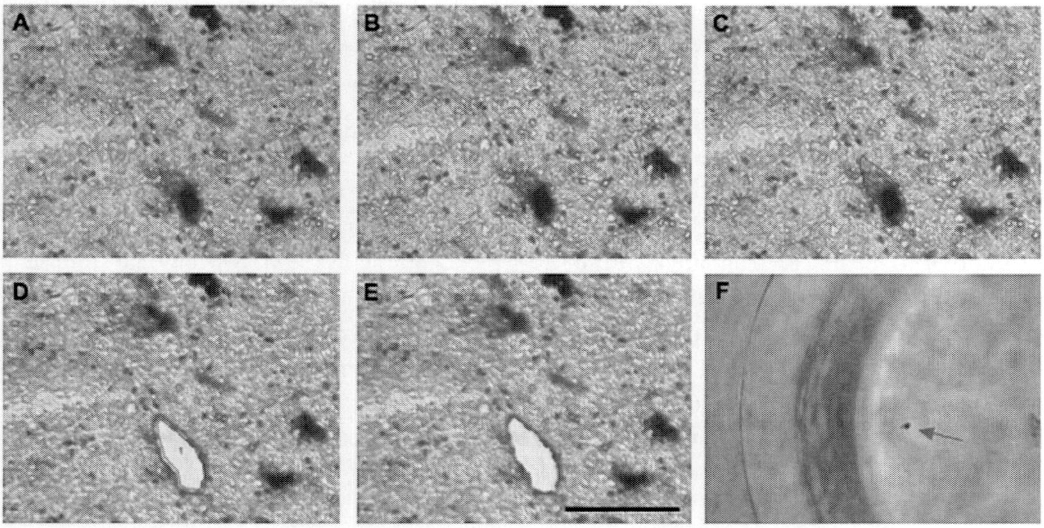

Reprinted from Analytical Biochemistry, Jun;425(2), Koob AO, Bruns L, Prassler C, Masliah E, Klopstock T, Bender A. Protein analysis through Western blot of cells excised individually from human brain and muscle tissue, p. 120-4, 2012, with permission from Elsevier.

Figure 1. Neurons visualized with Nissl stain in the human substantia nigra and captured with LCM. (A) the tissue without Eppendorf cap underneath, (B) the cell with Eppendorf cap underneath ready for capture of excised cells, (C) cell outlined for selection, (D) cell dissected out with laser, (E) selection line removed (scale bar A–E, 50 μm), (F) and the cell in the Eppendorf cap in lysis buffer with proteinase and phosphotase inhibitors (arrow).

In an attempt to modify ELISA for LCM for sensitive measurements, an experimental analysis attempted to infer the amount of prostate specific antigens in number of PSA molecules per microdissected tissue cell ranged from 2×10^4 to 6.3×10^6 in normal epithelium, prostate intraepithelial neoplasia (PIN), and invasive carcinoma (54). However, this assay was not able to achieve the reliability of Western blot for measuring protein expression, nor was it as sensitive as other assays now developed.

In proteomics research, because of the large tissue requirements for these techniques, they have been traditionally used in oncology research to study proteomics of tumorigenesis and qualities for malformation in comparison to normal tissue, because of the large amount of congruent tissue that can be excised for analysis [39]. Additionally, LCM has been a powerful tool to understand the proteome of structures within tissue organs. For example, immediately after the discovery of LCM, novel biomarkers in cancers were elucidated, due to the mass amount of tissue capable of being captured from tumorigenic tissue, as Emmert-Buck and colleagues identified cytokeratin 1 and annexin I after using two dimensional polyacrylamide get electrophoresis (2D PAGE) and mass spectrometry (MS) to compare tumor cells and normal controls from two patients with esophageal cancer [55]. Using 2D PAGE requires a large sample for analyze, and they dissected out 50,000 cells utilizing congruent epithelium [55]. The strategy of using 2D PAGE originally in combination with MS is now established as a method for biomarker discovery in several cancers [36]. For example, 50,000 microdissected adenocarcinoma cells and whole sections of pancreatic samples were analyzed with 2D PAGE and MS which identified S100A6 as a protein that was over-expressed in pancreatic cancer cells [56].

Other more sensitive methods such as ProteinChip followed by MS, identified differentially identified proteins such as annexin V in head and neck cancer [51] and heat shock protein 10 in colorectal cancer from 3000-5000 cells a sample [57]. In another study that demonstrates powerful applications of LCM with proteomics, researchers developed protein set enrichment analysis for a breast cancer study and identified 121 differentially expressed proteins at a 99% confidence level after excising 60,000 cells per condition and control with liquid chromatography and MS [58].

In order to locate cells and areas of interest for excision with LCM, histological stains can be done quickly on frozen sections with minimal degradation, as has been shown in these previous experiments, and in experiments for RNA analysis. However, in order to tag specific cells types for cell population enrichment, immunohistochemistry is required. Now rapid immunohistochemistry (Immuno-LCM) can also be performed before LCM take excise cells for downstream sensitive analysis and measurement without disruption of proteome integrity.

Immuno-LCM

With LCM, tissue sections need to be processed without being coverslipped, therefore allowing for the possibility for tissue degradation before cell capture and proteome analysis. However, because of the increasing amounts of tissue banks with snap frozen human tissue available, protein preservation without formalin or formaldehyde treatment has made proteomic analysis increasingly viable. Because of the unstable nature of RNA, Fend et al. developed a quick immunohistochemical staining method in order to minimize tissue degradation to perform RT-PCR to analyze gene expression [59] after LCM. They coined the method Immuno-LCM, and it has been employed successfully to study cell morphology and immunophenotype before microdissection [60]. This procedure has also been used before LCM for protein analysis [42, 61], and recent advances have also combined 3D imaging of immunohistochemically stained sections with LCM [62].

Normal histological stains such as the Nissl stain or hemotoxylin and eosin can be done quickly with little tissue degradation before cell dissection with LCM. For example, with a Nissl stain, the entire staining procedure can take 15 min. The tissue sections can be fixed for 3 min, rinsed in PBS, placed in 1% cresyl violet in acetic acid buffer for 5 min, then acetic acid buffer for 1 min followed by dehydration steps of 1 min in 70% ethanol and 1 min of 100% ethanol, then allowed to air dry. Once fixed and stained, cells can easily be visualized for LCM and remain stable enough for a few hours to extract RNA even without coverslipping. These original methods were used for LCM to study genome and gene expression after its development. However, because of the heterogeneity of human tissue, in order to study specific cell types, as well as cell subtypes, especially in nervous tissue, labeling through immunohistochemistry proves to be incredibly beneficial [11].

Fend et al. described 9 different antibodies optimized for rapid immunohistochemistry in their original study [59]. Frozen tissue sections were taken from snap frozen neoplastic and normal lymph node, breast and prostate tissue. After sectioning, the experimenters placed the slides in various fixatives for as little as 3 minutes. They chose the following different fixatives to test: acetone at -20, methanol at -20, 4% formaldehyde, 70% ethanol, an ethanol/formaldehyde mixture 3:1 and 15 seconds of ethanol followed by 5 minutes of -20 acetone. They found the precipitating agents acetone and methanol preserved the recovery of RNA to allow for analysis and conversion to cDNA and qPCR [11]. Cross-linking agents paraformaldehyde and ethanol recovered no perceptible analysis of RNA. For protein analysis, use of -20 acetone or methanol also likely allows for maximum analytic ability downstream [59].

Additionally, incubation in primary antibody is traditionally incubated at 4°C overnight or for up to 2 hours. The researchers discovered that with nine different antibodies, they could visualize the proteins of interest in as little as 5-10 min incubation times at room temperature with normal procedures. Secondary antibodies only required incubation times of 5 min. Conjugation steps and 4,4′-diaminodenzidine tetrahydrochloride (DAB) development also was less than 5 min. Sections were cleared with xylene before being allowed to air dry for processing. Interestingly, because of tissue hydration during the staining process, dehydration with alcohol after tissue immunohistochemistry does not disrupt analysis. Xylene is a necessary clearing step before use of IR-LCM because of the possibility of unwanted tissue adhering to the cap when acquiring tissue that may not be fully dehydrated. However, the xylene step can be omitted in staining for the UV-LCM method of catapulting. The dehydration step in this process uses ethanol for as little as 15 sec a step in 70% and 100% solutions [41, 63, 64]. All steps of the procedure are completed at room temperature and then the slides are allowed to air dry until LCM.

In the literature, primary and secondary antibody concentrations used have been as low as 1:50, and as high as 1:10 in the solution [63]. Blocking agents have been shown to be around 2% or 10% serum in PBS or TBS [41, 59]. Figure 2 shows glial cells in the human brain revealed for LCM with the rapid immunohistochemistry method originally developed by Fend [59, 63], and the associated Table gives a generalized outline for Immuno-LCM as seen in the literature.

DAB staining with light microscopy allows for stable visualization techniques. However, most LCM systems also employ the ability for fluorescent microscopy. A fluorescently conjugated secondary antibody would decrease immunostaining time due to the elimination of steps required for horseradish peroxidase and DAB development. However, photobleaching and background staining can be a concern while staining with fluorescent conjugates in order to visualize cells [60]. Although, double labeling of the same section would also be possible with rapid fluorescent immunohistochemistry, and be a significant benefit as enriched populations of two or three different cell types could be excised from the same section [65].

Regardless, these procedures have been successful maintaining protein and RNA integrity in human tissue and are therefore viable to use with any protein measurement assay. The following sections will discuss possible applications that could be reliable sensitive assays for protein measurements to enable the ability for analyzing specific cell types in situ.

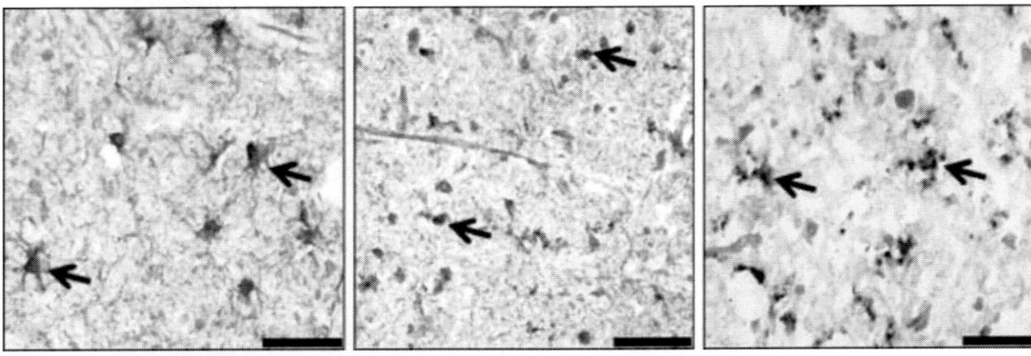

Reprinted from Journal of Neuroscience Methods, Jul 15; 208(2), Waller R, Woodroofe MN, Francese S, Heath PR, Wharton SB, Ince PG, et al. Isolation of enriched glial populations from post-mortem human CNS material by immuno-laser capture microdissection, p. 108-13, 2012, with permission from Elsevier.

Figure 2. Rapid immunohistochemistry for detection of glia cells in human CNS tissue. (A) Glial fibrillary acidic protein (GFAP) immunohistochemistry was used to identify astrocytes, (B) OSP to identify oligodendrocytes, and (C) CD68 to visualise microglia. Examples of immunopositive cells are indicated by the arrows. Scale bar represents 50μm.

Immuno-PCR

Immuno-PCR (IPCR) was developed by Sano et al. in the early 1990s [31]. The idea behind IPCR was to take the beneficial aspects of PCR and apply it to proteomics. The most accepted proteomics procedure at the time was ELISA, and capture ELISA techniques that had been developed increased sensitivity of original ELISA assays 100-10000 times.

IPCR was achieved by attaching biotinylated double-stranded dsDNA to an antibody through a protein A-streptavidin chimera. The original development of IPCR was with a direct ELISA format instead of a sandwich capture ELISA, as using a capture antibody would result in nonspecific binding of the chimera [66]. It had been known that streptavidin bound specifically with biotin and development of short dsDNA oligonucleotides that bound selectively to biotin had been developed [67]. The dsDNA linked with the antibody through streptavidin-protein A chimera binding with immunoglobulin G (IgG) – therefore biotinylated dsDNA could mark a protein through the antibody, and then be separated and analyzed through PCR similarly to how DNA and RNA had been studied. This allows for amplification of the recognition element for the protein, resulting in higher sensitivity and possibilities for quantification of minute amounts of protein.

For IPCR in this initial study, the limit of detection (LOD) for Bovine Serum Albumin (BSA) was 2 femtamolar, 10^5 more sensitive than normal ELISA for BSA [31]. There were several other beneficial applications for this original method, including an assay for carcinoembyonicgenic antigen (CEA), which is elevated in the blood serum of patients with various types of malignant carcinomas [68].

The next step was to eliminate the use of the protein A-streptavidin chimera to enable the use of a capture antibody. Zhou et al. achieved this soon after the original IPCR work. They took advantage of readily available biotinylated antibodies and linked biotinylated dsDNA through streptavidin. At the time, this was called 'universal IPCR' and is still the assay that is used the most today for antibody conjugation to an oligonucleotide [69] (Figure 3).

Table 1. Generalized Rapid Immunohistochemistry to Visualize Specific Cell Types for Microdissection and Proteomic Analysis from Human Frozen Tissue Sections

Steps (at room temperature)	Time
1. Place slides in -20 acetone	3 min
2. Wash in PBS or TBS	
3. Pretreatment with 1% Triton X-100, 3% hydrogen peroxide, protease and/or phosphotase inhibitors	3-5 min
4. Wash in PBS or TBS	
5. 5-10% blocking serum matching animal from secondary antibody	3-5 min
6. Primary antibody (concentration between 1:10 – 1:50)	3-10 min
7. Wash in PBS or TBS	
8. Secondary antibody (1:10 – 1:50)	3-5 min
9. Wash in PBS or TBS	
10. Conjugation (i.e., 4% horseradish peroxidase with ABC reagent from Vector Laboratories)	3-7 min
11. Wash in PBS or TBS	
12. 20% 4,4′-diaminodenzidine tetrahydrochloride	30 sec – 5 min
13. Double distilled water	30 sec
14. 70% alcohol	1 min
15. 100% alcohol	1 min
16. Air dry in air flow hood	
Total time	**30 min – 1 hr**

This procedure requires several steps where thermodynamic equilibrium needs to be achieved for each step in the incubation process for oligonucleotide conjugation. This can lead to incomplete streptavidin binding and loss of sensitivity if the experimenter is not experienced with these nuances [66]. However, the LOD for these experiments was 9.6 femtamolar for the Ets-1 protein, a level of sensitivity similar to what was reported by Sano et al. The Ets-1 protein is a transcription factor that is another prominent biomarker in ovarian cancer [70].

Specificity and background were still issues even with the use of a capture antibody [71]. IPCR strategies to combat this problem included creating supramolecules, or complexes of multiple bound streptavidin and biotinylated DNA molecules, that could be manipulated to only bind to the antibody of interest. These new formations are more specific, and were shown to detect rabbit IgG with a 30 femtomolar LOD, an order of magnitude more specific than universal IPCR [71, 72].

Other sources have bypassed the streptavidin-biotin affinity altogether and covalently bound dsDNA directly to the antibody for analysis [73, 74].

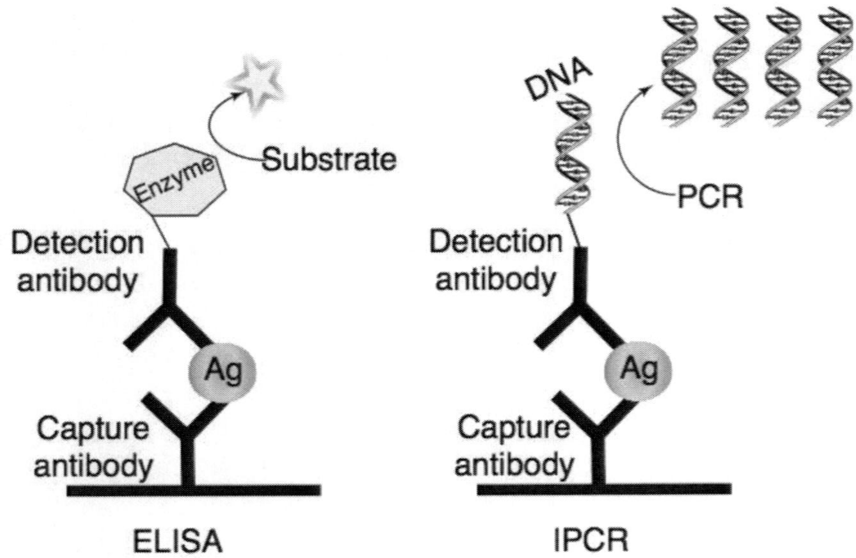

Reprinted from Trends in Biotechnology, Apr; 23(4), Niemeyer CM, Adler M, Wacker R. Lmmuno-PCR: high sensitivity detection of proteins by nucleic acid amplification, p. 208-16, 2005, with permission from Elsevier.

Figure 3. The general setup of IPCR is similar to that of antigen detection through ELISA. An antibody–enzyme conjugate in ELISA converts a substrate into a detectable product. In IPCR a conjugate comprising an antibody and a DNA marker is utilized. Nucleotides, specific primers and a polymerase are added and the marker is amplified by PCR for signal generation. The number of PCR amplicons produced is proportional to the initial quantity of antigen to be detected.

Although initially promising and easier to design multiplexed assays with covalently bound oligonucleotides that will remain bound to the antibody and retain sensitivity, the difficulty in replicating the process has likely led to these procedures not being widely adopted [75]. To date, universal IPCR methods are still the most commonly used bench to technique because of the replicable, if not overly efficient conjugation process [75].

Other variations have also arisen to make the assay more viable. A new novel way to conjugate the antibody to the oligonucleotide with a photocleavable linker has been reported, that could easily remove the oligonucleotide by simply subjecting the bound antibody to certain wavelengths of light [76]. Another prominent variation on the IPCR theme is Rolling Circle Amplification technology as a reporter [77]. After the initial binding steps, a circular ssDNA oligonucleotide is amplified through repeated signals, each hybridized with a fluorescent probe. As the ssDNA is amplified, the fluorescent emission can be quantified, simplifying the measurement of the captured protein to increase the sensitivity of the assay. This has proved flexible and simple for analysis and protein multiplexing [78].

Another sensitive variation of IPCR was the proximity ligation assay by Lundegren et al [79]. Only when the target analyte is bound by more than one antibody can a signal be achieved through ligation between both antibody-linked oligonucleotides. Subsequent fluorescent emission due to ligation can be quantified. The original strategy reported detection levels on the level of zeptamole sensitivity for platelet derived growth factor (PDGF) [79, 80]. Interestingly, this original strategy bound PDGF with aptamers (the oligonucleotides themselves binding the protein without an antibody), however, more recently, these assays have required antibodies typical with IPCR methods. The triple-binding PLA was developed afterwards, with an oligonucleotide that can be retrieved only if double linking occurs and then analyzed through qPCR in order to quantify small amounts of protein [81, 82]. The further development of this assay was the proximity extension assay, which adds a third oligonucleotide that binds to the first two, and when bound it can be extended and then used for qPCR [83, 84].

Although technically flexible because of its reliability on oligonucleotides, one of the concerns with immuno-PCR in a complex sample such as enriched cell population from LCM, is the possibility of high background through nonspecific binding of the oligonucleotide reporter. And widespread use of DNases is not possible because the reporter oligonucleotide would also be lost. One way around this was to encapsulate the reporter oligonucleotide in a liposome [85, 86]. This protects the oligonucleotide itself, and use of DNases are then possible in order to avoid background contamination (86). After labeling, the liposome is ruptured and then DNA collected for PCR analysis. In the initial study they revealed sensitivity several orders of magnitude better for cholera and botulism toxin than ELISAs [86].

IPCR has been used to measure levels of many different antigens, including viral markers such as the hepatitis B surface-antigen [87], and HIV core protein p24 [88], 100 fold and 25 fold more specific than ELISA respectively. In addition to the already mentioned CEA and ETS-1 proteins, IPCR has reliably detected Prostate-specific antigen with 1000 fold specificity over ELISA [77] and TNF-alpha [89], as well as other proteins that are known tumor markers. In addition to protein, IPCR has been used to probe for other molecules, showing a 1000 fold better detection of amines such as serotonin [71].

The use of IPCR with a complex sample of specific cells dissected with Laser Capture Microdissection is now a possibility. Hashimoto et al. acquired frozen sections of human brain from the hippocampus [32]. They dissected CA1 pyramidal neurons and detected them for amyloid-$\beta_{1\text{-}40}$ using the IPCR protocol modified from the sandwich ELISA with an antibody covalently bound directly to a ssDNA oligonucleotide. Their preliminary tests showed that the assay could detect amyloid-$\beta_{1\text{-}40}$ at 2 attomoles, and subsequently that CA1 pyramidal neurons have an average level of 10 attomolar amyloid-$\beta_{1\text{-}40}$, therefore at those concentrations, they only needed three neurons for detection level [32] in sample dilutions. At this level, this technique qualifies as a viable option to measure protein levels reliably and more specific than Western blot.

IPCR has been a beneficial advance for sensitivity detection, but has not yet been used as a widespread bench top technique analysis of protein measurements from small complex samples, and therefore had been impractical for measurements of protein levels from individual cell types. One issue with these techniques is antibody development, which requires expertise in order to be able to develop an antibody bound with the oligonucleotide properly each time analysis needs to be conducted [71]. Many of these as of yet are not commercially available, although as the use of LCM increases and researchers become interested in protein analysis on the cellular level, this could become more commonplace.

The technology continues to be developing towards that goal. And encouraging results from human blood detection with proximity extension assays, and detection from human neurons indicate that as this technology advances it could be very beneficial for use with Laser Capture Microdissection to understand the proteomics at the cellular level in human tissue.

Biobarcoding and Nanoparticles

In order to avoid the cumbersome procedure of uniquely binding antibody conjugates to oligonucleotides, a variation of Immuno-PCR was developed 10 years ago called biobarcoding. The term biobarcoding has been used to indicate any use of nucleic acid reporting of a structure or molecule. And therefore, under this definition, IPCR is a type of biobarcoding. However, for our purposes here, biobarcoding will consider applications using nanoparticles bound to ssDNA or RNA. The principle remains the same as IPCR, as a capture antibody is used in a sandwich assay.

Originally reported by Mirkin and colleagues 10 years ago, biobarcoding to recognize proteins involved the use of magnetic microparticles functionalized with monoclonal antibodies for prostate specific antigen (PSA) that acted as the capture antibody [90, 91]. The magnetic microparticle is 1μm in diameter with an iron oxide core. After binding with PSA by the monoclonal antibody, the solution was washed, and then gold nanoparticles functionalized with a polyclonal antibody for PSA were added. The gold nanoparticle is also hybridized through thiol-gold chemistry with oligonucleotides that can act as the reporter molecule. Researchers then utilized a magnet to the wall of the tube where the reaction took place; the magnetic microparticles immediately migrate to the magnet. After washing, the researchers could collect the oligonucleotide "barcodes" from the gold nanoparticle indicative of the presence of the protein, then measure the protein levels through gel electrophoresis, qPCR or through their own devised scanometric method [92, 93].

The scanometric method consists of a chip with DNA for the binding of oligonucleotide barcodes. Then a separate gold nanoparticle binds to the DNA as well as acting as a nucleation site for silver deposition after the chemical reduction of silver from solution [94]. The silver deposits can then be captured by a simple desktop scanner for quantification [92, 93]. They reported that the assay was more sensitive for PSA, a marker for prostate and breast cancer, on six orders of magnitude than normal ELISA procedures and two orders of magnitude more sensitive than optimal IPCR. Scanometric detection was able to visualize the protein at 30 attomolar LOD and PCR at 3 attomolar [90]. This assay avoids the complex conjugation issues with antibodies and DNA oligonucleotides that is problematic in IPCR, as well as the issues that occur with the heterogeneous nature of the target-capture procedure, resulting in high background through nonspecific binding [95].

After the development of this assay, the researchers attempted to use it for a beneficial diagnostic. Amyloid-β diffusible ligands (ADDLs) were known to be present in the soluble extract of the frontal cortex of brains from Alzheimer's disease patients [96]. However, although this indicated the likely presence of ADDLs in the cerebral spinal fluid, and therefore a diagnostic tool for people with the disease and disease progression, it wasn't quantifiable with ELISA or immunoblot. Mirkin and colleagues showed that using their biobarcode assay, with 15 people with the disease, and 15 normal controls, the median concentration for the Alzheimer's patients was 1.7 femtomolar, and around 200 attomolar concentration (8.5 times on average lower) for the normal controls [94, 97] (Figure 4).

Additionally, in another study, three different antibody magnetic microparticle probes for three different protein markers were detected in buffer and serum media. The markers were PSA, human chorionic gonadotropin (HCG) and alpha-fetoprotein (AFP) [98]. This demonstrated the advantageous nature for the biobarcoding method to avoid problems with cross-reactivity between probes and reporters. HCG is a testicular cancer marker, PSA is a breast and prostate cancer marker and AFP is an effective marker for hepatocellular carcinoma [99]. More sensitive assays for these proteins of interest, especially in a multiplexing format, could be helpful for the diagnostics of cancer.

After these first experiments, researchers reported detection of HIV-1 p24 Gag occurred with traditional ELISA in 23 out of 112 people tested, while 111 out of 112 tested positive for the protein with the biobarcode assay, demonstrating its specificity [100]. More recently, PCR and gel electrophoresis for the p24 antigen from HIV infected patients was recognized after biobarcoding [101]. Also, in a slightly modified version, researchers used biobarcoding technology to detect respiratory syncytial viruses [102].

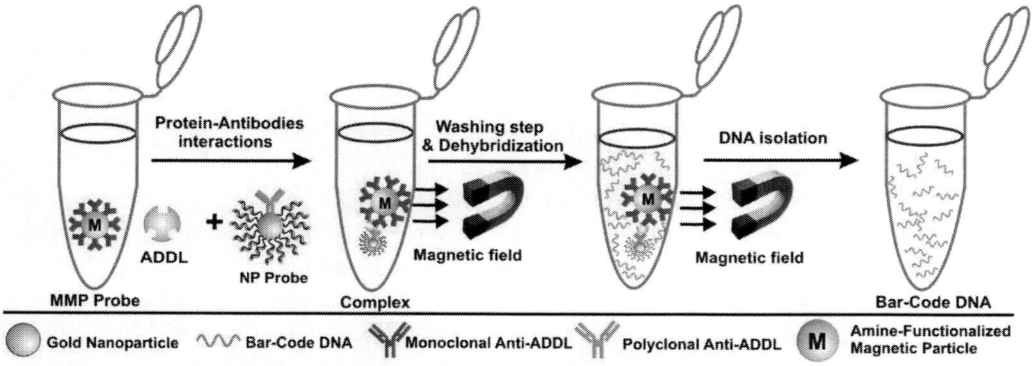

Reprinted with permission from Proceedings of the National Academy of Sciences of the United States of America, Feb;102(7), Georganopoulou DG, Chang L, Nam JM, Thaxton CS, Mufson EJ, Klein WL, et al. Nanoparticle-based detection in cerebral spinal fluid of a soluble pathogenic biomarker for Alzheimer's disease, p. 2273-6, Copyright (2005) National Academy of Sciences, U.S.A.

Figure 4. The bio-barcode amplification assay. The assay uses magnetic microparticles functionalized with monoclonal antibodies that recognize and bind Amyloid-β **diffusible ligands** (ADDLs) from human serum samples in this example. The ADDLs are then sandwiched with a gold nanoparticle probe that has been functionalized with double-stranded DNA and an anti-ADDL polyclonal antibody. After repeated washing while using a magnet to immobilize the magnetic microparticles, a dehybridization step releases hundreds of barcode DNA strands for each antigen-binding event.

One medically beneficial use of this assay was to determine after radical prostatectomy whether cancer would recur. PSA is normally undetectable in the blood of patients who have undergone this procedure. Using the biobarcode assay, researchers were able to detect levels of PSA, as reported by Nam et al. in the original paper, and then if the levels started to rise, noted that recurrence of cancer happened in those cases, where if the levels didn't rise, the cancer did not recur [103].

Another variation was even more specific and reported an LOD of 400/500 femtomolar of prostate specific antigen from patients blood serum levels in an assay that only took 14 minutes (104), confirming the useful diagnostic of the original methods report by Nam et al.

Biobarcoding has also been modified for use in an ELISA format with a gold nanoparticle functionalized with oligonucleotides and recognition secondary antibody, but with a primary antibody on the surface as a capture antibody, a hybrid technique modified from IPCR and the biobarcoding technique, the so called nano-IPCR, to detect the Hantaan virus nucleocapsid protein [105].

Researchers compared nano-IPCR with IPCR and ELISA for the cytokines interleukin-3 (IL-3) and stem cell factor (SCF) produced by cells in culture. They found that detection in serum was much more rapid and specific with the nano-IPCR method [64].

The simplicity of biobarcoding makes it an ideal candidate for routine use with LCM to determine cellular differences in protein measurements in human tissue, although it hasn't been used for this to date. Recently, it has become more cost effective as protocols for labeling gold nanoparticles with antibodies and oligonucleotides, as well as labeling magnetic microparticles have become more available. These protocols are much simpler and easier to achieve in house than development of an antibody/oligonucleotide complex win IPCR.

Just as gold nanoparticles can be used to bind with antibodies and functionalized with oligonucleotides, they can also be used to bind and recognize proteins themselves and then quantified with other reporter molecules [106]. Gold nanoparticles with a fluorescently bound GFP molecule were used for competitive expression (chemical noise) of specific proteins, whereby when the protein bound to the nanoparticle, GFP was removed as the indicator. Researchers could detect and identify proteins in human serum [107, 108].

Colloidal gold nanoparticles have also increased the sensitivity of plasmon surface resonance dramatically, making this technique another possibility for analyzing proteins reliably in small quantities [109]. These strategies have resulted in more sensitive and reliable protein detection than traditional plasmon surface resonance strategies. However, instead of using oligonucleotides as the reporter, the amount of gold nanoparticles that had bound to protein can be reported though an electrostatic conjugate that amplified the gold nanoparticles. This resulted in picomolar detection of human IgG, and had also ben used to detected metalloproteinases-2 and antiglutamic acid decarboxylase antibody [110, 111].

Another method to use gold nanoparticles for reporting either through simple UV spectrophotometry or plasmon surface resonance was similar to the biobarcoding method, where magnetomicroparticles and gold nanoparticles are functionalized with antibodies to a protein of interest. However the gold nanoparticles were then 'grew' for reporting in the presence of ascorbic acid, as ascorbic acid would be reduced in the presence of gold ions until the reaction was exhausted, effectively changing the color of the gold nanoparticles [112].

Interestingly, when gold nanoparticles aggregate, they no longer emit a red color, but the color turns purple or blue and simple colorimetric analysis can be used [113, 114]. Researchers reported using aptamer functionalized gold nanoparticles, with aptamers designed for PDGF, and PDGF receptors. They found they could detect PDGF receptor beta at 3.2 nM [113].

Aptamers themselves without gold nanoparticle association can act as the recognition element and the reporter molecule when binding to protein. This could also be a possible detection system going forward with Laser Capture Microdissection.

Aptamers

An oligonucleotide of single-stranded DNA or RNA molecule that can act as a recognition element is termed an aptamer. When aptamers were first described in 1990 [115, 116], the idea that binding affinity could be achieved with oligonucleotides themselves, eliminating the need for an antibody soon followed [117, 118]. Additionally, the molecule could also then act as its own reporter as well, as a part of the oligonucleotide, or a ligand to the nucleotide, could then be amplified through PCR. Since aptamers are defined as an oligonucleotide that can recognize a specific target, many of the strategies used in biobarcoding and IPCR are utilizing aptamers. However, for the purposes of this section, aptamers will be referred to as an assay utilizing oligonucleotides as the recognition element for the protein of interest and not just as the reporter molecule. There are many strategies that can be employed for aptamer binding and reporting, including manipulation of the secondary and tertiary structure of the aptamer to facilitate protein recognition and binding [119].

To develop an aptamer that will selectively bind to a particular protein, a process called systematic evolution of ligands by exponential enrichment, or SELEX is used [115, 116, 118, 120, 121]. In order to 'evolve' a DNA or RNA sequence to bind to a protein of interest (or any molecule for that matter), a random set of oligonucleotide sequences, for example, random sequences of 25 bases long (the possibility for 4^{25} different oligonucleotide species) are added to a tube with a protein immobilized on beads, then the sequences that do not bind are washed out. The binding oligonucleotide sequences are amplified, and then the procedure is repeated again. This repetition is needed anywhere between 5-20 times. With RNA, as was originally reported in 1990, the amplification step also requires reverse transcription, PCR and then in vitro transcription [115, 116], whereas DNA binding sequences can be resolved in fewer steps. An aptamer database on aptamer sequences generated in vitro, and their binding targets, is available and updated monthly [122].

Since RNA aptamers are believed to occur more commonly in vivo, they are of specific interest, but for the purposes of recognizing protein in basic research, DNA aptamers can be cheaper to make and more stable [123]. DNA aptamers were initially used to bind thrombin [117]. Thrombin is not naturally bound to a nucleic acid sequence, so it was ideally suited to artificial aptamer development. When the SELEX procedure was used, a specific region that was unique to all sequences was determined to be the region of the DNA or RNA that had special affinity for the thrombin protein. Many of the protein binding developments with aptamers use thrombin as a their proof of principle [123].

For example, researchers combined aptamer binding for thrombin and IgE with capillary electrophoresis and discovered a LOD of 40 nM [124]. Other assays followed – including the use of an aptamer bound to a fluorescent reporter that fluoresced upon binding to its target, or also called a molecular beacon. This was originally used to study thrombin and realized a 112 picomolar LOD [125-127].

Another strategy to increase sensitivity was to eliminate unbound aptamers with exonucleases. Aptamers were designed specifically to be digested by the enzymes when not bound to thrombin, which demonstrated much higher affinity for the protein [128].

Also, notably, aptamers were successfully used to create an ELISA based assay, with aptamer capture and recognition elements. However, the original level of detection was no more specific than ELISA [129].

Aptamers have been shown to have an affinity for proteins they are designed to recognize that rivals antibodies, and in some cases is more specific [130]. Because aptamers can then easily be labeled with nanoparticles, fluorescent probes, or be the reporter themselves and amplified through PCR, the sensitivity can be vastly superior to antibodies [131, 132].

Although aptamer technology has yet to be used with Laser Capture Microdissection to recognize specific proteins in tissue, the advancement of the technology could be a promising avenue for study. Two main issues with the use of aptamers as recognition elements for protein within complex tissue samples have arisen. The first is aptamer non-specific binding to other nucleic acids in the cell that could decrease reliability [133]. Treatment of cells with RNases still resulted in non-specific binding by an aptamer designed to detect green fluorescent protein in vitro [133]. Also, secondly, there appears to be a susceptibility of aptamers to enzyme breakdown and lack of stability for some aptamers [119].

One of the ways these issues were addressed was with the proximity ligation assay. The assay originally identified platelet-derived growth factor (PDGF) through the ligation of two ssDNA aptamer sequences developed through SELEX to recognize different PDGF complexes [79]. This increased specificity for the protein and decreased background, could be detected in human serum and cerebrospinal fluid, but was not achieved with a tissue sample [134]. The PLA assay moved on to use antibody intermediates through IPCR for increased specificity, because at the time there was a very small aptamer library for protein detection [135] and two distinct complexes of the protein needed to be recognized by the two different aptamers, decreasing its flexibility.

Other advances have shown that aptamers coupled with a molecular electronic device, can analyze thrombin on the level of a single molecule [136]. Additionally, nicking enzymes such as endonucleases have been used to detach part of the aptamer recognition element for qPCR, but can also be bound to gold nanoparticles that can be analyzed through colorimetric methods when they aggregate [137]. Another variation of these methods uses fluorescence on the nicked probes, that would then accumulate and fluoresce for analysis (138).

However, although promising, these applications have not yet become widespread for the analysis of complex tissue samples. As these technical advances progress, aptamers could be the most useful and universal strategy for detection, but difficulties can arise without the biochemical knowledge and specific SELEX training to develop these strategies (131). Additionally, non-specific binding in analysis of complex tissue samples is a real concern because of the loss of the assay's reliability and reproducibility. Strategies that address this will be the most successful with LCM.

Perhaps the most interesting application to measure protein levels from a lysate of single cells isolated from LCM would be the use of aptamers after a traditional and established Western blotting technique.

Tagging aptamers with a fluorescent marker such as quantum dots, or finding ways to sensitize the measurements of proteins after they have been separated through electrophoresis, could be simple and routine [132], and take advantage of Western blot's superior recognition of protein for quantification based on molecular weight. Using an aptamer recognition element after Western blot technique is one of the many that has taken the moniker of the so-called 'Eastern blot' [139].

In one instance, researchers attached quantum dots through a His-tag that could then be analyzed after binding. The procedure dramatically reduced the time required to do the assay, as no blocking was required or probing of the transfer membrane with primary and secondary antibodies [140]. (Figure 5) Strategies along these lines that are more sensitive than a normal Western blot could possibly be a reliable and universal tool for measurement of protein after LCM of specific cell types, as well as verifying protein size after separation with electrophoresis, something that is not possible with IPCR or biobarcoding.

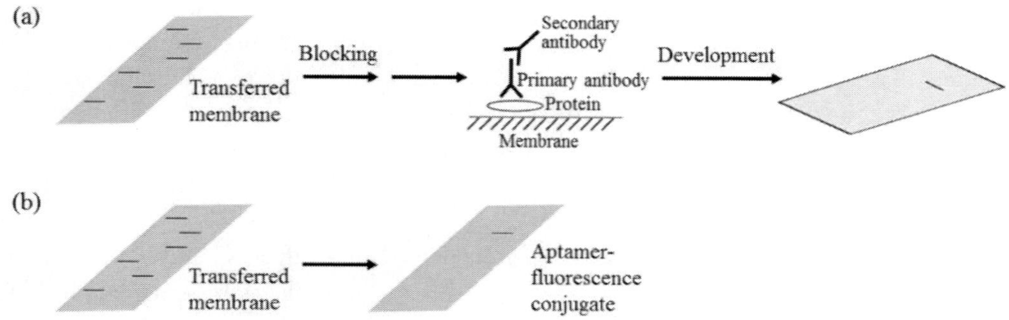

Reprinted from an open source, Sensors, Jan;12(1), Song KM, Lee S, Ban C. Aptamers and Their Biological Applications, p. 612-3, 2012.

Figure 5. (a) The protein binding steps required in conventional Western blot analysis, (b) The aptamer-based Western blot analysis requiring no blocking and antibody binding or recognition steps. An aptamer recognizes the protein of interest and is associated with quantum dots or fluorescent probe as a reporter.

Mass Spectrometry

Although Mass Spectrometry (MS) is not necessarily a benchtop and cost effective technique like the other assays discussed, the advancement of this technology is making it pertinent to consider in terms of protein quantification and measurement from small cell populations. 10,000 to 100,000 cells are more routinely being analyzed by Mass Spectrometry with LCM; however, that cell population has decreased as technologies have advanced. One early method showed the reliable discrimination of proteins between different cell types from placenta with only 125 cells [141]. Although currently not considered quantitative, this LOD has been improved upon with more creative techniques. Originally, Mass Spectrometry has been associated with the discovery and characterization of proteins, and quantification has been considered unreliable, especially with small cell populations.

However, quantification of protein has been attempted by using different sample concentrations after isotope labeling [142-144]. Although with this strategy, protein adsorption may vary results and it is still not as reliable as other assays, as well as requiring a large sample that cannot be utilized with LCM [145]. Stable isotope labeling by amino acids in cell culture (SILAC) [146] before MS or in vivo [147] provides a unique approach that is more reliable, but is not a factor in quantitative analysis of protein from complex human tissue samples. The major benefit of mass spectrometry is still the ability to analyze proteins processed from complex tissue without specific recognition elements, but as the techniques advance, quantification from small samples could be possible.

The two common methods are electrospray ionization (ESI) and matrix-assisted laser desorption/ionization (MALDI). With MALDI, a laser light causes ionization of sample molecules after it is pulsed onto the sample for analysis into a gaseous phase [6]. ESI utilizes electrical energy to ionize analytes and transfers them into the gaseous phase. Separation techniques such as of liquid chromatography and capillary electrophoresis have made mass spectrometry much more sensitive [148, 149]. Amino acid sequences can then be determined through tandem MS (MS/MS) and compared with a database for protein determination.

Traditionally, MALDI wasn't reliable quantitatively, and could only accurately analyze proteins less than 20 kDa in size as well [150]. MALDI was also reliant initially on techniques like 2D Electrophoresis for separation of samples before MALDI analysis. 2D-E requires a large sample for reliable quantification. Despite the large sample requirements, this was initially attempted with laser capture microdissection in 2002, where they compared mammary carcinoma with normal breast epithelium and found 40 peak differences in protein markers [151].

Because of the unreliability of quantification of protein with MS, most often what is discovered is verified with Western blot or ELISA, and therefore not practical for measurement with Laser Capture Microdissection, especially with minute samples. However, researchers were recently able to detect 70 differentially expressed proteins in stromal cells from colon adenocarcinoma compared with non-neoplastic colon mucosa with MS/MS using isobaric tags for relative and absolute quantification (iTRAQ) after LCM; 8 of the 70 were subsequently verified with Western blot as being significantly different in expression [152]. Additionally, the variety of techniques for protein isolation have made the prospect interesting as of late. Researchers have reported quantitative measurements of proteins in only 1000 cells [153]. They combined PLA and IPCR with MALDI-TOF to analyze the DNA that was used to tag the antibodies in the samples. The amplified DNA was then analyzed through MALDI quantitatively with femtomolar LODs. Additionally, up to 8 proteins from serum could be analyzed at once [153].

Also, stable isotopes aren't needed in some MS procedures, resulting in a label free approach that is more sensitive and gives fewer false positives [154, 155]. Original liquid chromatography and electrospray ionization (ESI) with tandem mass spectrometry (MS/MS) resulted in protein LOD at the zeptamolar concentration [156].

In another study, researchers were able to determine 10 proteins uniquely expressed in renal amyloidosis after LCM and MS/MS proteomic analysis [157]. Interestingly, they only required two glomeruli for each of the 127 cases they studied in order to have enough material for protein recognition. This amounted to an aggregate surface area of ~50,000-60,000 μm^2 [157]. This was not a quantitative study, but the small amount of tissue needed demonstrates the advancement of the sensitivity of these techniques.

Mass spectrometry has also been used in association with plasmon surface resonance to increase the sensitivity and resolution of both techniques [158-160].

Nanoparticles can also be used as matrices for Laser desorption ionization in order to provide a more enhanced technique, that could offset the problems with reproduction and resolution [7]. For instance, gold nanoparticles functionalized with a cationic ligand were used, and subsequent cellular uptake analyzed by LDI-MS [161-164]. Strategies to functionalize nanostructured surfaces with avidin-biotin conjugates and through antibody immobilization have been shown using these techniques [165, 166]. If nanoparticles are functionalized with aptamers to recognize specific proteins of interest, the detection and measurement between enriched cell populations could also be achieved with MS [167].

Although verification of protein measurement with MS at this time would also need to be confirmed with other methods already discussed, the importance of mass spectrometry with LCM in terms of understanding the proteomics of particular cell types in situ cannot be underestimated. These techniques are improving at a rapid rate as researchers are more creative with the increasing developments.

Discussion

The use of LCM to dissect specific cell types from human tissue will give a better understanding of the human proteome. Currently, new biomarker discovery for disease states with mass spectrometry and LCM can be routinely completed, as well as rapid immunohistochemistry before LCM for identification of specific cell types for excision. As cost-effective techniques that are reliable and trusted for measurement of protein expression levels from specific cell types aggregated through LCM become universal, the understanding of the levels of biomarkers in early disease states in different cell types will help in determining better therapeutics and prognosis.

For example, in neurodegenerative diseases, the pathological hallmark is the accumulation of proteins such as tau, α-synuclein and amyloid-β in tissue [168-171]. These biomarkers have been useful for the study of the disease but have led to no relevant treatments, and an assumption is that expression of a protein upstream of their accumulation or the proteins themselves are the cause of the disease [172]. However, the cell type origin that results in the accumulation of these pathological hallmarks and downstream byproducts could be among any cell type in the brain. Developments with LCM will help to further analyze specific cell types in human tissue to elucidate problems such as these.

At this time it appears that the most promising candidates for assays that could be cost-effective and reliable for measurement of protein from complex tissue samples are the IPCR technique developed by Sano et al., the biobarcoding technique developed by Mirkin et al., and possibly the use of aptamers as the recognition element after Western blot. Further studies of these assays on different cell types from complex tissue samples and focusing on optimization of protein level measurement instead of simple LOD determinations need to be conducted. Studies have shown that LCM is sufficiently specific for the cell type of interest [42], especially in a 10 μm section, where a cell type with an average diameter of 20 μm, would be obtained at > 80% enriched for cell specific protein after excising ~100 cells from LCM. Also, most protein concentrations in tissue are well above the LOD of these assays if 100 cells are taken. Excising 100 cells after rapid immunohistochemistry with LCM is a dissecting procedure that can take under an hour. Abundant proteins, such as tyrosine hydroxylase in dopaminergic neurons can be detected reliably with Western blot with a cell aggregate surface area of 360,000 μm^2, or around 200 neurons and 1700 astrocytes. If IPCR can reliably detect protein in three neurons, then accuracy after optimizing IPCR or biobarcoding for measurement of protein levels could easily be achieved with ~100 cells, of any type [90].

At the moment, it appears that background staining and contamination with IPCR is an issue when using complex tissue samples, although this may be minimal with the biobarcoding technique. Additionally, new assays with aptamers, including the use of aptamer recognition elements after Western blotting, could also provide sensitive, efficient, reliable and cost effective bench top techniques for the analysis of proteins from single cells in heterogeneous tissue.

As these techniques become more reliable and trusted, and commonly produced, vast avenues will be on the horizon for researchers to travel towards the understanding of differential protein expression at the cellular level in biomedicine.

Acknowledgments

I would like to thank Andreas Bender for inviting me to the University of Munich through the Bavarian-California Technological Center (BaCaTeC) Exchange to work on initial experiments that led to this review, and would also like to acknowledge Amber Qureshi and Fred Bonilla for useful comments and suggestions. Additionally, I'd like to thank Irene and Luccia Koob, and François-Marie Arouet for help with manuscript preparation.

References

[1] Yalow RS, Berson SA. Immunoassay of endogenous plasma insulin in man. The Journal of clinical investigation. 1960 Jul;39:1157-75. PubMed PMID: 13846364. Pubmed Central PMCID: PMC441860. Epub 1960/07/01. eng.

[2] Engvall E, Perlmann P. Enzyme-linked immunosorbent assay, Elisa. 3. Quantitation of specific antibodies by enzyme-labeled anti-immunoglobulin in antigen-coated tubes. *Journal of immunology* (Baltimore, Md : 1950). 1972 Jul;109(1):129-35. PubMed PMID: 4113792. Epub 1972/07/01. eng.

[3] Yolken RH, Stopa PJ. Enzyme-linked fluorescence assay: Ultrasensitive solid-phase assay for detection of human rotavirus. *J Clin Microbiol.* 1979 Sep;10(3):317-21. PubMed PMID: 226564. Pubmed Central PMCID: PMC273160. Epub 1979/09/01. eng.

[4] Burnette WN. "Western blotting": electrophoretic transfer of proteins from sodium dodecyl sulfate--polyacrylamide gels to unmodified nitrocellulose and radiographic detection with antibody and radioiodinated protein A. *Anal Biochem.* 1981 Apr;112(2):195-203. PubMed PMID: 6266278. Epub 1981/04/01. eng.

[5] Giepmans BNG, Adams SR, Ellisman MH, Tsien RY. Review - The fluorescent toolbox for assessing protein location and function. Science. 2006 Apr;312(5771):217-24. PubMed PMID: WOS:000236765300035. English.

[6] Karas M, Hillenkamp F. Laser Desorption Ionization Of Proteins With Molecular Masses Exceeding 10000 Daltons. *Analytical Chemistry.* 1988 Oct;60(20):2299-301. PubMed PMID: WOS:A1988Q513900029. English.

[7] Chiang CK, Chen WT, Chang HT. Nanoparticle-based mass spectrometry for the analysis of biomolecules. *Chem Soc Rev.* 2011;40(3):1269-81. PubMed PMID: WOS:000287585000009. English.

[8] Ma XJ, Dahiya S, Richardson E, Erlander M, Sgroi DC. Gene expression profiling of the tumor microenvironment during breast cancer progression. *Breast cancer research : BCR.* 2009;11(1):R7. PubMed PMID: 19187537. Pubmed Central PMCID: PMC2687710. Epub 2009/02/04. eng.

[9] Nakazono M, Qiu F, Borsuk LA, Schnable PS. Laser-capture microdissection, a tool for the global analysis of gene expression in specific plant cell types: identification of genes expressed differentially in epidermal cells or vascular tissues of maize. *The Plant cell.* 2003 Mar;15(3):583-96. PubMed PMID: 12615934. Pubmed Central PMCID: PMC150015. Epub 2003/03/05. eng.

[10] Petricoin EF, 3rd, Bichsel VE, Calvert VS, Espina V, Winters M, Young L, et al. Mapping molecular networks using proteomics: a vision for patient-tailored

combination therapy. *Journal of clinical oncology: official journal of the American Society of Clinical Oncology*. 2005 May 20;23(15):3614-21. PubMed PMID: 15908672. Epub 2005/05/24. eng.

[11] Buckanovich RJ, Sasaroli D, O'Brien-Jenkins A, Botbyl J, Conejo-Garcia JR, Benencia F, et al. Use of immuno-LCM to identify the in situ expression profile of cellular constituents of the tumor microenvironment. *Cancer biology & therapy*. 2006 Jun;5(6):635-42. PubMed PMID: 16627987. Epub 2006/04/22. eng.

[12] Sugiyama Y, Sugiyama K, Hirai Y, Akiyama F, Hasumi K. Microdissection is essential for gene expression profiling of clinically resected cancer tissues. *Am J Clin Pathol.* 2002 Jan;117(1):109-16. PubMed PMID: 11789716. Epub 2002/01/16. eng.

[13] Bonner RF, EmmertBuck M, Cole K, Pohida T, Chuaqui R, Goldstein S, et al. Cell sampling - Laser capture microdissection: Molecular analysis of tissue. *Science.* 1997 Nov;278(5342):1481-&. PubMed PMID: WOS:A1997YG85800052.

[14] EmmertBuck MR, Bonner RF, Smith PD, Chuaqui RF, Zhuang ZP, Goldstein SR, et al. Laser capture microdissection. *Science.* 1996 Nov;274(5289):998-1001. PubMed PMID: WOS:A1996VR79200054.

[15] Bota M, Dong HW, Swanson LW. From gene networks to brain networks. *Nature neuroscience*. 2003 Aug;6(8):795-9. PubMed PMID: 12886225. Epub 2003/07/30. eng.

[16] Boone DR, Sell SL, Hellmich HL. Laser capture microdissection of enriched populations of neurons or single neurons for gene expression analysis after traumatic brain injury. Journal of visualized experiments: *JoVE.* 2013 (74). PubMed PMID: 23603738. Pubmed Central PMCID: PMC3653564. Epub 2013/04/23. eng.

[17] Mullis K, Faloona F, Scharf S, Saiki R, Horn G, Erlich H. Specific enzymatic amplification of DNA in vitro: the polymerase chain reaction. Cold Spring Harbor symposia on quantitative biology. 1986;51 Pt 1:263-73. PubMed PMID: 3472723. Epub 1986/01/01. eng.

[18] Saiki RK, Gelfand DH, Stoffel S, Scharf SJ, Higuchi R, Horn GT, et al. Primer-Directed Enzymatic Amplification Of DNA With A Thermostable Dna-Polymerase. Science. 1988 Jan;239(4839):487-91. PubMed PMID: WOS:A1988L802900034. English.

[19] Livak KJ, Schmittgen TD. Analysis of relative gene expression data using real-time quantitative PCR and the 2(T)(-Delta Delta C) method. Methods. 2001 Dec;25(4):402-8. PubMed PMID: WOS:000173949500003.

[20] Heid CA, Stevens J, Livak KJ, Williams PM. Real time quantitative PCR. *Genome Research.* 1996 Oct;6(10):986-94. PubMed PMID: WOS:A1996VN30500009.

[21] Suarez-Quian CA, Goldstein SR, Pohida T, Smith PD, Peterson JI, Wellner E, et al. Laser capture microdissection of single cells from complex tissues. B*ioTechniques.* 1999 Feb;26(2):328-35. PubMed PMID: 10023545. Epub 1999/02/19. eng.

[22] Simone NL, Bonner RF, Gillespie JW, Emmert-Buck MR, Liotta LA. Laser-capture microdissection: opening the microscopic frontier to molecular analysis. *Trends in genetics : TIG.* 1998 Jul;14(7):272-6. PubMed PMID: 9676529. Epub 1998/07/24. eng.

[23] Schutze K, Lahr G. Identification of expressed genes by laser-mediated manipulation of single cells. *Nat Biotechnol.* 1998 Aug;16(8):737-42. PubMed PMID: 9702771. Epub 1998/08/14. eng.

[24] Luo L, Salunga RC, Guo HQ, Bittner A, Joy KC, Galindo JE, et al. Gene expression profiles of laser-captured adjacent neuronal subtypes. *Nat Med.* 1999 Jan;5(1):117-22. PubMed PMID: WOS:000077885000040. English.

[25] Weis S, Llenos IC, Dulay JR, Elashoff M, Martinez-Murillo F, Miller CL. Quality control for microarray analysis of human brain samples: The impact of postmortem factors, RNA characteristics, and histopathology. *J Neurosci Methods.* 2007 Sep 30;165(2):198-209. PubMed PMID: 17628689. Epub 2007/07/14. eng.

[26] Chu TT, Liu Y, Kemether E. Thalamic transcriptome screening in three psychiatric states. *Journal of human genetics.* 2009 Nov;54(11):665-75. PubMed PMID: 19834500. Epub 2009/10/17. eng.

[27] Cunnea P, McMahon J, O'Connell E, Mashayekhi K, Fitzgerald U, McQuaid S. Gene expression analysis of the microvascular compartment in multiple sclerosis using laser microdissected blood vessels. *Acta neuropathologica.* 2010 May;119(5):601-15. PubMed PMID: 19967542. Epub 2009/12/08. eng.

[28] Keeney PM, Bennett JP, Jr. ALS spinal neurons show varied and reduced mtDNA gene copy numbers and increased mtDNA gene deletions. *Molecular neurodegeneration.* 2010;5:21. PubMed PMID: 20504367. Pubmed Central PMCID: PMC2889994. Epub 2010/05/28. eng.

[29] Simpson JE, Ince PG, Shaw PJ, Heath PR, Raman R, Garwood CJ, et al. Microarray analysis of the astrocyte transcriptome in the aging brain: relationship to Alzheimer's pathology and APOE genotype. *Neurobiology of aging.* 2011 Oct;32(10):1795-807. PubMed PMID: 21705112. Epub 2011/06/28. eng.

[30] Humerick M, Hanson J, Rodriguez-Canales J, Lubelski D, Rashid OM, Salinas YD, et al. Analysis of Transcription Factor mRNAs in Identified Oxytocin and Vasopressin Magnocellular Neurons Isolated by Laser Capture Microdissection. PLoS One. 2013;8(7):e69407. PubMed PMID: 23894472. Pubmed Central PMCID: PMC3722287. Epub 2013/07/31. eng.

[31] Sano T, Smith CL, Cantor CR. IMMUNO-PCR - VERY SENSITIVE ANTIGEN-DETECTION BY MEANS OF SPECIFIC ANTIBODY-DNA CONJUGATES. Science. 1992 Oct;258(5079):120-2. PubMed PMID: WOS:A1992JQ61900032.

[32] Hashimoto M, Aoki M, Winblad B, Tjernberg LO. A novel approach for A beta(1-40) quantification using immuno-PCR. *J Neurosci Methods.* 2012 Apr;205(2):364-7. PubMed PMID: WOS:000302435400016. English.

[33] Mouledous L, Hunt S, Harcourt R, Harry JL, Williams KL, Gutstein HB. Proteomic analysis of immunostained, laser-capture microdissected brain samples. *Electrophoresis.* 2003 Jan;24(1-2):296-302. PubMed PMID: WOS:000180637300036.

[34] He H, Sun G, Ping FY. Laser-capture Microdissection and Protein Extraction for Protein Fingerprint of OSCC and OLK. Artificial Cells Blood Substitutes and Biotechnology. 2009;37(5):208-13. PubMed PMID: WOS:000269812700004.

[35] Zhang DH, Lim SG, Koay ESC. Proteomic identification of down-regulation of oncoprotein DJ-1 and proteasome activator subunit 1 in hepatitis B virus-infected well-differentiated hepatocellular carcinoma. *International Journal of Oncology.* 2007 Sep;31(3):577-84. PubMed PMID: WOS:000248695800013.

[36] Ornstein DK, Gillespie JW, Paweletz CP, Duray PH, Herring J, Vocke CD, et al. Proteomic analysis of laser capture microdissected human prostate cancer and in vitro

prostate cell lines. *Electrophoresis.* 2000 Jun;21(11):2235-42. PubMed PMID: WOS:000087975100016.

[37] Martinet W, Abbeloos V, Van Acker N, De Meyer GR, Herman AG, Kockx MM. Western blot analysis of a limited number of cells: a valuable adjunct to proteome analysis of paraffin wax-embedded, alcohol-fixed tissue after laser capture microdissection. *Journal of Pathology.* 2004 Mar;202(3):382-8. PubMed PMID: WOS:000220031100015.

[38] Braakman RB, Luider TM, Martens JW, Foekens JA, Umar A. Laser capture microdissection applications in breast cancer proteomics. *Methods in molecular biology* (Clifton, NJ). 2011;755:143-54. PubMed PMID: 21761300. Epub 2011/07/16. eng.

[39] Mukherjee S, Rodriguez-Canales J, Hanson J, Emmert-Buck MR, Tangrea MA, Prieto DA, et al. Proteomic analysis of frozen tissue samples using laser capture microdissection. *Methods in molecular biology* (Clifton, NJ). 2013;1002:71-83. PubMed PMID: 23625395. Epub 2013/04/30. eng.

[40] Lu QZ, Munugesan N, Macdonald JA, Wu SL, Pachter JS, Hancock WS. Analysis of mouse brain microvascular endothelium using immuno-laser capture microdissection coupled to a hybrid linear ion trap with Fourier transform-mass spectrometry proteomics platform. *Electrophoresis.* 2008 Jun;29(12):2689-95. PubMed PMID: WOS:000257481300026.

[41] Murugesan N, Macdonald JA, Lu Q, Wu SL, Hancock WS, Pachter JS. Analysis of mouse brain microvascular endothelium using laser capture microdissection coupled with proteomics. *Methods in molecular* biology (Clifton, NJ). 2011;686:297-311. PubMed PMID: 21082378. Epub 2010/11/18. eng.

[42] Koob AO, Bruns L, Prassler C, Masliah E, Klopstock T, Bender A. Protein analysis through Western blot of cells excised individually from human brain and muscle tissue. *Anal Biochem.* 2012 Jun;425(2):120-4. PubMed PMID: WOS:000304691900007. English.

[43] Isenberg G, Bielser W, Meier-Ruge W, Remy E. Cell surgery by laser micro-dissection: a preparative method. *Journal of microscopy.* 1976 May;107(1):19-24. PubMed PMID: 781257. Epub 1976/05/01. eng.

[44] Domazet B, Maclennan GT, Lopez-Beltran A, Montironi R, Cheng L. Laser capture microdissection in the genomic and proteomic era: targeting the genetic basis of cancer. *International journal of clinical and experimental pathology.* 2008;1(6):475-88. PubMed PMID: 18787684. Pubmed Central PMCID: PMC2480591. Epub 2008/09/13. eng.

[45] Simone NL, Bonner RF, Gillespie JW, Emmert-Buck MR, Liotta LA. Laser-capture microdissection: opening the microscopic frontier to molecular analysis. *Trends in Genetics.* 1998 Jul;14(7):272-6. PubMed PMID: WOS:000074742000008.

[46] Kolble K. The LEICA microdissection system: design and applications. Journal of molecular medicine (Berlin, Germany). 2000;78(7):B24-5. PubMed PMID: 11043387. Epub 2000/10/24. eng.

[47] Micke P, Ostman A, Lundeberg J, Ponten F. Laser-assisted cell microdissection using the PALM system. *Methods in molecular biology* (Clifton, NJ). 2005;293:151-66. PubMed PMID: 16028418. Epub 2005/07/21. eng.

[48] Palinauskas V, Dolnik OV, Valkiunas G, Bensch S. Laser microdissection microscopy and single cell PCR of avian hemosporidians. *The Journal of parasitology*. 2010 Apr;96(2):420-4. PubMed PMID: 20001098. Epub 2009/12/17. eng.

[49] Schermelleh L, Thalhammer S, Heckl W, Posl H, Cremer T, Schutze K, et al. Laser microdissection and laser pressure catapulting for the generation of chromosome-specific paint probes. *BioTechniques.* 1999 Aug;27(2):362-7. PubMed PMID: 10457845. Epub 1999/08/24. eng.

[50] Schutze K, Becker I, Becker KF, Thalhammer S, Stark R, Heckl WM, et al. Cut out or poke in--the key to the world of single genes: laser micromanipulation as a valuable tool on the look-out for the origin of disease. *Genetic analysis : biomolecular engineering*. 1997 Mar;14(1):1-8. PubMed PMID: 9158958. Epub 1997/03/01. eng.

[51] Gillespie JW, Gannot G, Tangrea MA, Ahram M, Best CJ, Bichsel VE, et al. Molecular profiling of cancer. *Toxicologic pathology*. 2004 Mar-Apr;32 Suppl 1:67-71. PubMed PMID: 15209405. Epub 2004/06/24. eng.

[52] Maitra A, Gazdar AF. Tissue microdissection and processing. *Cancer treatment and research.* 2001;106:63-84. PubMed PMID: 11225009. Epub 2001/02/28. eng.

[53] Vandewoestyne M, Goossens K, Burvenich C, Van Soom A, Peelman L, Deforce D. Laser capture microdissection: Should an ultraviolet or infrared laser be used? *Anal Biochem*. 2013 Aug 15;439(2):88-98. PubMed PMID: 23643622. Epub 2013/05/07. eng.

[54] Simone NL, Remaley AT, Charboneau L, Petricoin EF, 3rd, Glickman JW, Emmert-Buck MR, et al. Sensitive immunoassay of tissue cell proteins procured by laser capture microdissection. *The American journal of pathology*. 2000 Feb;156(2):445-52. PubMed PMID: 10666374. Pubmed Central PMCID: PMC1850045. Epub 2000/02/10. eng.

[55] Emmert-Buck MR, Gillespie JW, Paweletz CP, Ornstein DK, Basrur V, Appella E, et al. An approach to proteomic analysis of human tumors. *Molecular carcinogenesis*. 2000 Mar;27(3):158-65. PubMed PMID: 10708477. Epub 2000/03/09. eng.

[56] Shekouh AR, Thompson CC, Prime W, Campbell F, Hamlett J, Herrington CS, et al. Application of laser capture microdissection combined with two-dimensional electrophoresis for the discovery of differentially regulated proteins in pancreatic ductal adenocarcinoma. *Proteomics.* 2003 Oct;3(10):1988-2001. PubMed PMID: 14625861. Epub 2003/11/20. eng.

[57] Melle C, Bogumil R, Ernst G, Schimmel B, Bleul A, von Eggeling F. Detection and identification of heat shock protein 10 as a biomarker in colorectal cancer by protein profiling. *Proteomics.* 2006 Apr;6(8):2600-8. PubMed PMID: 16502466. Epub 2006/02/28. eng.

[58] Cha S, Imielinski MB, Rejtar T, Richardson EA, Thakur D, Sgroi DC, et al. In situ proteomic analysis of human breast cancer epithelial cells using laser capture microdissection: annotation by protein set enrichment analysis and gene ontology. *Molecular & cellular proteomics*: MCP. 2010 Nov;9(11):2529-44. PubMed PMID: 20739354. Pubmed Central PMCID: PMC2984225. Epub 2010/08/27. eng.

[59] Fend F, Emmert-Buck MR, Chuaqui R, Cole K, Lee J, Liotta LA, et al. Immuno-LCM: Laser capture microdissection of immunostained frozen sections for mRNA analysis. *American Journal of Pathology.* 1999 Jan;154(1):61-6. PubMed PMID: WOS:000077973800010.

[60] Murakami H, Liotta L, Star RA. IF-LCM: laser capture microdissection of immunofluorescently defined cells for mRNA analysis rapid communication. *Kidney Int.* 2000 Sep;58(3):1346-53. PubMed PMID: 10972700. Epub 2000/09/06. eng.

[61] Drew DA, Devers T, Horelik N, Yang S, O'Brien M, Wu R, et al. Nanoproteomic analysis of extracellular receptor kinase-1/2 post-translational activation in microdissected human hyperplastic colon lesions. *Proteomics.* 2013 May;13(9):1428-36. PubMed PMID: 23467982. Epub 2013/03/08. eng.

[62] Paschon V, Higa GS, Walter LT, de Sousa E, Zuzarte FC, Weber VR, et al. A new and reliable guide for studies of neuronal loss based on focal lesions and combinations of in vivo and in vitro approaches. *PLoS One.* 2013;8(4):e60486. PubMed PMID: 23585836. Pubmed Central PMCID: PMC3622006. Epub 2013/04/16. eng.

[63] Waller R, Woodroofe MN, Francese S, Heath PR, Wharton SB, Ince PG, et al. Isolation of enriched glial populations from post-mortem human CNS material by immuno-laser capture microdissection. *J Neurosci Methods.* 2012 Jul 15;208(2):108-13. PubMed PMID: 22609336. Epub 2012/05/23. eng.

[64] Potuckova L, Franko F, Bambouskova M, Draber P. Rapid and sensitive detection of cytokines using functionalized gold nanoparticle-based immuno-PCR, comparison with immuno-PCR and ELISA. *Journal of Immunological Methods.* 2011 Aug;371(1-2):38-47. PubMed PMID: WOS:000294518100005. English.

[65] von Smolinski D, Blessenohl M, Neubauer C, Kalies K, Gebert A. Validation of a novel ultra-short immunolabeling method for high-quality mRNA preservation in laser microdissection and real-time reverse transcriptase-polymerase chain reaction. The Journal of molecular diagnostics: *JMD.* 2006 May;8(2):246-53. PubMed PMID: 16645212. Pubmed Central PMCID: PMC1867592. Epub 2006/04/29. eng.

[66] Niemeyer CM, Wacker R, Adler M. Combination of DNA-directed immobilization and immuno-PCR: very sensitive antigen detection by means of self-assembled DNA-protein conjugates. *Nucleic Acids Research.* 2003 Aug;31(16). PubMed PMID: WOS:000184783000037.

[67] Weber PC, Ohlendorf DH, Wendoloski JJ, Salemme FR. STRUCTURAL ORIGINS OF HIGH-AFFINITY BIOTIN BINDING TO STREPTAVIDIN. Science. 1989 Jan;243(4887):85-8. PubMed PMID: WOS:A1989R646900037. English.

[68] Ren J, Ge L, Li YQ, Bai J, Liu WC, Si XM. Detection of circulating CEA molecules in human sera and leukopheresis of peripheral blood stem cells with E. coli expressed bispecific CEAScFv-streptavidin fusion protein-based immuno-PCR technique. In: Lo YMD, Chiu RWK, Johnson PJ, editors. Circulating Nucleic Acids in Plasma or Serum II. *Annals of the New York Academy of Sciences*. 9452001. p. 116-8.

[69] Zhou H, Fisher RJ, Papas TS. UNIVERSAL IMMUNO-PCR FOR ULTRA-SENSITIVE TARGET PROTEIN-DETECTION. Nucleic Acids Research. 1993 Dec;21(25):6038-9. PubMed PMID: WOS:A1993MR19900034.

[70] Verschoor ML, Wilson LA, Verschoor CP, Singh G. Ets-1 Regulates Energy Metabolism in Cancer Cells. *PLoS One.* 2010 Oct;5(10):9. PubMed PMID: WOS:000283419100009. English.

[71] Niemeyer CM, Adler M, Wacker R. Lmmuno-PCR: high sensitivity detection of proteins by nucleic acid amplification. *Trends in Biotechnology.* 2005 Apr;23(4):208-16. PubMed PMID: WOS:000228286000008.

[72] Niemeyer CM, Adler M, Pignataro B, Lenhert S, Gao S, Chi LF, et al. Self-assembly of DNA-streptavidin nanostructures and their use as reagents in immuno-PCR. *Nucleic Acids Research.* 1999 Dec;27(23):4553-61. PubMed PMID: WOS:000084033500004.

[73] Hendrickson ER, Truby TMH, Joerger RD, Majarian WR, Ebersole RC. A SENSITIVITY MULTIANALYTE IMMUNOASSAY USING COVALENT DNA-LABELED ANTIBODIES AND POLYMERASE CHAIN-REACTION. *Nucleic Acids Research.* 1995 Feb;23(3):522-9. PubMed PMID: WOS:A1995QK64600030.

[74] Sims PW, Vasser M, Wong WL, Williams PM, Meng YG. Immunopolymerase chain reaction using real-time polymerase chain reaction for detection. *Anal Biochem.* 2000 Jun;281(2):230-2. PubMed PMID: WOS:000087391100013. English.

[75] Janssen KPF, Knez K, Spasic D, Lammertyn J. Nucleic Acids for Ultra-Sensitive Protein Detection. *Sensors.* 2013 Jan;13(1):1353-84. PubMed PMID: WOS:000314024800077. English.

[76] Agasti SS, Liong M, Peterson VM, Lee H, Weissleder R. Photocleavable DNA Barcode-Antibody Conjugates Allow Sensitive and Multiplexed Protein Analysis in Single Cells. Journal of the American Chemical Society. 2012 Nov;134(45):18499-502. PubMed PMID: WOS:000311192100006. English.

[77] Schweitzer B, Wiltshire S, Lambert J, O'Malley S, Kukanskis K, Zhu ZR, et al. Immunoassays with rolling circle DNA amplification: A versatile platform for ultrasensitive antigen detection. *Proceedings of the National Academy of Sciences of the United States of America.* 2000 Aug;97(18):10113-9. PubMed PMID: WOS:000089067500055.

[78] Schweitzer B, Roberts S, Grimwade B, Shao WP, Wang MJ, Fu Q, et al. Multiplexed protein profiling on microarrays by rolling-circle amplification. *Nature Biotechnology.* 2002 Apr;20(4):359-65. PubMed PMID: WOS:000174681900022.

[79] Fredriksson S, Gullberg M, Jarvius J, Olsson C, Pietras K, Gustafsdottir SM, et al. Protein detection using proximity-dependent DNA ligation assays. *Nature Biotechnology.* 2002 May;20(5):473-7. PubMed PMID: WOS:000175364800029. English.

[80] Gustafsdottir SM, Schallmeiner E, Fredriksson S, Gullberg M, Soderberg O, Jarvius M, et al. Proximity ligation assays for sensitive and specific protein analyses. *Anal Biochem.* 2005 Oct;345(1):2-9. PubMed PMID: WOS:000232431700002. English.

[81] Darmanis S, Nong RY, Vanelid J, Siegbahn A, Ericsson O, Fredriksson S, et al. ProteinSeq: High-Performance Proteomic Analyses by Proximity Ligation and Next Generation Sequencing. *PLoS One.* 2011 Sep;6(9):10. PubMed PMID: WOS:000295939600036. English.

[82] Schallmeiner E, Oksanen E, Ericsson O, Spangberg L, Eriksson S, Stenman UH, et al. Sensitive protein detection via triple-binder proximity ligation assays. *Nature Methods.* 2007 Feb;4(2):135-7. PubMed PMID: WOS:000243998100014. English.

[83] Di Giusto DA, Wlassoff WA, Gooding JJ, Messerle BA, King GC. Proximity extension of circular DNA aptamers with real-time protein detection. *Nucleic Acids Research.* 2005 2005;33(6). PubMed PMID: WOS:000228398400012.

[84] Lundberg M, Eriksson A, Tran B, Assarsson E, Fredriksson S. Homogeneous antibody-based proximity extension assays provide sensitive and specific detection of low-abundant proteins in human blood. *Nucleic Acids Research.* 2011 Aug;39(15):8. PubMed PMID: WOS:000294555800005. English.

[85] He J, Evers DL, O'Leary TJ, Mason JT. Immunoliposome-PCR: a generic ultrasensitive quantitative antigen detection system. *Journal of nanobiotechnology*. 2012;10(1):26. PubMed PMID: 22726242. Pubmed Central PMCID: PMC3466442. Epub 2012/06/26. eng.

[86] Mason JT, Xu LX, Sheng ZM, O'Leary TJ. A liposome-PCR assay for the ultrasensitive detection of biological toxins. *Nature Biotechnology*. 2006 May;24(5):555-7. PubMed PMID: WOS:000237331300030. English.

[87] Maia M, Takahashi H, Adler K, Garlick RK, Wands JR. DEVELOPMENT OF A 2-SITE IMMUNO-PCR ASSAY FOR HEPATITIS-B SURFACE-ANTIGEN. *J Virol Methods*. 1995 Apr;52(3):273-86. PubMed PMID: WOS:A1995QR24600004.

[88] Barletta JM, Edelman DC, Constantine NT. Lowering the detection limits of HIV-1 viral load using real-time Immuno-PCR for HIV-1 p24 antigen. American Journal of Clinical Pathology. 2004 Jul;122(1):20-7. PubMed PMID: WOS:000222326900003.

[89] Saito K, Kobayashi D, Sasaki M, Araake H, Kida T, Yagihashi A, et al. Detection of human serum tumor necrosis factor-alpha in healthy donors, using a highly sensitive immuno-PCR assay. *Clinical Chemistry*. 1999 May;45(5):665-9. PubMed PMID: WOS:000080209900009.

[90] Nam JM, Thaxton CS, Mirkin CA. Nanoparticle-based bio-bar codes for the ultrasensitive detection of proteins. *Science*. 2003 Sep;301(5641):1884-6. PubMed PMID: WOS:000185536700044.

[91] Nam JM, Park SJ, Mirkin CA. Bio-barcodes based on oligonucleotide-modified nanoparticles. *Journal of the American Chemical Society*. 2002 Apr 17;124(15):3820-1. PubMed PMID: WOS:000175014000010.

[92] Nam JM, Stoeva SI, Mirkin CA. Bio-bar-code-based DNA detection with PCR-like sensitivity. *Journal of the American Chemical Society*. 2004 May 19;126(19):5932-3. PubMed PMID: WOS:000221416700002.

[93] Taton TA, Mirkin CA, Letsinger RL. Scanometric DNA array detection with nanoparticle probes. *Science*. 2000 Sep;289(5485):1757-60. PubMed PMID: WOS:000089195200048.

[94] Keating CD. Nanoscience enables ultrasensitive detection of Alzheimer's biomarker. *Proceedings of the National Academy of Sciences of the United States of America*. 2005 Feb;102(7):2263-4. PubMed PMID: WOS:000227073100001. English.

[95] Muller UR. Protein detection using biobarcodes. Molecular Biosystems. 2006 Oct;2(10):470-6. PubMed PMID: WOS:000241101100010.

[96] Gong YS, Chang L, Viola KL, Lacor PN, Lambert MP, Finch CE, et al. Alzheimer's disease-affected brain: Presence of oligomeric A beta ligands (ADDLs) suggests a molecular basis for reversible memory loss. *Proceedings of the National Academy of Sciences of the United States of America*. 2003 Sep;100(18):10417-22. PubMed PMID: WOS:000185119300052. English.

[97] Georganopoulou DG, Chang L, Nam JM, Thaxton CS, Mufson EJ, Klein WL, et al. Nanoparticle-based detection in cerebral spinal fluid of a soluble pathogenic biomarker for Alzheimer's disease. *Proceedings of the National Academy of Sciences of the United States of America*. 2005 Feb;102(7):2273-6. PubMed PMID: WOS:000227073100006. English.

[98] Stoeva SI, Lee JS, Smith JE, Rosen ST, Mirkin CA. Multiplexed detection of protein cancer markers with biobarcoded nanoparticle probes. *Journal of the American*

Chemical Society. 2006 Jul;128(26):8378-9. PubMed PMID: WOS:000238590000006. English.

[99] Bidart JM, Thuillier F, Augereau C, Chalas J, Daver A, Jacob N, et al. Kinetics of serum tumor marker concentrations and usefulness in clinical monitoring. *Clinical Chemistry*. 1999 Oct;45(10):1695-707. PubMed PMID: WOS:000083038200001. English.

[100] Kim EY, Stanton J, Korber BTM, Krebs K, Bogdan D, Kunstman K, et al. Detection of HIV-1 p24 Gag in plasma by a nanoparticle-based bio-barcode-amplification method. *Nanomedicine*. 2008 Jun;3(3):293-303. PubMed PMID: WOS:000256789400007. English.

[101] Dong HH, Liu JL, Zhu H, Ou CY, Xing WG, Qiu MF, et al. Two types of nanoparticle-based bio-barcode amplification assays to detect HIV-1 p24 antigen. *Virol J*. 2012 Aug;9:7. PubMed PMID: WOS:000311124700001. English.

[102] Perez JW, Vargis EA, Russ PK, Haselton FR, Wright DW. Detection of respiratory syncytial virus using nanoparticle amplified immuno-polymerase chain reaction. *Anal Biochem*. 2011 Mar;410(1):141-8. PubMed PMID: WOS:000286711300020. English.

[103] Thaxton CS, Elghanian R, Thomas AD, Stoeva SI, Lee JS, Smith ND, et al. Nanoparticle-based bio-barcode assay redefines "undetectable" PSA and biochemical recurrence after radical prostatectomy. *Proceedings of the National Academy of Sciences of the United States of America*. 2009 Nov;106(44):18437-42. PubMed PMID: WOS:000271429800006. English.

[104] Ranzoni A, Sabatte G, van Ijzendoorn LJ, Prins MWJ. One-Step Homogeneous Magnetic Nanoparticle Immunoassay for Biomarker Detection Directly in Blood Plasma. *ACS Nano*. 2012 Apr;6(4):3134-41. PubMed PMID: WOS:000303099300032. English.

[105] Chen LY, Wei HP, Guo YC, Cui ZQ, Zhang ZP, Zhang XE. Gold nanoparticle enhanced immuno-PCR for ultrasensitive detection of Hantaan virus nucleocapsid protein. *Journal of Immunological Methods*. 2009 Jul;346(1-2):64-70. PubMed PMID: WOS:000268199900008. English.

[106] Saha K, Agasti SS, Kim C, Li XN, Rotello VM. Gold Nanoparticles in Chemical and Biological Sensing. *Chemical Reviews*. 2012 May;112(5):2739-79. PubMed PMID: WOS:000303698000006. English.

[107] De M, Rana S, Akpinar H, Miranda OR, Arvizo RR, Bunz UHF, et al. Sensing of proteins in human serum using conjugates of nanoparticles and green fluorescent protein. *Nat Chem*. 2009 Sep;1(6):461-5. PubMed PMID: WOS:000269215100012. English.

[108] You CC, Miranda OR, Gider B, Ghosh PS, Kim IB, Erdogan B, et al. Detection and identification of proteins using nanoparticle-fluorescent polymer 'chemical nose' sensors. *Nat Nanotechnol*. 2007 May;2(5):318-23. PubMed PMID: WOS:000246688900017. English.

[109] Lyon LA, Musick MD, Natan MJ. Colloidal Au-enhanced surface plasmon resonance immunosensing. *Analytical Chemistry*. 1998 Dec;70(24):5177-83. PubMed PMID: WOS:000077550100011. English.

[110] Pieper-Furst U, Stocklein WFM, Warsinke A. Gold nanoparticle-enhanced surface plasmon resonance measurement with a highly sensitive quantification for human tissue

inhibitor of metalloproteinases-2. *Anal Chim Acta*. 2005 Sep;550(1-2):69-76. PubMed PMID: WOS:000232356200010. English.

[111] Cao C, Sim SJ. Signal enhancement of surface plasmon resonance immunoassay using enzyme precipitation-functionalized gold nanoparticles: A femto molar level measurement of anti-glutamic acid decarboxylase antibody. Biosens Bioelectron. 2007 Apr;22(9-10):1874-80. PubMed PMID: WOS:000245541600006. English.

[112] Cao C, Li X, Lee J, Sim SJ. Homogenous growth of gold nanocrystals for quantification of PSA protein biomarker. *Biosens Bioelectron*. 2009 Jan;24(5):1292-7. PubMed PMID: WOS:000263199800039. English.

[113] Huang CC, Huang YF, Cao ZH, Tan WH, Chang HT. Aptamer-modified gold nanoparticles for colorimetric determination of platelet-derived growth factors and their receptors. *Analytical Chemistry*. 2005 Sep;77(17):5735-41. PubMed PMID: WOS:000231652300053. English.

[114] Wei H, Li BL, Li J, Wang EK, Dong SJ. Simple and sensitive aptamer-based colorimetric sensing of protein using unmodified gold nanoparticle probes. *Chem Commun*. 2007 (36):3735-7. PubMed PMID: WOS:000249437900011. English.

[115] Tuerk C, Gold L. SYSTEMATIC EVOLUTION OF LIGANDS BY EXPONENTIAL ENRICHMENT - RNA LIGANDS TO BACTERIOPHAGE-T4 DNA-POLYMERASE. *Science*. 1990 Aug;249(4968):505-10. PubMed PMID: WOS:A1990DR81100032.

[116] Ellington AD, Szostak JW. INVITRO SELECTION OF RNA MOLECULES THAT BIND SPECIFIC LIGANDS. *Nature*. 1990 Aug;346(6287):818-22. PubMed PMID: WOS:A1990DW48800045.

[117] Bock LC, Griffin LC, Latham JA, Vermaas EH, Toole JJ. SELECTION OF SINGLE-STRANDED-DNA MOLECULES THAT BIND AND INHIBIT HUMAN THROMBIN. *Nature*. 1992 Feb;355(6360):564-6. PubMed PMID: WOS:A1992HC52600065. English.

[118] Gold L, Polisky B, Uhlenbeck O, Yarus M. DIVERSITY OF OLIGONUCLEOTIDE FUNCTIONS. *Annu Rev Biochem*. 1995;64:763-97. PubMed PMID: WOS:A1995RH35300026.

[119] Liu JW, Cao ZH, Lu Y. Functional Nucleic Acid Sensors. *Chemical Reviews*. 2009 May;109(5):1948-98. PubMed PMID: WOS:000266096800007. English.

[120] Robertson DL, Joyce GF. SELECTION INVITRO OF AN RNA ENZYME THAT SPECIFICALLY CLEAVES SINGLE-STRANDED-DNA. *Nature*. 1990 Mar;344(6265):467-8. PubMed PMID: WOS:A1990CW47800075. English.

[121] Wilson DS, Szostak JW. In vitro selection of functional nucleic acids. *Annu Rev Biochem*. 1999;68:611-47. PubMed PMID: WOS:000082693200021. English.

[122] Lee JF, Hesselberth JR, Meyers LA, Ellington AD. Aptamer database. Nucleic Acids Research. 2004 Jan;32:D95-D100. PubMed PMID: WOS:000188079000019. English.

[123] Baird GS. Where Are All the Aptamers? *American Journal of Clinical Pathology*. 2010 Oct;134(4):529-31. PubMed PMID: WOS:000282013900001.

[124] German I, Buchanan DD, Kennedy RT. Aptamers as ligands in affinity probe capillary electrophoresis. *Analytical Chemistry*. 1998 Nov;70(21):4540-5. PubMed PMID: WOS:000076839000022. English.

[125] Li BL, Wei H, Dong SJ. Sensitive detection of protein by an aptamer-based label-free fluorescing molecular switch. *Chem Commun*. 2007 (1):73-5. PubMed PMID: WOS:000243380800014. English.

[126] Fang XH, Cao ZH, Beck T, Tan WH. Molecular aptamer for real-time oncoprotein platelet-derived growth factor monitoring by fluorescence anisotropy. *Analytical Chemistry*. 2001 Dec;73(23):5752-7. PubMed PMID: WOS:000172541200031.

[127] Fang XH, Mi YM, Li JWJ, Beck T, Schuster S, Tan WH. Molecular beacons - Fluorogenic probes for living cell study. *Cell Biochemistry and Biophysics*. 2002;37(2):71-81. PubMed PMID: WOS:000179416800001.

[128] Wang XL, Li F, Su YH, Sun X, Li XB, Schluesener HJ, et al. Ultrasensitive detection of protein using an aptamer-based exonuclease protection assay. *Analytical Chemistry*. 2004 Oct 1;76(19):5605-10. PubMed PMID: WOS:000224217400019.

[129] Drolet DW, MoonMcDermott L, Romig TS. An enzyme-linked oligonucleotide assay. Nature Biotechnology. 1996 Aug;14(8):1021-5. PubMed PMID: WOS:A1996UZ19900038. English.

[130] Yoshida Y, Waga I, Horii K. Quantitative and sensitive protein detection strategies based on aptamers. *Proteom Clin Appl*. 2012 Dec;6(11-12):574-80. PubMed PMID: WOS:000312650700004. English.

[131] Cho EJ, Lee JW, Ellington AD. Applications of Aptamers as Sensors. Annual Review of Analytical Chemistry. *Annual Review of Analytical Chemistry*. 2. Palo Alto: Annual Reviews; 2009. p. 241-64.

[132] Song KM, Lee S, Ban C. Aptamers and Their Biological Applications. *Sensors*. 2012 Jan;12(1):612-31. PubMed PMID: WOS:000299537100034. English.

[133] Stanlis KKH, McIntosh JR. Single-strand DNA aptamers as probes for protein localization in cells. *Journal of Histochemistry & Cytochemistry*. 2003 Jun;51(6):797-808. PubMed PMID: WOS:000183098100011. English.

[134] Fredriksson S, Dixon W, Ji H, Koong AC, Mindrinos M, Davis RW. Multiplexed protein detection by proximity ligation for cancer biomarker validation. *Nature Methods*. 2007 Apr;4(4):327-9. PubMed PMID: WOS:000245584900013.

[135] Gullberg M, Gustafsdottir SM, Schallmeiner E, Jarvius J, Bjarnegard M, Betsholtz C, et al. Cytokine detection by antibody-based proximity ligation. *Proceedings of the National Academy of Sciences of the United States of America*. 2004 Jun 1;101(22):8420-4. PubMed PMID: WOS:000221831800031.

[136] Liu S, Zhang XY, Luo WX, Wang ZX, Guo XF, Steigerwald ML, et al. Single-Molecule Detection of Proteins Using Aptamer-Functionalized Molecular Electronic Devices. Angewandte Chemie-International Edition. 2011;50(11):2496-502. PubMed PMID: WOS:000288037600007.

[137] Li J, Fu HE, Wu LJ, Zheng AX, Chen GN, Yang HH. General Colorimetric Detection of Proteins and Small Molecules Based on Cyclic Enzymatic Signal Amplification and Hairpin Aptamer Probe. *Analytical Chemistry*. 2012 Jun;84(12):5309-15. PubMed PMID: WOS:000305320500023. English.

[138] Xue LY, Zhou XM, Xing D. Sensitive and Homogeneous Protein Detection Based on Target-Triggered Aptamer Hairpin Switch and Nicking Enzyme Assisted Fluorescence Signal Amplification. *Analytical Chemistry*. 2012 Apr;84(8):3507-13. PubMed PMID: WOS:000302838500005. English.

[139] Lin JS, McNatty KP. Aptamer-based regionally protected PCR for protein detection. *Clin Chem.* 2009 Sep;55(9):1686-93. PubMed PMID: 19589846. Epub 2009/07/11. eng.

[140] Shin S, Kim I-H, Kang W, Yang JK, Hah SS. An alternative to Western blot analysis using RNA aptamer-functionalized quantum dots. *Bioorganic & Medicinal Chemistry Letters.* 2010 Jun 1;20(11):3322-5. PubMed PMID: WOS:000277788900027.

[141] de Groot CJ, Steegers-Theunissen RP, Guzel C, Steegers EA, Luider TM. Peptide patterns of laser dissected human trophoblasts analyzed by matrix-assisted laser desorption/ionisation-time of flight mass spectrometry. *Proteomics.* 2005 Feb;5(2):597-607. PubMed PMID: 15627953. Epub 2005/01/04. eng.

[142] Kirkpatrick DS, Gerber SA, Gygi SP. The absolute quantification strategy: a general procedure for the quantification of proteins and post-translational modifications. *Methods.* 2005 Mar;35(3):265-73. PubMed PMID: 15722223. Epub 2005/02/22. eng.

[143] Ross PL, Huang YN, Marchese JN, Williamson B, Parker K, Hattan S, et al. Multiplexed protein quantitation in Saccharomyces cerevisiae using amine-reactive isobaric tagging reagents. *Molecular & cellular proteomics* : MCP. 2004 Dec;3(12):1154-69. PubMed PMID: 15385600. Epub 2004/09/24. eng.

[144] Liu H, Sadygov RG, Yates JR, 3rd. A model for random sampling and estimation of relative protein abundance in shotgun proteomics. *Anal Chem.* 2004 Jul 15;76(14):4193-201. PubMed PMID: 15253663. Epub 2004/07/16. eng.

[145] Warwood S, Byron A, Humphries MJ, Knight D. The effect of peptide adsorption on signal linearity and a simple approach to improve reliability of quantification. *Journal of proteomics.* 2013 May 9. PubMed PMID: 23665148. Epub 2013/05/15. Eng.

[146] Ong SE, Blagoev B, Kratchmarova I, Kristensen DB, Steen H, Pandey A, et al. Stable isotope labeling by amino acids in cell culture, SILAC, as a simple and accurate approach to expression proteomics. *Molecular & cellular proteomics: MCP.* 2002 May;1(5):376-86. PubMed PMID: 12118079. Epub 2002/07/16. eng.

[147] Zanivan S, Krueger M, Mann M. In vivo quantitative proteomics: the SILAC mouse. *Methods in molecular biology* (Clifton, NJ). 2012;757:435-50. PubMed PMID: 21909926. Epub 2011/09/13. eng.

[148] Lovgren U, Johansson S, Jensen LS, Ekstrom C, Carlshaf A. Quantitative determination of peptide drug in human plasma samples at low pg/ml levels using coupled column liquid chromatography-tandem mass spectrometry. *J Pharm Biomed* Anal. 2010 Nov;53(3):537-45. PubMed PMID: WOS:000280435800042. English.

[149] Lapainis T, Rubakhin SS, Sweedler JV. Capillary Electrophoresis with Electrospray Ionization Mass Spectrometric Detection for Single-Cell Metabolomics. *Analytical Chemistry.* 2009 Jul;81(14):5858-64. PubMed PMID: WOS:000268135000037. English.

[150] von Eggeling F, Melle C, Ernst G. Microdissecting the proteome. *Proteomics.* 2007 Aug;7(16):2729-37. PubMed PMID: WOS:000249295500005. English.

[151] Xu BGJ, Caprioli RM. Direct analysis of laser capture microdissected cells by MALDI mass spectrometry. *J Am Soc Mass Spectrom.* 2002 Nov;13(11):1292-7. PubMed PMID: WOS:000179082200005. English.

[152] Mu Y, Chen Y, Zhang G, Zhan X, Li Y, Liu T, et al. Identification of stromal differentially expressed proteins in the colon carcinoma by quantitative proteomics.

Electrophoresis. 2013 Jun;34(11):1679-92. PubMed PMID: 23737015. Epub 2013/06/06. eng.

[153] Flanigon J, Kamali-Moghaddam M, Burbulis I, Annink C, Steffen M, Oeth P, et al. Multiplex protein detection with DNA readout via mass spectrometry. *New Biotech.* 2013 Jan;30(2):153-8. PubMed PMID: WOS:000313786400008. English.

[154] America AHP, Cordewener JHG. Comparative LC-MS: A landscape of peaks and valleys. *Proteomics.* 2008 Feb;8(4):731-49. PubMed PMID: WOS:000253791600009. English.

[155] Wiener MC, Sachs JR, Deyanova EG, Yates NA. Differential mass spectrometry: A label-free LC-MS method for finding significant differences in complex peptide and protein mixtures. *Analytical Chemistry.* 2004 Oct;76(20):6085-96. PubMed PMID: WOS:000224565200013. English.

[156] Shen Y, Tolic N, Masselon C, Pasa-Tolic L, Camp DG, Hixson KK, et al. Ultrasensitive proteomics using high-efficiency on-line micro-SPE-NanoLC-NanoESI MS and MS/MS. *Analytical Chemistry.* 2004 Jan;76(1):144-54. PubMed PMID: WOS:000187806100024. English.

[157] Sethi S, Vrana JA, Theis JD, Leung N, Sethi A, Nasr SH, et al. Laser microdissection and mass spectrometry-based proteomics aids the diagnosis and typing of renal amyloidosis. *Kidney Int.* 2012 Jul;82(2):226-34. PubMed PMID: WOS: 000305968800016. English.

[158] Homola J. Surface plasmon resonance sensors for detection of chemical and biological species. *Chemical Reviews.* 2008 Feb;108(2):462-93. PubMed PMID: WOS:000253165500007. English.

[159] Madeira A, Ohman E, Nilsson A, Sjogren B, Andren PE, Svenningsson P. Coupling surface plasmon resonance to mass spectrometry to discover novel protein-protein interactions. *Nature Protocols.* 2009;4(7):1023-37. PubMed PMID: WOS: 000268858700003. English.

[160] Krone JR, Nelson RW, Dogruel D, Williams P, Granzow R. BIA/MS: Interfacing biomolecular interaction analysis with mass spectrometry. *Anal Biochem.* 1997 Jan;244(1):124-32. PubMed PMID: WOS:A1997WC06000019. English.

[161] Zhu ZJ, Rotello VM, Vachet RW. Engineered nanoparticle surfaces for improved mass spectrometric analyses. *Analyst.* 2009;134(11):2183-8. PubMed PMID: WOS:000270941900001. English.

[162] Zhu ZJ, Tang R, Yeh YC, Miranda OR, Rotello VM, Vachet RW. Determination of the Intracellular Stability of Gold Nanoparticle Monolayers Using Mass Spectrometry. *Analytical Chemistry.* 2012 May;84(10):4321-6. PubMed PMID: WOS: 000303965500014.

[163] Zhu ZJ, Ghosh PS, Miranda OR, Vachet RW, Rotello VM. Multiplexed Screening of Cellular Uptake of Gold Nanoparticles Using Laser Desorption/Ionization Mass Spectrometry. *Journal of the American Chemical Society.* 2008 Oct;130(43):14139-43. PubMed PMID: WOS:000260301700049.

[164] Creran B, Yan B, Moyano DF, Gilbert MM, Vachet RW, Rotello VM. Laser desorption ionization mass spectrometric imaging of mass barcoded gold nanoparticles for security applications. *Chem Commun.* 2012;48(38):4543-5. PubMed PMID: WOS: 000302743000004.

[165] Meng JC, Siuzdak G, Finn MG. Affinity mass spectrometry from a tailored porous silicon surface. *Chem Commun.* 2004 (18):2108-9. PubMed PMID: WOS: 000223873000047. English.

[166] Lowe RD, Szili EJ, Kirkbride P, Thissen H, Siuzdak G, Voelcker NH. Combined Immunocapture and Laser Desorption/Ionization Mass Spectrometry on Porous Silicon. *Analytical Chemistry.* 2010 May;82(10):4201-8. PubMed PMID: WOS: 000277531500034. English.

[167] Ocsoy I, Gulbakan B, Shukoor MI, Xiong XL, Chen T, Powell DH, et al. Aptamer-Conjugated Multifunctional Nanoflowers as a Platform for Targeting, Capture, and Detection in Laser Desorption Ionization Mass Spectrometry. *ACS Nano.* 2013 Jan;7(1):417-27. PubMed PMID: WOS:000314082800045. English.

[168] Goedert M, Spillantini MG, Del Tredici K, Braak H. 100 years of Lewy pathology. *Nature reviews Neurology.* 2013 Jan;9(1):13-24. PubMed PMID: 23183883. Epub 2012/11/28. eng.

[169] Wolozin BL, Pruchnicki A, Dickson DW, Davies P. A neuronal antigen in the brains of Alzheimer patients. *Science.* 1986 May 2;232(4750):648-50. PubMed PMID: 3083509. Epub 1986/05/02. eng.

[170] Kirschner DA, Abraham C, Selkoe DJ. X-ray diffraction from intraneuronal paired helical filaments and extraneuronal amyloid fibers in Alzheimer disease indicates cross-beta conformation. *Proc Natl Acad Sci USA.* 1986 Jan;83(2):503-7. PubMed PMID: 3455785. Pubmed Central PMCID: PMC322888. Epub 1986/01/01. eng.

[171] Spillantini MG, Schmidt ML, Lee VM, Trojanowski JQ, Jakes R, Goedert M. Alpha-synuclein in Lewy bodies. *Nature.* 1997 Aug 28;388(6645):839-40. PubMed PMID: 9278044. Epub 1997/08/28. eng.

[172] Crump CJ, Johnson DS, Li YM. Development and Mechanism of gamma-Secretase Modulators for Alzheimer's Disease. *Biochemistry.* 2013 May 2. PubMed PMID: 23614767. Epub 2013/04/26. Eng.

In: Recent Advances in Proteomics Research
Editor: Ming D. Li

ISBN: 978-1-62948-218-7

Chapter 2

Neuroproteomics and Its Application in Drug Addiction Study

Ju Wang[1*] *and Ming D. Li*[2]
[1]School of Biomedical Engineering, Tianjin Medical University, China
[2]Department of Psychiatry and Neurobehavioral Sciences, University of Virginia, US

Abstract

Drug addiction is a chronic neurological disease. Because of its multigenic nature and the anatomic and cellular complexity of the brain, the underlying biochemical mechanisms are unclear. In recent years, proteomics technology has been applied to assess the changes in protein expression in the brains of animals and humans exposed to addictive drugs. The technique shows great promise in characterizing protein function, biochemical pathways, and networks to clarify the mechanisms of drug dependence. In this chapter, we first provide an overview of proteomics technologies and the bioinformatics tools available to analyze the data. Then, we summarize the recent application of proteomics to profiling protein expression changes in response to several drugs of abuse. Many biochemical pathways and biological processes are enriched among the proteins associated with exposure to one or more of these drugs, including those involved in protein degradation and modification, oxidative stress response, energy metabolism, and synaptic transmission and signaling pathways involved in neuronal functions.

Keywords: Proteomics, addiction, brain, pathways, nicotine

* Correspondence to: Ju Wang, PhD, School of Biomedical Engineering, Tianjin Medical University, Tianjin, China. Email: wangjujw@gmail.com

1. Introduction

Drug addiction is marked by compulsive use of a particular drug even in the face of serious negative consequences (Potenza et al. 2011). These behavioral abnormalities appear progressively during repeated exposure to a drug and often persist for a long time after discontinuation of drug consumption (Nestler 2004). Addiction can be attributed to multiple genetic, neurobiological, and psychosocial factors. Biologically, the addictive drugs function as reinforcing stimuli that contribute to their abuse liability. They exert their maladaptive effects in the brain, leading to further intake, drug craving, binge administration, and relapse.

The neurobiological substrates and the biochemical mechanisms underlying drug addiction are complex. The mesolimbic and mesocortical pathways of the reward system involving the ventral tegmental area (VTA), nucleus accumbens (NAc), amygdala, hippocampus, and frontal cortex are the key structures in addiction. Exposure to an addictive drug can alter the expression of various genes/proteins in these brain regions, which further mediates the functions of individual neurons and the related neural circuits (Wong et al. 2011; Renthal and Nestler 2009; Nestler 2001). Acute administration of drugs activates the mesolimbic dopaminergic system; the physiological effects are additionally mediated by the serotonergic-, opioid- and c-aminobutyric acid (GABA) systems in these brain areas, directly or indirectly. By contrast, chronic drug exposure leads to a decrease in neurotransmitter concentrations and receptor density in the same neuronal networks, eventually causing a decline in function. Besides the alterations of multiple neurotransmitter systems, a series of receptor-meditated signal transduction pathways are modulated. At the cellular level, drug addiction can be considered a form of synaptic dysregulation and pathology (Russo et al. 2010; Tan et al. 2011; Luscher and Malenka 2011). For example, significant morphological and electrophysiological disturbances are found in mesolimbic brain structures after cocaine administration.

Much has been learned about the genes and pathways involved in drug addiction, but the detailed cellular and molecular mechanisms, as well as their interactions that predispose to drug abuse/dependence and neurotoxicity in the central nervous system (CNS), are far from clear. Thus, identification of the genes/proteins related to the development and maintenance of drug dependence is not only critical to the understanding of the molecular mechanisms underlying addiction, but also imperative for the development of new pharmacological means to reverse the addictive state, prevent relapse, or reduce the use of these drugs (Larsson and Engel 2004).

Recent advances in genomics technology, especially the advent of high-throughput screening technologies, along with the development of bioinformatics, are making it possible to analyze simultaneously a large set of genes or proteins expressed in a particular cell line or tissue, which has greatly improved our capability to detect and identify molecular mechanisms related to a disease. These technical advances have facilitated the identification and characterization of the three major genetic units: the genome, the transcriptome, and the proteome. Whereas the genome is the entire set of genes encoded by the DNA of an organism, the transcriptome and the proteome consist of the ensemble of mRNA transcripts and protein forms expressed in a biological sample at a given time and in a particular situation. Both the transcriptome and the proteome differ from cell to cell and fluctuate in response to physiological signals, including developmental cues, stress, drugs, changes in the extracellular environment, and disease.

In transcriptome research, functional genomics methods, such as quantitative RT-PCR, large-scale microarray analysis, and next-generation sequencing technologies, provide sensitive and high-throughput approaches to quantitating the mRNA (and/or miRNA) species present in the cells of the nervous system and other tissues (DeRisi et al. 1996; Southern et al. 1999). These methods present an efficient way to map complex biological pathways and to identify novel genes under an experimental condition of interest. In recent years, numerous microarray studies have been conducted focusing on the effects of nicotine and other addictive drugs (e.g., alcohol, cocaine, morphine) on gene expression profiles (Dunckley and Lukas 2003; Konu et al. 2001; Konu et al. 2004; Li et al. 2002; Li et al. 2004; Zhang et al. 2001; Kerns et al. 2005; Bahi and Dreyer 2005; McClung et al. 2005; Rhodes and Crabbe 2005). There is no doubt that studying RNA expression changes in response to these addictive drugs have greatly helped illuminate the genes involved in the development of behaviors related to drug abuse and provided significant insight into the molecular mechanisms underlying the dependence-inducing properties of these substances (Li et al. 2002; Rhodes and Crabbe 2005; Pollock 2002; Yuferov et al. 2005). However, it is protein, not mRNA, that ultimately works as the molecular machine for cellular functions. Although protein expression is often correlated with mRNA abundance, the amounts of a protein in a cell cannot be predicted easily by the amount of mRNA due to protein stability and translation regulation. Also, analyzing mRNA expression cannot provide information on post-translational modifications or subcellular localization of the protein product. Thus, analyzing the mRNA and protein species in the same tissue provides related but different information, and both have unique value. Many pharmacologic and physiological effects of nicotine and other addictive substances in humans and animals are mediated, not only at the level of RNA, but also at the level of protein and/or post-translation. Although some genes have been demonstrated to be regulated by drug exposure at the mRNA level, it is difficult to infer the differences in protein expression. At the same time, there are numerous reports on the effect of addictive drugs on the regulation of an individual protein, but it is difficult to obtain a global protein expression pattern in a specific cell line or tissue based on this information alone. The recently developed proteomic techniques thus have been essential tools for drug addiction study.

Proteomics is a multidisciplinary and technology-driven science concerned with systematic, large-scale analysis of the amount, structure, and function of proteins in a biological system. The roots of proteomics go back to the 1980s, when the usual methods of protein identification were immunoblotting and co-migration with purified proteins in one-dimensional gel electrophoresis (Thrift et al. 1986). So far, two types of approaches have been employed in proteomics research to characterize proteins in large-scale production in the sample of interest: two-dimensional (2D) gel electrophoresis/mass spectrometry (MS) (Lewis et al. 2000; Gerner et al. 2000; Kanamoto et al. 2002; Predic et al. 2002) and protein arrays (Haab et al. 2001; Sreekumar et al. 2001; Miller et al. 2003; Nielsen et al. 2003).

Compared with conventional approaches, proteomic analysis allows the simultaneous assessment of a much larger number of proteins, creating a global view of the physiological condition investigated. Broadly speaking, proteomics has four main objectives: 1) to identify all proteins from a proteome, creating a catalogue of information; 2) to analyze differential protein expression associated with a disease, different cell states, and different treatments; 3) to characterize proteins by discovering their function, cellular localization, post-translational modifications (PTMs), etc.; and 4) to describe and understand protein-interaction networks.

Although proteomics is still an evolving field, proteomic approaches have been proved useful for elucidating the molecular effects of addictive drugs such as amphetamine, morphine, cocaine, nicotine, and alcohol. Thus, proteomic studies offer great promise for increasing understanding of the biochemical basis of addiction. However, given the multigenic nature of drug addiction, the anatomical and cellular complexity of the brain, as well as the relatively limited knowledge of the molecular pathology of these disorders, there are still many challenges. With more and more investigators intending to take advantage of proteomic techniques in drug addiction study, an understanding of the capabilities and limitations of proteomic analysis is essential. In the following sections, we briefly introduce the technologies used in addiction proteomics studies, and then summarize the information obtained from related analyses.

2. Proteome Dynamics and Complexity in Substance Abuse Research

The rapidly evolving field of proteomics aims to depict the characteristics and activity of every cellular protein. The proteome is clearly more complicated than the genome, as a single gene can encode multiple protein variants through alternative splicing of the mRNA transcript, use of alternative translational start and/or stop sites, and frame-shifting.

A surprising discovery of the genome projects for human, mouse, and rat was the small number of protein-coding genes (Waterston et al. 2002; Gibbs et al. 2004; Consortium 2004). Between 20,000 and 25,000 human protein-coding genes have been identified or predicted, and this figure represents less than 2% of the total genome sequence (Consortium 2004). Expression analysis using tiling arrays and the comprehensive characterization of transcriptional start and stop sites added new facets to this apparently low degree of complexity (Claverie 2005; Mendes Soares and Valcarcel 2006). Analysis of 1,000,000 expressed sequences of mouse genome revealed more than 181,047 transcripts, surpassing the number of predicted mouse genes by a factor of nearly 10 (Carninci et al. 2005). More than 56,000 protein-coding transcripts could be assigned gene names, including many that had previously been undetected. At least 65% of the transcriptional units were modified by splicing, and numerous of new splice variants were detected.

The highly dynamic proteome will require similarly dynamic quantitative models of protein pathways to capture an integrated cellular response to a stimulus. A cellular phenotype involves several dynamic processes that are influenced by environmental signals. Transitions in the transcriptome and proteome caused by substance abuse involve several modifications that include transcriptional cues and post-translational processes.

The complexity of the cellular proteome, depicted in its temporal and spatial dynamic nature, results from the multiple isoforms of many genes. An estimated ten protein isoforms can be generated from a single gene (Liebler 2002; Kim et al. 2004). Approximately 300 types of PTMs exist, examples being amino- and carboxy-terminal cleavage, phosphorylation, glycosylation, and myristoylation (Huber 2003; Garavelli 2004). The human proteome contains more than 100,000 putative phosphorylation sites, and about 50% of all proteins are believed to be phosphorylated by one of the more than 500 protein kinases (Manning et al. 2002). Considerable complexity also results from glycosylation. More than 2700 unique glycan structures are already known that reflect variations in the type, number, and position of individual sugar residues, the degree of branching, and the extent of acetylation, methylation, phosphorylation, and sulfation (Cooper et al. 2003). On average, a protein probably undergoes 2 to 20 PTMs (Fountoulakis 2004). Taking into account the much larger number of protein-coding transcripts than the predicted number of genes and the complexity of the brain transcriptome, several hundred thousand protein species are expected to exist in the nervous system. After drug treatment, the transcriptome is subject to several modifications, and modified mRNA transcripts encode different sets of proteins that are subject to further modification, such as phosphorylation/dephosphorylation by kinases and phosphatases, proteolytic processing, acetylation, and glycosylation. Proteins can also be cross-linked by transglutaminase or conjugated to small tags such as ubiquitin or SUMO. Post-translational modifications are important ways in which proteins acquire new functions or states in response to a specific cellular condition such as activation, turnover, downregulation, conformation, or localization (Morrison et al. 2002).

3. Proteomic Techniques

Efforts to make functional proteomics into a more useful and popular practice have led to a number of platforms for protein concentration assessment or functional characterization in high-throughput assays (Stoevesandt et al. 2009; Lee et al. 2013).

In most proteomic analysis, the procedure involves a two-stage strategy: protein separation and identification. Various workflows can be used for protein separation prior to identification with bioinformatics analysis. The most widely adopted separation method is two-dimensional electrophoresis (2DE), whereas the central platform for protein identification is tandem mass spectrometry (MS) (Roepstorff 2012). The predominant approaches are 2DE and subsequent MS, liquid chromatography-MS (LC-MS), and *gel-enhanced* liquid chromatography-MS (GeLC-MS). Multidimensional separations are required to minimize the number of peptides introduced into the mass spectrometer. Two-dimensional gel electrophoresis (2D-PAGE), first described in 1975 (O'Farrell 1975), is still the most common technique for monitoring changes in expression in protein mixtures. This method produces high-resolution separations according to the molecular mass and p*I* of the protein, resulting in the display of potentially thousands of spots (O'Farrell 1975). The following is a summary of key steps in the application of such proteomics technologies to substance abuse research.

3.1. Overview of 2D-PAGE Analysis

During the past few years, many technical issues that initially limited proteomics have been addressed. Important technologic advances include the use of immobilized pH gradients for isoelectric focusing and fluorescent stains for protein imaging [for reviews, see (Alban et al. 2003; Gorg et al. 2000)]. Similarly, sophisticated and high-throughput procedures are now available for protein identification. Peptide mass fingerprinting by MS and tandem MS (MS/MS) produces partial amino acid sequences of chosen proteins that can be used to identify protein spots from 2D gels through database searching [reviewed in (Link et al. 1999)]. These techniques can be optimized by focusing on a specific pH or size range (zoom gels) and separating proteins of interest from bulk proteins before electrophoresis. However, 2D-PAGE can be limited by detection sensitivity and protein solubility. High-abundance proteins generally greater than 120 kD are favored, whereas integral membrane or basic proteins may be difficult to resolve (Resing and Ahn 2005). Currently, 2D gel electrophoresis is limited to analysis of a few hundred to approximately 2,000 of the most abundant proteins in a given sample (Fountoulakis 2004; Williams et al. 2004; Witzmann and Strother 2004), which is only a small fraction of the probable size of a proteome.

Two-dimensional protein maps have been constructed for the whole brain of mouse (Gauss et al. 1999), rat (Fountoulakis et al. 1999), and human (Langen et al. 1999) to analyze the entire brain proteome systematically. More than 200 discrete proteins have been identified. The significant degree of overlap in the proteins identified in various studies implies that the techniques are reliable and reproducible, at least with respect to the more abundant protein species.

3.2. Biological Samples

Brain tissue is the most direct means to investigate protein changes associated with drug dependence. Proteomic analysis of postmortem human brain tissue from drug-abusing subjects provides valuable translational information, and several excellent studies have been published (Lull et al. 2010; Kashem et al. 2009; Kashem et al. 2008; Kashem et al. 2007; Tannu et al. 2007). However, such studies are technically difficult because of small amount of tissue available and factors such as variations in genetic background, drug abuse history, and comorbid conditions. Also, tissue deteriorates rapidly after death (Franzen et al. 2003; Lehrmann et al. 2008) and subject to delayed retrieval (Whitehouse et al. 1984). Thus, optimization of sample preparation can be difficult.

The second sample source is cell culture. The main advantage of using cultures is the ability to observe responses to stimuli in the absence of the complexity of the entire organism. Furthermore, every replicate sample from subculture will be virtually identical; therefore, examination of the influences induced by some stimuli and comparison with the control culture will be relevant and trustworthy. Moreover, reagents can be administered directly at a defined concentration, avoiding one of the major problems with *in vivo* experiments. Also, culture techniques ensure significant control of the environment. The entire model can be well controlled, and thus, observation of dynamic changes in protein fingerprints over time or after addition of a drug is much easier than in whole tissues (Bodzon-Kulakowska et al. 2009; Suder et al. 2009).

However, cell culture may not reflect processes as they occur in an intact animal. Because of the limitations of post-mortem brain and cell culture, animal models have been used extensively in drug-dependence studies. Here, brain tissue can be collected in a controlled environment and processed immediately to minimize protein degradation (Lull et al. 2010). Furthermore, a large number of samples can be harvested reproducibly from defined anatomic regions. In the case of addiction research, brain regions associated with these phenomena are of primary interest. With animal models, the complexity of the whole system is maintained, while the protein content and its abundance and diversity are extended. However, problems persist. For example, low-abundance proteins may not be detectable, and prefractionation and enrichment procedures might be necessary.

3.3. Biological Variability

Quantitative proteomics investigates molecular physiology by measuring differences in protein expression in samples subjected to different experimental conditions or different external stimuli (e.g., control vs. drug-treated samples). A major obstacle in determining quantitative changes in protein expression reliably is error imposed by technical variation (e.g., “procedure noise”) and biological variation (e.g., genetic and epigenetic variation). In gaining insight into mechanisms of action of drugs and profiling biomarkers, reproducibly measuring the myriad protein-expression changes across numerous multivariate experiments is a key challenge (e.g., multiple animals, doses, drugs, and time points).

Understanding the degree of experimental variation is central to all quantitative proteomic investigations. Two-dimensional gel electrophoresis is currently the most robust high-throughout method for this purpose. Replicate 2D gels of the same sample show a spot volume correlation coefficient of about 0.85 (Molloy et al. 2003). For the five sample types investigated, the average coefficient of variation (CV = SD/mean × 100%) for spot volumes from replicate gels is in the range of 18-27%. A previous study by Blomberg et al. (1995) showed average CVs from three laboratories for 470 matched proteins from the same sample to be 20-28% in replicate experiments using 2D gels with immobilized pH gradients (IPGs) and autoradiography.

A study using an assortment of samples, methods, and formats indicated an average baseline technical variation from the 2-DE separation and detection process of 20-30%. Thus, a sample size of three or four would be sufficient to detect a two-fold change in protein expression with statistical confidence of 80% power and a P value of 0.05.

3.4. Experimental Design and Sample Size

Experimental neuroproteomics design must account for biological variability, whatever the type of sample. However, there currently is a lack of structured procedures to deduce sample quality and sample size or a means to calculate confidence intervals, as is routinely done in genomics experiments (Boguski and McIntosh 2003). Biological variance is determined largely by sample heterogeneity; and biological heterogeneity is usually measured by CV, which increases from 20-26% for cultures of mammalian cell lines to 40-50% for primary cell cultures (Molloy et al. 2003). Animal samples and biofluids would show even more sample variance because of individual differences between animals, which must be minimized by controlling such factors as the treatment, animal lineage, age, sex, environmental exposure, and diet.

3.5. Protein Preparation and Separation

Protein must be extracted before analysis, which begins with homogenization of tissue or cells. A solubilization buffer, usually with surfactants, is added to lyse cell membranes and extract the proteins. Samples must then be cleansed of contaminants such as cellular debris through centrifugation and filtration (Ueyama et al. 2003). If one uses protein extracted from plasma, serum, or cerebrospinal fluid, high-abundance housekeeping proteins such as albumin (>50%) and immunoglobulins (>15%) account for 95% of the sample and often mask the less-abundant proteins of interest (Anderson and Anderson 2002; Ahmed et al. 2003; Zhang et al. 2005). Several methods have been described for removing housekeeping proteins without affecting the proteins of interest, including membrane filtration, precipitation, affinity resins, and chromatography. The highly selective affinity methods are the most promising, although they tend to remove also some less-abundant or albumin-bound proteins that are desired (Ahmed et al. 2003; Mehta et al. 2003). New technologies are needed to address these issues. Also, salts, ionic detergents, lipids, and nucleic acids can interfere with proteomic analysis (Jiang et al. 2004). High salt concentrations should be avoided because they disrupt gel electrophoresis and MS. Because sodium dodecyl sulfate (SDS) and other ionic detergents are generally incompatible with or tolerated only in a small amount by isoelectric focusing in 2D-PAGE and MS, non-ionic detergents (e.g., Triton X-100 or Nonidet P-40) at sufficiently low concentrations (1% is common) should be used as extraction buffers.

Gel electrophoresis continues to be a powerful tool for protein separations. One-dimensional SDS-PAGE is often used for immunoblotting, a highly selective technique that uses protein-specific antibodies to detect the proteins of interest. However, immunoblotting is limited to proteins against which antibodies are available and generally is reserved for a focused study of a particular protein. Multi-dimensional separations can resolve complex biological mixtures more effectively than any 1D separation. Two-dimensional PAGE was the first multi-dimensional protein separation method (O'Farrell 1975) and can resolve thousands of proteins to enable large-scale protein identification and quantification through MS (Lilley et al. 2002).

In 2D-PAGE, proteins are first separated by p*I* through isoelectric focusing (IEF) and then resolved by SDS-PAGE. Immobilized pH gradient strips have become the standard for IEF, in which weak acids and bases are copolymerized in a polyacrylamide gel. In an IPG strip (commercially available in broad or narrow pH ranges), proteins migrate in an applied electric field until they reach a zone where their net charge is neutral, revealing their p*I* value (Freeman and Hemby 2004). In the second separation, the IPG strip is equilibrated with SDS to provide a uniform protein charge and then mounted on a polyacrylamide gel for mass resolution.

Through 2D-PAGE, considerable information can be obtained about a protein, including the isoelectric point, relative molecular mass (*Mr*), relative quantity, and PTMs. The p*I* and *Mr* values can identify the protein by searching 2D brain protein map databases (Fountoulakis 2004). Despite acceptance of 2D-PAGE as the benchmark in protein separations, it is impossible to visualize an entire proteome in this way without a supplemental technique because of the difficulty it has in resolving hydrophobic, alkaline, high- or low-mass, and low-abundance proteins (Herbert et al. 2001).

3.6. Protein Identification

Protein identification for neuroproteomics has been done primarily with 2D-PAGE brain protein map databases, which are available for different species, notably the rat (Krapfenbauer et al. 2003; Taoka et al. 2000), mouse (Tsugita et al. 2000; Beranova-Giorgianni et al. 2002), and human (Yang et al. 2004). Recently, 2D-PAGE has been coupled with matrix-assisted laser desorption/ionization time-of-flight mass spectrometry (MALDI-TOFMS) in drug-dependence research. Although MALDI has been available for 20 years, it became popular for protein analysis only in the late 1990s (Nguyen et al. 1995; Fenselau 1997). Peptide fragmentation patterns can augment peptide mass mapping for protein identification. With tandem MS, peptides of interest can be isolated and fragmented into often-predictable pieces that are used for sequence correlation. Although this approach has been employed widely in proteomics research, it is still plagued by difficulties with sample complexity and dynamic change. The most elaborate separation schemes cannot handle the immense complexity of biological samples and identify only a subset of the predominant proteins.

3.7. Protein Quantification

Quantitative analysis in neuroproteomics has been used primarily for differential expression profiling by 2D-PAGE for biomarker discovery (Fountoulakis 2004; Brewis and Brennan 2010). For example, control and drug-treated samples from animal models, clinical material, or cell cultures can be resolved by 2D-PAGE and then visually compared using different staining methods. However, a major limitation of 2D-PAGE for protein quantification is gel-to-gel variability (Choe and Lee 2003). To overcome this limitation, 2D differential gel electrophoresis (2D-DIGE) was developed in which two samples are distinguished by different cyanine fluorescent labels (Tonge et al. 2001; Lilley and Friedman 2004). The 2D-DIGE method minimizes gel-to-gel reproducibility concerns by resolving two samples in the same gel. Gels can be compared by using a third dye as an internal control and pooling equal amounts of all the samples. Every spot thus would be represented in the standard, improving quantification and the power of statistical analysis for various gels. Despite the advantages of 2D-DIGE over 2D-PAGE, there are some disadvantages; e.g., the expression of cytosolic protein is favored above membrane proteins by 2D electrophoresis (Wu and Yates 2003), and gel-to-gel variation cannot be totally eliminated.

4. Bioinformatics

Protein separation and MS data must be assembled and processed to provide biologically relevant information. In bottom-up proteomic methods, peptides and their fragment mass-to-charge values must be matched with theoretical peptides from protein- or gene-sequence databases. Protein data are collated in non-redundant public archives and can be organized into tissue- and species-specific databases to reduce analysis time and false-positives/negative results. Bioinformatics tools for proteomics span a large panel of diverse applications, from simple tools to compare amino acid compositions to sophisticated software for large-scale protein-structure determination. Herein, we concentrate on describing bioinformatics tools for 2-DE analysis, protein identification by peptide mass fingerprinting (PMF) and peptide fragment fingerprinting (PFF), and protein quantitation with MS data.

4.1. 2DE Gel Image Analysis

Various software programs are available for 2DE gel image analysis, which include DeCyder and ImageMaster 2D Platinum from GE Healthcare, Delta2D from Decodon, PDQuest from BioRad, Progenesis from Nonlinear Dynamics, and Proteomweaver from BioRad, to name a few (Table 1). Although each software package has its own strengths and weaknesses, depending on the gel type and experimental conditions, almost all have the same basic operations and functionalities necessary to carry out a complete gel study, which should end up highlighting differentially expressed spots in a population of gels (Millioni et al. 2010).

Besides basic visualization properties, the major functions of software for a 2DE gel image analysis system include: 1) the detection and quantification of protein spots on the gel; 2) the matching of corresponding spots across the gels; and 3) the localization of significant changes in protein expression (Dowsey et al. 2003).

Other features such as data management, integration, and analysis may or may not be included. Given that different packages utilize different sets of user-adjustable parameters plus a complicated technique of image analysis, only a few studies have been conducted to evaluate the performance of each package. For example, the software PDQuest, Melanie 3 (called ImageMaster 2D Platinum since version 3), Phoretx, Progenesis, and Z3 performed well in all evaluations (Rogers et al. 2003). Three other comparative studies showed equivalent results for PDQuest and Progenesis (Rosengren et al. 2003), PDQuest and Phoretix (Wheelock and Buckpitt 2005), and Melanie and Z3 (Raman et al. 2002). Thus, the choice of a software program depends on the availability of the 2DE system and software and the preference of each researcher or research group.

Table 1. Tools available for 2DE gel image analysis

Software name	Developer	Web link
DeCyder 2D	GE Healthcare	http://www.gelifesciences.com
ImageMaster 2D Platinum	GE Healthcare	http://www.gelifesciences.com
Delta2D	Decodon	*https://www.decodon.com/delta2d.html*
PDQuest	Bio-Rad	http://www.bio-rad.com/
Proteomweaver	Bio-Rad	http://www.bio-rad.com/
Progenesis Samespots	Nonlinear Dynamics	http://www.nonlinear.com/
BioNumerics 2D	Applied Maths	www.applied-maths.com
Melanie	Geneva Bioinformatics	http://www.genebio.com
REDFIN	Ludesi	www.ludesi.com/redfin

4.2. PMF and PFF Analysis

One of the main objectives of proteomics is to separate and identify particular proteins. Typically, one protein or a mixture is digested enzymatically, and the resulting peptide masses are measured, producing an MS spectrum (called a PMF). Peptides also can be isolated and fragmented with the mass spectrometer, leading to MS/MS spectra (also called PFF). In PMF analysis, the experimental spectrum is compared with theoretical ones computed from protein sequences stored in databases and *in silico* digested using the cleavage specificity of the protease employed in the experiment. In a simplistic view, the procedure roughly counts masses overlapping between the experimental and theoretical spectra, leading to a "similarity score" for each candidate protein. The candidate proteins are then stored according to their scores, and the top-ranked protein is considered the identification of the spectrum. The key step lies in the scoring function, and various scoring schemes have been implemented in different algorithms. Table 2 provides a summary of a number of PMF analysis tools. A detailed description of these search tools and their corresponding algorithms is beyond the scope of this chapter but is available in recent reviews (Hernandez et al. 2006; Gras and Muller 2001).

The PFF approach is similar to the PMF technique but is applied to MS/MS spectra and hence correlates peptide spectra with theoretical peptides from a database. Except for this fundamental difference, the search algorithm for the PFF analysis is almost identical to the PMF approach; i.e., the theoretical MS/MS spectra are computed from the theoretical peptide sequences and correlated with the experimental MS/MS spectrum in order to find the candidate peptide with the highest score. Compared with PMF analysis, the MS/MS-based PFF approach offers several advantages: 1) it provides significant peptide coverage, with detailed information about the peptide sequence and possible modifications; 2) it is possible to work with complex peptide mixtures or to search homologous databases; and 3) it does not require all the peptide of a given protein to be confirmed to achieve identification. Some representative PFF analysis tools are listed in Table 3 in the hope that this information can aid researchers in their choice of methods.

Regardless of whether PMF or PFF analysis is used in protein identification, the following parameters should be considered. There is no question that the similarity score and the corresponding *p* value obtained with classic criteria can serve as useful clues in protein identification. However, they cannot substitute for the careful interpretive analysis necessary to distinguish a false- from a true-positive result. The second parameter is the percent coverage (i.e., the proportion of a theoretical protein covered by the MS data of the protein spot). According to many proteomics researchers, more than 20% coverage is likely to be significant (Gasteiger et al. 2005; Barrett et al. 2005). Even so, this criterion is not enough to avoid the risk of false-positive identifications. For most programs, the percent coverage is given mainly as a criterion to help researchers find the best theoretical protein-matching with the observed MS data. Finally, the p*I* and MW should be relatively close to the theoretical values for species with complete genome sequences in databases. For cross-species identification, an M_r variation of ± 30 and a p*I* variation of ± 2.0 are generally used (Gasteiger et al. 2005; Barrett et al. 2005; Habermann et al. 2004).

Table 2. List of protein identification software for PMF analysis available on Internet

Software mame	Parameters available for search	Databases available	Web link
Aldente	p*I*, MW, taxon, digestion, modifications (variable and/or fixed), missed cleavages, thresholds of spectrometer, and peptide scorning	Swiss-Prot, TrEMBL	http://www.genebio.com/products/phenyx/aldente/
MASCOT	MW, taxon, digestion, modifications (variable and/or fixed), missed cleavages, thresholds of spectrometer, peptide scoring	MSDB, NCBI, Swiss-Prot, Random	http://www.matrixscience.com/
Prospector	p*I*, MW, taxon, digestion, modifications (variable and/or fixed), missed cleavages, thresholds of spectrometer, peptide scoring	Genpept, NCBI, Swiss-Prot, EST_mouse, EST_human, EST_others	http://prospector.ucsf.edu
PeptideSearch	p*I*, MW, taxon, digestion, modifications (variable and/or fixed), missed cleavage, thresholds of spectrometer, peptide scoring	NRDB	http://newt-omics.mpi-bn.mpg.de/Peptidesearch.php
ProFound	p*I*, MW, taxon, digestion, modifications (variable and/or fixed), missed cleavage, thresholds of spectrometer, peptide scoring	NCBI	http://prowl.rockefeller.edu/prowl-cgi/profound.exe

Table 3. List of protein identification software for PFF analysis available on Internet

Software name	Parameters available for search	Databases available	Web link
MASCOT	Taxon, digestion, modifications (variable and/or fixed), missed cleavages, thresholds of spectrometer, type of spectrometer, peptide scoring	MSDB, NCBI, Swiss-Prot, Random, EST_mouse, EST_human, EST_others	http://www.matrixscience.com/
Phenyx	Taxon, digestion, modifications (variable and/or fixed), missed cleavages, thresholds of spectrometer, type of spectrometer, peptide scoring	TrEMBL, NCBI, Swiss-Prot, MSDB, EST_mouse, EST_human, EST_others	http://www.genebio.com/products/phenyx/
OMSSA	Taxon, digestion, modifications (variable and/or fixed), missed cleavages, thresholds of spectrometer, type of spectrometer, peptide scoring	MSDB, NCBI, Swiss-Prot, Random, EST_mouse, EST_human, EST_others	http://www.ionsource.com/functional_reviews/OMSSA/OMSSA_Rev.htm
PeptideSearch	MW, taxon, digestion, modifications (variable and/or fixed), missed cleavage, thresholds of spectrometer, peptide scoring	NRDB	http://newt-omics.mpi-bn.mpg.de/Peptidesearch.php
PepFrag	Taxon, digestion, modifications (variable and/or fixed), missed cleavage, thresholds of spectrometer, type of spectrometer, peptide scoring	NCBI, dbEST	http://prowl.rockefeller.edu/prowl/pepfrag.html

4.3. Quantitative Analysis

Several strategies are available for relative quantitation of proteins in different samples. The labeling of proteins or peptides followed by MS is a rapidly advancing approach. Within this category, the isotope-coded affinity tags (ICATs) and isotope tags for relative and absolute quantification (iTRAQ) are commonly used in practice (Gygi et al. 1999; Ross et al. 2004). For the ICATs, samples from a control and an experimental condition are each tagged with a stable isotope, combined, digested, and subjected to MS analysis. By labeling each sample with its own isotope, the relative ratio of the peptides of both conditions can be compared in the same MS run. On the other hand, rather than labeling cysteine-containing peptides, primary amines are labeled in the iTRAQ method, which allows tagging of most tryptic peptides. In addition, whereas ICAT labels only two conditions, iTRAQ allows as many as four conditions, making it ideal for experiments in which more than two experimental groups need to be compared. Although the labeling step differs in the two approaches, the overall experimental procedures are similar and can be summarized as: 1) isotopic labeling of separate protein mixtures; 2) combined digestion of the labeled proteins followed by multi-dimensional liquid chromatographic separation; 3) automated MS/MS of the separate peptides; and 4) automated database searching to identify the peptide sequences and quantify the relative protein abundance based on the MS/MS.

5. Challenges and Limitations of Neuroproteomics Approaches

There are many things that can be done with proteomics in substance abuse research: detection of the brain proteome, comparison of protein expression profiles, PTM profiling, and mapping protein-protein interactions. All of them have their strengths and limitations, and one of the major challenges ahead is to determine the most appropriate application of proteomics technology to the system. Although the 2DE-based approach is one of the most reliable in quantitative proteomics, it can routinely identify only several hundred to a few thousand protein species (Freeman and Hemby 2004; Gorg et al. 2004). The upper limit of 2DE resolution might be close to 10,000 individual spots per gel (Gorg et al. 2004; Klose et al. 2002), but no method is available to annotate such a large number of spots on a single gel. In addition to this discrepancy between proteome size, resolution power, and identified proteins, 2DE has some severe limitations on its use in neuroproteomics. For example, it requires large amounts of protein and fails to separate the important class of hydrophobic membrane proteins (Santoni et al. 2000). Furthermore, high- and low-molecular weight proteins are hard to display (Gorg et al. 2004; Lubec et al. 2003). Of all the rat proteins stored in the Swiss-Prot database, simply because of their MW and *pI*, roughly 13% would not appear on a gel with a pH from 3.0 to 10.0 in the first dimension and from 14 to 200 kDa in the second dimension (Becker et al. 2006). Another limitation of 2DE is the low dynamic range of the amount of protein that can be displayed. Most often, only a small and very similar subset of the proteome is detected by 2DE. This corresponds to high- to medium-abundance proteins: low-abundance proteins are not amenable to the analysis (Gygi et al. 2000). To overcome this limitation, the use of narrow pH-gradient gels has been suggested

(Westbrook et al. 2001). However, this technique requires huge amounts of sample. Finally, an emerging critical issue is the presence of spot fusion and comigration of different protein species to create a single gel spot.

6. Protein arrays

Although the proteomic investigation mainly relies on electrophoresis, liquid chromatography, and mass spectrometry, the protein array has emerging as a powerful approach (Lee et al. 2013). Compared with a 2D gel electrophoresis platform, the protein array has a relatively short history. The development of protein arrays benefited from DNA microarray technology, in which hundreds or thousands of transcripts can be measured in a single experiment, and the equipment needed to manufacture microarrays became more accessible.

Although the protein arrays were not applied in biomedical research until the late 1990s, the concept of the protein array is not new. Miniaturized, multiplexed, solid-phase immunoassays were developed in the late 1980s using protein microdots spotted manually onto nitrocellulose sheets or other solid support (Ekins et al. 1990). One of the earliest reports demonstrating the feasibility of multianalyte, microarray-based immunoassays was published by Silzel and associates in 1998 (Silzel et al. 1998). Since then, several studies have been published, with most concentrated on the analysis of human cytokines (Wiese et al. 2001; Jenison et al. 2001; Tam et al. 2002; Schweitzer et al. 2002). There are two reasons for this focus. First, matched pairs of high-affinity, high-specificity antibodies have been commercialized. The availability of such reagents circumvents the largest hurdle in developing microarray-based assays. Second, cytokines are secreted from cells; thus, analyses are done on simpler mixtures than cellular lysates.

Although protein arrays present an innovative and versatile approach to studying protein abundance and function on an unprecedented scale, the development of the protein array has been more challenging than building DNA arrays. There is no doubt that protein function arrays will benefit from better methods of fabrication, processing, and analysis, but at present, the greatest obstacle to their widespread use is the lack of large collections of pure recombinant proteins.

Most available antibodies do not function well in the array format: it is estimated that only 30% of commercially available antibodies can be used qualitatively and even fewer (ca. 20%) can be used quantitatively (Haab et al. 2001). Thus, it is not surprising that successful protein arrays have been restricted to cytokines. Furthermore, even if the limited availability of antibodies is not an issue, protein arrays will not be applicable to all kinds of samples. For example, if one is interested in identifying proteins with altered abundance in an experimental sample relative to a reference sample, the 2D gel electrophoresis platform should be used, because this approach is unbiased, and the investigators do not impose their knowledge of biology on the experimental design. On the other hand, if one is interested in studying a particular biological process, only those proteins involved in that process need to be examined. Under such a circumstance, a protein array may work better if appropriate affinity reagents are available. Therefore, the protein array is best suited to system-oriented investigations.

7. Recent Applications of Neuroproteomics in Research on Nicotine and Other Substances of Abuse

At present, a comprehensive genome-wide proteome analysis in drug addiction remains beyond our capability because of its extreme complexity and the limits of the current technology. A more feasible approach is to focus on substructures, such as the brain and its components. During recent years, proteomics has been used extensively to profile the protein expression pattern in cultures of neurons or in different brain regions of animals or humans in response to substances of abuse such as alcohol, amphetamine, cocaine, morphine, and nicotine. On the other hand, the application of proteomics technology to the study of a specific addictive drug is still limited, which makes a thorough and detailed summary for a single substance unlikely. Thus, we attempt here a collective summary of the application of proteomics techniques to studies of different abused substances so that we can have a better understanding of the proteins and pathways commonly modified by abused substances. In the following sections, we first discuss the current reports on the application of proteomics in studies of several addictive substances and then focus on the key proteins identified in these studies and discuss their relevance to drug addiction.

7.1. Nicotine

Although only a few proteomic studies on nicotine addiction have been reported, much information on the biochemistry and biological processes underlying nicotine dependence has been obtained. Chronic nicotine administration (7 days) significantly regulated the expression of seven proteins in the striatum of male Holtzman rats (Yeom et al. (2005)). Of these proteins, three were induced by nicotine; i.e., zinc-finger binding protein-89 (ZBP-89), cyclic nucleotide phosphodiesterase (CNP1), and deoxyribonuclease 1-like 3 (DNASE1L3); whereas the other four proteins; i.e., tandem pore domain halothane-inhibited potassium channel 2 (THIK2), brain-specific hyaluronan-binding protein (BRAL1), death effecter domain-containing protein (DEDD), and brain-derived neurotrophic factor (BDNF), were suppressed. The expression patterns of most proteins were coincident with their mRNAs. This work demonstrated the potential ability of the proteomic approach to identify novel markers related to nicotine addiction. In a more comprehensive study (Hwang and Li 2006), protein expression profiles were analyzed in five brain regions: the amygdala, NAc, prefrontal cortex (PFC), striatum, and ventral tegmental area (VTA) of rats that received nicotine for 7 days. The expression patterns seen were region specific. Several proteins (e.g., dynamin1, laminin receptors, aldolase A, SNAP-beta, and *N*-ethylmaleimide-sensitive fusion protein) were differentially expressed in multiple brain regions. These proteins can be grouped into several biological process categories, such as energy metabolism, synaptic transmission, and oxidative stress response. Although nicotine addiction is caused primarily by neural adaptations to nicotine in the CNS, nicotine causes alterations in other tissues also (Benowitz 1996). Because of their ready availability, peripheral tissues have been used in proteomic studies to infer the status of the CNS. In one study, 15 proteins were identified in serum and polymorphonuclear and mononuclear leukocytes from rats receiving chronic nicotine (14 days) (Piubelli et al. 2005). The regulation of these proteins was suggested to be related to activities of nicotine in the CNS. In a study of extinction and relapse, serum proteomic analysis in rats self administering nicotine revealed specific protein patterns (Cecconi et al. 2008).

7.2. Alcohol

Proteomic approaches have been used extensively to study alcohol addiction in both animal models and humans. Differences in protein expression were identified in two regions (amygdala and NAc) of alcohol-naïve inbred alcohol-preferring and -non-preferring rats (Witzmann and Strother 2004; Witzmann et al. 2003; Bell et al. 2006). Breeding for specific drinking behaviors produced alterations in the expression of several proteins that could influence neuronal function within the NAc and amygdala. The data imply differences in the mechanisms of synaptic transmission in the NAc of alcohol-preferring and -non-preferring rats. Changes in protein expression in the amygdala and NAc of alcohol-preferring rats given either continuous or scheduled access to ethanol over 6 weeks suggested that proteins related to intracellular signal transduction, cytoskeletal structure, metabolism, response to stress, and synaptic transmission were associated with ethanol self-administration in these two brain structures (Bell et al. (2006)). The effects of ethanol on protein expression in the NAc shell of alcohol-preferring, alcohol-non-preferring, and Wistar rats (McBride et al. 2009) indicated a unique response in the non-preferring animals. Moreover, chronic alcohol treatment for 4 weeks alters protein expression in the brains of zebrafish (Damodaran et al. 2006). Those investigators found eight proteins to be differentially regulated by alcohol, namely, voltage-dependent anion channel proteins (VDAC1 and VDAC2), heat shock protein 70 (HSP70), alpha subunit of G_o (GNAO1), and subunit A of the catalytic domain of H^+ transporting ATPase (ATP6V1A1). For Park et al. (2004), the different responses to alcohol seen in inbred mouse strains C57BL/6J and DBA/2J may reflect differences in the response rates and interactions of variants of alcohol-responsive proteins. Sari et al. (2010) analyzed the protein expression pattern in the brains of fetal C57BL/6J mice exposed to ethanol and found significant downregulation of several proteins important for energy generation, apoptosis, cellular signaling, and neuronal migration.

Some proteomic studies have been conducted on alcoholic human brain tissues. Lewhol et al (2004) applied such techniques to autopsy tissue and saw 182 significant changes in the superior frontal cortex, most of them reflecting lower expression in alcoholic brains. Also, more than 60 altered proteins were identified, confirming the potential of the technique.

Other regions involved in neurocognitive function have been studied by proteomic methods. Two thiamine-dependent proteins, transketolase and pyruvate dehydrogenase E1 beta-subunit (PDHB), proved to be differentially expressed in the dorsolateral PFC in alcoholics (Alexander-Kaufman et al. 2006; Alexander-Kaufman et al. 2007). A large number of metabolism-related enzymes, in particular, ones involved in energy transduction, also are differentially expressed in this region. In the cerebellar vermis of uncomplicated and cirrhotic alcoholics (Alexander-Kaufman et al. 2007), many proteins were found to be differentially expressed in the cirrhotics, implying that liver dysfunction strikingly affects protein expression (Matsumoto 2009). Changes also were seen in the splenium, genu, and body of the corpus callosum in the brains of alcoholics. Proteins with differential expression were found in all of these subregions, among them glutamate carboxypeptidase (GCP1) and phospholipase D (PLD) (Kashem et al. 2009; Kashem et al. 2008; Kashem et al. 2007). Expression profiles also identified the region-specific proteins in these areas. About 20 differentially expressed proteins were seen in the hippocampus of alcoholics (Matsuda-Matsumoto et al. 2007), among them glutamine synthetase, predominantly an astrocyte protein.

7.3. Morphine/Heroin/Butorphanol

Morphine and similar drugs exert their diverse actions through interaction with opioid receptors. Most of the proteomics investigations on these drugs were made with animal models or cell lines. Because of their importance in drug dependence, brain regions such as the NAc (Li et al. 2006), frontal cortex (Kim et al. 2005; Bierczynska-Krzysik et al. 2006), hippocampus (Bierczynska-Krzysik et al. 2006), and striatum (Bodzon-Kulakowska et al. 2009; Bierczynska-Krzysik et al. 2006) are most often examined, although whole brain (Bierczynska-Krzysik et al. 2006), spinal cord (Shui et al. 2007), dorsal root ganglia (Li et al. 2009), or cultured neuronal cells (Suder et al. 2009; Neasta et al. 2006) also have been studied. Moreover, proteins in the synapses have been analyzed directly (Moron et al. 2007; Van den Oever et al. 2010; Prokai et al. 2005; Yang et al. 2007).

Given the availability of comprehensive discussions of the proteins associated with morphine use (Bodzon-Kulakowska et al. 2011; Bodzon-Kulakowska et al. 2005), we will confine our discussion to an overview of the subject. Most of these proteins fall into one of several major functional categories, namely protein modification, energy metabolism, cell cytoskeleton, signal transduction, and vesicular trafficking; many others are involved in synaptic plasticity, learning, and memory. Similar patterns are observed among the phosphotyrosyl proteins in the frontal cortex of morphine- or butorphanol-dependent rats (Kim et al. 2005; Kim et al. 2004).

7.4. Amphetamine/Methamphetamine

All proteomic analyses of the effects of amphetamine/methamphetamine on the central nervous system (CNS) were done with animal models. Freeman et al. (2005) analyzed the amphetamine self-administration transitional states in the hippocampus of rats. Actin- and cytoskeletal-related proteins were over-represented among the proteins changed by abstinence from the drugs. After acute methamphetamine administration in rats, there were significant alterations in proteins involved in energy metabolism, oxidative stress response, or the ubiquitin-proteasome pathway (Iwazaki et al. 2006). The same investigators analyzed the amygdala and striatum of rats with methamphetamine-induced behavioral sensitization (Iwazaki et al. 2007; Iwazaki et al. 2008), finding that the proteins of the amygdala react differently to acute and chronic methamphetamine administration. Synaptic, cytoskeletal, oxidative stress, apoptosis, and mitochondria-related proteins were differentially expressed in the amygdala of sensitized animals. In the striatum, 31 such proteins were identified in methamphetamine-sensitized compared with control rats, such as synapsin II, synaptosomal-associated protein 25 (SNAP25), and adenylyl cyclase-associated protein 1 (CAP1). Acute methamphetamine exposure caused differential expression of 82 proteins, with 40, such as collapsin response mediator protein-2 (CRMP-2), superoxide dismutase 1 (SOD 1), phosphatidylethanolamine-binding protein-1 (PEBP-1), and mitogen activated kinase kinase-1 (MKK-1), being reduced and the other 42, including authophagy-linked microtubule-associated protein light chain 3 (LC3), synapsin-1, and Parkinsonism-linked ubiquitin carboxy-terminal hydroxylase-L1 (UCHL1), being induced (Kobeissy et al. 2008). After eight methamphetamine injections, 47 altered cytosolic and 42 altered membrane proteins were identified in the rat frontal cortex, the altered proteins being associated with protein degradation, redox regulation, energy metabolism, cellular growth, cytoskeletal modifications, and synaptic function (Faure et al. 2009). Li et al. (2008) studied global protein profiles associated with methamphetamine-induced neurotoxicity in the striatum, hippocampus, and frontal cortex. Alpha-synuclein was induced, and mitochondrial ATP synthase D chain had reduced expression in all brain regions. Two proteins, Cu/Zn SOD 1 and subunit of mitochondrial H-ATP synthase, also were reduced. The results suggested that methamphetamine neurotoxicity involves the same pathways in different brain regions. In an exploration of the effect of methamphetamine in the mesolimbic dopamine system of rats, Yang et al. (2008) found that a conditioned place preference was induced by the drug. Protein profiles in the NAc, striatum, PFC, cingulate cortex, and hippocampus were compared. Further screening identified 27 proteins that were differentially expressed, including those involved in the cytoskeleton, transport/endocytosis or exocytosis, and signal transduction.

7.5. Cocaine

Comprehensive reviews of the findings of proteomics and other high-throughput studies of cocaine addiction are available (Hemby 2010; Lull et al. 2008), so we provide only a brief review. Hemby and colleagues (Tannu et al. 2007; Hemby 2006) analyzed the protein profile in the NAc of cocaine overdose victims. Among the proteins differentially expressed in the overdose and control groups were beta-tubulin, liprin-alpha3, and neuronal enolase, with decreases in parvalbumin, ATP synthase beta-chain, and peroxiredoxin 2. In Rhesus monkeys using cocaine chronically by self-administration, the proteome and phosphoproteome of the NAc showed differential expression of proteins involved in cytoskeleton structure, mitochondrial function, energy metabolism, and cell signaling (Tannu et al. 2010). del Castillo et al. (2009) identified five proteins responsible for the different rate of extinction of cocaine-induced conditioned place preference in the NAc of Sprague Dawley rats. A single dose of cocaine could alter the expression of multiple proteins in this brain region, and the drug enhanced the replication of HIV-1 in normal human astrocytes as a result of alterations in multiple proteins from different pathways (Reynolds et al. 2006).

7.6. Marijuana

Central nervous system dysregulation and altered immune responses follow interaction of cannabinoids, the main psychoactive component of marijuana, with specific receptors. Among the proteins expressed differentially in normal human astrocytes exposed to marijuana are those involved in cell structure (e.g., chromobox homolog 1 [CBX1], RAN binding protein 1 [RANBP1], tubulin alpha 6 [TUBA6]), energy metabolism (e.g., aldolase fructose-bisphosphate [ALDOA], creatine kinase B chain [CKB], enolase 1 alpha [ENO1], M2 pyruvate kinase [PKM2]), signal transduction (e.g., prohibitin [PHB], rho GDP-dissociation inhibitor beta [ARHGDIB]), protein synthesis or modification (e.g., cathepsin D [CTSD], heat shock 60 kDa protein 1 [HSPD1], ribosomal protein S4 [RPS4]), as well as transport (e.g., chloride channel [CLCN1] and heterogeneous nuclear ribonucleoprotein F [HNRNPF]) (Bindukumar et al. 2008). Some studies have looked at protein expression in brain tissues of animals exposed to delta-9- tetrahydrocannabinol (THC), a particularly important cannabinoid. By comparing the protein expression profiles in the hippocampus of adolescent and adult rats subjected to repeated THC injections, Quinn et al. (2008) uncovered 17 unique differentially expressed proteins in the adolescent rats and 9 in the adult rats, with many of the affected proteins being involved in oxidative stress/mitochondrial function and cytoarchitecture. In another study, Rubino et al. (2009) identified 11 proteins differentially expressed in the PFC synaptosomes of THC-treated female rats. Of these, 10 were suppressed by THC and only one was induced, suggesting reduced synaptic efficiency in response to THC treatment. Colombo et al. (2009) identified 24 differentially expressed proteins in the cerebellum of chronically THC-treated mice. Expression of several alpha and beta subunits of guanine nucleotide-binding proteins, calcium-binding proteins (calretinin and hippocalcin-like protein 1), and dynamin-1 was induced.

8. Proteomics Provides Insights into Mechanisms Underlying Drug Addiction

Similar to other high-throughput approaches such as microarray technology, the true power of the proteomic approach is its ability to provide a comprehensive perspective on the protein families and pathways related to a specific condition such as exposure to a substance of abuse. As the two most widely used high-throughput technologies in the functional genomics field, both microarray and proteomics can provide RNA/protein expression information on a genomic scale. Whereas microarray analysis yields expression profiles at the mRNA level, proteomic analysis provides profiles at the protein level. Thus, the two approaches should be considered complementary. Proteomics provides an efficient way to map the expression of complex biological pathways and identifies proteins regulated under an experimental condition of interest. Furthermore, it enables comparison of the expression of the pathways or function-related proteins under different but related experimental conditions. This is especially valuable for the study of drug addiction, because such a comparison can help researchers not only to understand the specificity of physiological effects of different drugs, but also to gain insight into the similarity between the molecular and cellular mechanisms related to addiction to these drugs.

Because of the diversity of addictive substances, it would be difficult to cover all the proteins affected by all the drugs. A better way is to concentrate on a few relatively common drugs; i.e., alcohol, amphetamine/methamphetamine, cocaine, morphine, heroin, butorphanol, marijuana, and nicotine. In an earlier study, we compiled 497 proteins reported to be directly or indirectly related to the biological response to addictive drugs (Wang et al. 2011). Among them, there were 160 proteins related to two or more drugs. These proteins may be among the most important molecules related to exposure to drugs of abuse. At the same time, because they have been identified for different drugs by various research laboratories, this information is considered to be more reliable.

The proteins were further grouped into multiple categories according to their functional relevance to biological processes in neurons based on their Gene Ontology (GO) categories (Figure 1). Of the proteins, more than one third (56/160) are related to energy metabolism, some of which play important roles in cellular energy generation and are identified for multiple addictive drugs; e.g., ALDOA, M2 pyruvate kinase (PKM2), ATP synthase ATP5G1, ENO1, and triosephosphate isomerase 1 (TPI1). Oxidative stress response proteins (14; 8.75%) that function in the detoxification of hydrogen peroxide and other reactive oxidative species also have been found to be affected multiple addictive drugs, examples being glutathione peroxidase 1 (GPX1), glutathione S-transferase pi 1 (GSTP1), peroxiredoxin 2 (PRDX2), and SOD 1. A number of proteins involved in protein modification and degradation (15; 9.38%) are associated with addictive drugs, among which are several chaperones such as heat shock 70kD protein 8 (HSPA8), chaperonin 60 (HSPD1), chaperonin containing t-complex polypeptide 1 subunit 6a (CCT6A), peptidylprolyl isomerase A (PPIA), as well as proteins of the ubiquitin pathway; i.e., ubiquitin carboxy-terminal hydrolase L1 (UCHL1) and ubiquitin-conjugating enzyme E2N (UBE2N). The identification of a large number of proteins involved in energy metabolism, oxidative stress response, and protein modification and degradation for each additive drug clearly demonstrates the importance of these biological processes in the development and maintenance of drug dependence (Li and Wang 2007). The correlation of these biological processes with each addictive drug also has been discussed in some proteomics papers (Bodzon-Kulakowska et al. 2009; Hwang and Li 2006; Matsumoto 2009; Iwazaki et al. 2006; Tannu et al. 2010; Bindukumar et al. 2008).

Besides those involved in energy metabolism, oxidative stress response, and protein modification and degradation, proteomics analyses have identified proteins associated with other biological processes, such as synaptic function, signal transduction, neuronal function (other than synaptic transmission), cell structure and organization, and transport (Wang et al. 2011). Most of the proteins related to synaptic function are involved in vesicle trafficking. Multiple proteins involved in vesicle budding, docking, and fusion in cellular vesicular transport have been associated with addictive substances in proteomic studies; e.g., dynamin 1 (DNM1), NSF, NSF attachment proteins (NAPA and NAPB), syntaxin binding protein 1 (STXBP1), and synapsin IIb (SYN2). These proteins are critical for the regulation of highly specific and coupled intracellular vesicle trafficking, as well as the cellular endocytotic and exocytotic pathways. Several other proteins associated with exposure to addictive drugs are connected with the SNARE complex. For example, NSF, NAPA, NAPB, STXBP1, and SYN2 are associated with nicotine treatment, whereas NSF, NAPA, STXBP1 and SYN2 are associated with morphine exposure and NSF, STXBP1, SYN2, NSFL1 cofactor (NSFL1C or p47), synaptosomal-associated proteins (SNAP25 and SNAP29), and syntaxin 1B2 (STX1B) are associated with alcohol exposure.

A few other proteins involved in synaptic transmission are closely related to signal transduction, e.g., 4-aminobutyrate aminotransferase (ABAT), glutamate-ammonia ligase (GLUL; also known as L-glutamine synthetase), GTP-binding protein alpha o (GNAO), guanine nucleotide-binding protein beta-1 subunit (GNB1), and protein phosphatase 3 catalytic subunit alpha isoform (PPP3CA).

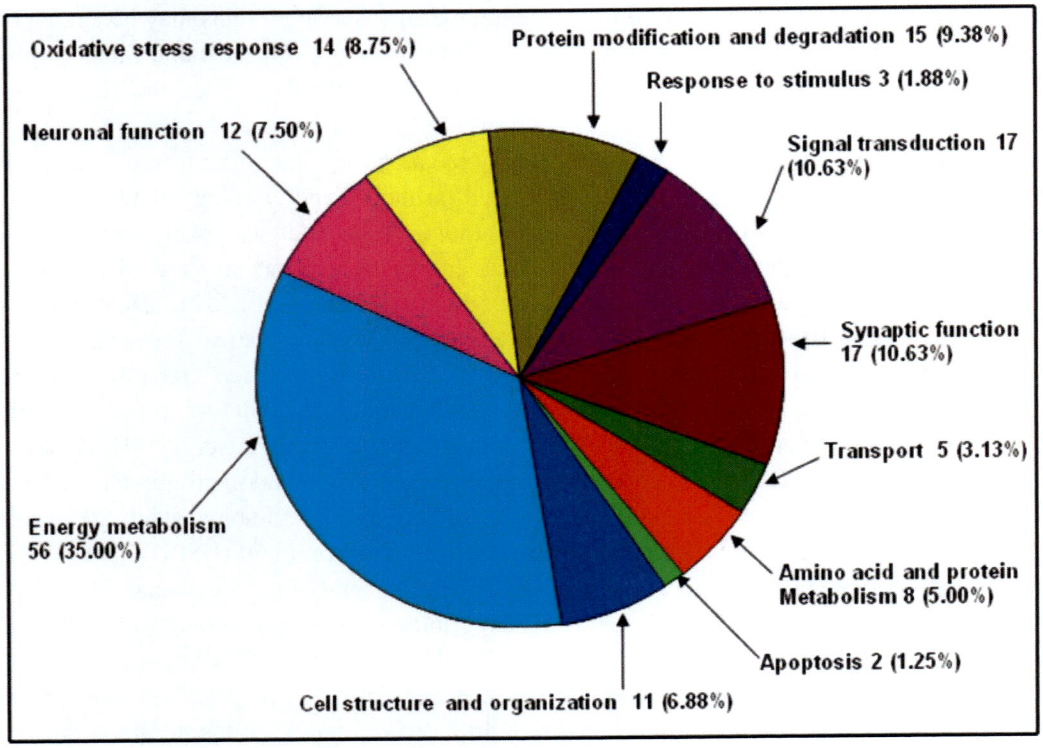

Figure 1. Functional categories of proteins associated with exposure to two or more addictive drugs. Numbers following the category names indicate the number of proteins in the category, and those in parentheses are the corresponding percentages.

Because of the diversity of the function of the proteins involved in signal transduction, we used an analysis tool called Ingenuity Pathway Analysis (IPA; www.ingenuity.com) to uncover the major biological pathways enriched among the 160 proteins.

Among the identified pathways are CREB signaling in neurons, GABA receptor signaling, glutamate receptor signaling, synaptic long-term potentiation, and calcium signaling. These pathways are involved in diverse neuronal functions. They are also among the central hubs of the molecular mechanisms underlying addiction to different drugs (Nestler 2004; Nestler 2005; Kauer and Malenka 2007). The involvement of these pathways in addiction has been investigated extensively (Hemby 2010; Wang and Li 2010; Mukherjee et al. 2008; Nestler 1997).

Whereas earlier studies usually focused on one or a few individual targets in these pathways, the proteomics approach has the power to analyze multiple molecular targets, even multiple pathways, simultaneously, thus providing more direct evidence on the interaction between the addictive drugs and brain signaling pathways and between different signaling pathways. Taken together, results from proteomic studies on drug addiction, as well as information accumulated from other approaches, may suggest a possible mechanism underlying the alteration in neuronal synaptic plasticity in the response to drug exposure (Figure 2).

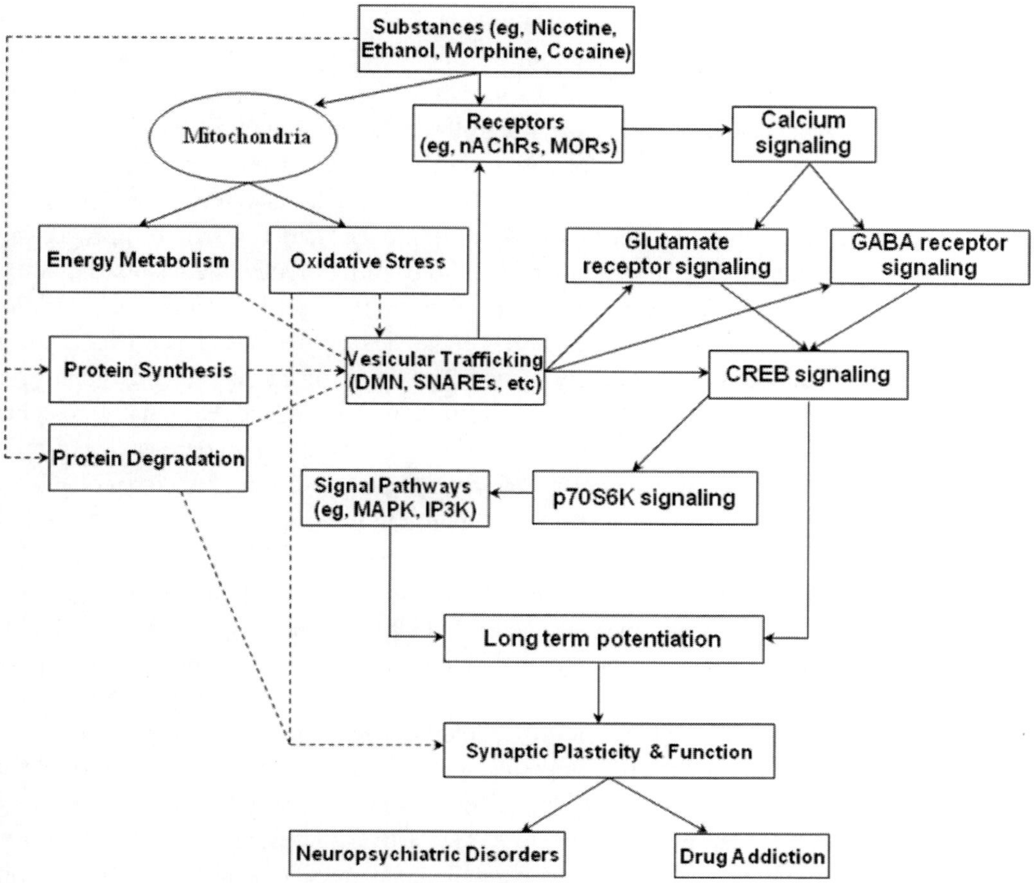

Figure 2. A schematic model for the biological effects of addictive drugs. By interacting with certain receptors, substances of abuse modulate multiple signaling pathways. At the same time, some other biological processes, including energy supply, oxidative stress, and protein synthesis, are also disturbed, which may evoke an alteration in the intracellular environment, forcing the neurons to adapt. These events may result in regulation of the vesicular trafficking system that is closely related to synaptic transmission and modulation of signaling pathways related to cell fate and neuron plasticity. The regulation of these pathways can evoke changes in neuron viability and structure and, ultimately, drug dependence and other neuronal disease.

Summary and Conclusion

By using proteomics technologies, changes in protein expression in the nervous system in response to treatment with substances of abuse can be profiled on a large scale. Even some of the proteins identified have been suggested to be associated with substances of abuse by traditional molecular technologies, but the neuroproteomics approach can provide a much broader and comprehensive view of the function and interaction of these proteins. Although at this time, it is difficult to assign a precise biological function to all the proteins identified simply on the basis of the proteomics results, determination of differentially expressed proteins and their corresponding biological pathways in response to a drug treatment should provide important insight into the mechanism underlying drug-neuron interaction.

Much of our knowledge of drug-neuron interactions has been accumulated through relatively traditional approaches that focus on the function or interaction of one or a few genes/proteins. Such approaches are critical in our exploration of the mechanisms underlying drug addiction. However, the newly developed high-throughput protein profiling technologies such as neuroproteomics can provide information regarding protein function and interaction on a much larger scale. In addition to its ability to obtain a relatively global view of metabolism and cell signaling, the neuroproteomics approach does not require preselecting target proteins in relation to the neural actions of addictive drugs. Thus, the results should be more objective, which is critical for its promise of identifying novel proteins involved in the neural effects of addictive drugs. Even at this early stage, neuroproteomics has proved to be a powerful tool for examining and identifying the dynamics of protein changes and the biological processes they are involved in that underlie drug dependence. Eventually, insights from neuroproteomics coupled with other conventional molecular technologies will not only help us to elucidate the mechanisms used by the neurons with respect to their architecture, synaptic transmission, and signaling cascades in response to treatment by drugs of abuse, but also can accelerate identification of new therapeutic targets for drug dependence.

Proteomics has already become a powerful tool for examining and identifying the dynamics of proteins and the biological processes underlying drug dependence. However, because of the high complexity of the proteome and the limitations of current technology, not all the proteins associated with exposure to addictive drugs can be detected by proteomics approaches. In most proteomics studies on drug addiction, the majority of identified proteins fall into a relatively small number of categories, such as energy metabolism, cytoskeletal proteins, and protein modification and degradation, whereas proteins critical for drug dependence, such as neurotransmitter receptors, GTPases, kinases, and phosphatases are under-represented. Taking the receptors as an example, some receptors, such as nAChRs and opioid and cannabinoid receptors, serve as gateways in the interactions between drugs and the neurons (Ferre et al. 2010; Contet et al. 2004; Changeux 2010), whereas receptors such as those for dopamine, glutamate, and GABA are important to the control of neurotransmitter release and thus play a fundamental role in the development of drug dependence. However, no such receptor has been identified in proteomic study so far, although these genes are known to be associated with drug exposure at both the mRNA and protein levels. The major cause underlying such observation is that the most commonly detected proteins belong to abundant and easily separable species, whereas the underrepresented proteins usually are expressed in low abundance and may also be difficult to separate by classical 2D-PAGE and mass spectrometry, often being obscured by more abundant proteins in proteomic studies. As suggested earlier, advances in technology, appropriate experimental model selection, and improvement in sample preparation will enhance the sensitivity and power of proteomic analysis and help us maximize the insights gained from proteomic studies of drug addiction (Huber 2003; Lull et al. 2010; Li and Wang 2007; Garbis et al. 2005).

Acknowledgments

The preparation of this chapter was supported in part by National Institutes of Health grants DA-12844 and DA-13783 to Ming D Li and National Natural Science Foundation of

China grant No. 31271411 to Ju Wang. We thank Dr. David L. Bronson for his excellent editing of the manuscript.

References

Ahmed N, Barker G, Oliva K, Garfin D, Talmadge K, Georgiou H, Quinn M, Rice G (2003) An approach to remove albumin for the proteomic analysis of low abundance biomarkers in human serum. *Proteomics* 3(10):1980-1987.

Alban A, David SO, Bjorkesten L, Andersson C, Sloge E, Lewis S, Currie I (2003) A novel experimental design for comparative two-dimensional gel analysis: two-dimensional difference gel electrophoresis incorporating a pooled internal standard. *Proteomics* 3(1):36-44.

Alexander-Kaufman K, Cordwell S, Harper C, Matsumoto I (2007) A proteome analysis of the dorsolateral prefrontal cortex in human alcoholic patients. *Proteomics Clin Appl* 1(1):62-72.

Alexander-Kaufman K, Harper C, Wilce P, Matsumoto I (2007) Cerebellar vermis proteome of chronic alcoholic individuals. *Alcohol Clin Exp Res* 31(8):1286-1296.

Alexander-Kaufman K, James G, Sheedy D, Harper C, Matsumoto I (2006) Differential protein expression in the prefrontal white matter of human alcoholics: a proteomics study. *Mol Psychiatry* 11(1):56-65.

Anderson NL, Anderson NG (2002) The human plasma proteome: history, character, and diagnostic prospects. *Mol Cell Proteomics* 1(11):845-867.

Bahi A, Dreyer JL (2005) Cocaine-induced expression changes of axon guidance molecules in the adult rat brain. *Mol Cell Neurosci* 28(2):275-291.

Barrett J, Brophy PM, Hamilton JV (2005) Analysing proteomic data. *Int J Parasitol* 35(5):543-553.

Becker M, Schindler J, Nothwang HG (2006) Neuroproteomics - the tasks lying ahead. *Electrophoresis* 27(13):2819-2829.

Bell RL, Kimpel MW, Rodd ZA, Strother WN, Bai F, Peper CL, Mayfield RD, Lumeng L, Crabb DW, McBride WJ, Witzmann FA (2006) Protein expression changes in the nucleus accumbens and amygdala of inbred alcohol-preferring rats given either continuous or scheduled access to ethanol. *Alcohol* 40(1):3-17.

Benowitz NL (1996) Pharmacology of nicotine: addiction and therapeutics. *Annu Rev Pharmacol Toxicol* 36597-613.

Beranova-Giorgianni S, Pabst MJ, Russell TM, Giorgianni F, Goldowitz D, Desiderio DM (2002) Preliminary analysis of the mouse cerebellum proteome. *Brain Res Mol Brain Res* 98(1-2):135-140.

Bierczynska-Krzysik A, Bonar E, Drabik A, Noga M, Suder P, Dylag T, Dubin A, Kotlinska J, Silberring J (2006) Rat brain proteome in morphine dependence. *Neurochem Int* 49(4):401-406.

Bierczynska-Krzysik A, Pradeep John JP, Silberring J, Kotlinska J, Dylag T, Cabatic M, Lubec G (2006) Proteomic analysis of rat cerebral cortex, hippocampus and striatum after exposure to morphine. *Int J Mol Med* 18(4):775-784.

Bindukumar B, Mahajan SD, Reynolds JL, Hu Z, Sykes DE, Aalinkeel R, Schwartz SA (2008) Genomic and proteomic analysis of the effects of cannabinoids on normal human astrocytes. *Brain Res* 11911-11.

Blomberg A, Blomberg L, Norbeck J, Fey SJ, Larsen PM, Larsen M, Roepstorff P, Degand H, Boutry M, Posch A, et al. (1995) Interlaboratory reproducibility of yeast protein patterns analyzed by immobilized pH gradient two-dimensional gel electrophoresis. *Electrophoresis* 16(10):1935-1945.

Bodzon-Kulakowska A, Bierczynska-Krzysik A, Drabik A, Noga M, Kraj A, Suder P, Silberring J (2005) Morphinome--proteome of the nervous system after morphine treatment. *Amino Acids* 28(1):13-19.

Bodzon-Kulakowska A, Kulakowski K, Drabik A, Moszczynski A, Silberring J, Suder P (2011) Morphinome--a meta-analysis applied to proteomics studies in morphine dependence. *Proteomics* 11(1):5-21.

Bodzon-Kulakowska A, Suder P, Mak P, Bierczynska-Krzysik A, Lubec G, Walczak B, Kotlinska J, Silberring J (2009) Proteomic analysis of striatal neuronal cell cultures after morphine administration. *J Sep Sci* 32(8):1200-1210.

Boguski MS, McIntosh MW (2003) Biomedical informatics for proteomics. *Nature* 422(6928):233-237.

Brewis IA, Brennan P (2010) Proteomics technologies for the global identification and quantification of proteins. *Adv Protein Chem Struct Biol* 801-44.

Carninci P, Kasukawa T, Katayama S, Gough J, Frith MC, Maeda N, Oyama R, Ravasi T, Lenhard B, Wells C, Kodzius R, Shimokawa K, Bajic VB, Brenner SE, Batalov S, Forrest AR, Zavolan M, Davis MJ, Wilming LG, Aidinis V, Allen JE, Ambesi-Impiombato A, Apweiler R, Aturaliya RN, Bailey TL, Bansal M, Baxter L, Beisel KW, Bersano T, Bono H, Chalk AM, Chiu KP, Choudhary V, Christoffels A, Clutterbuck DR, Crowe ML, Dalla E, Dalrymple BP, de Bono B, Della Gatta G, di Bernardo D, Down T, Engstrom P, Fagiolini M, Faulkner G, Fletcher CF, Fukushima T, Furuno M, Futaki S, Gariboldi M, Georgii-Hemming P, Gingeras TR, Gojobori T, Green RE, Gustincich S, Harbers M, Hayashi Y, Hensch TK, Hirokawa N, Hill D, Huminiecki L, Iacono M, Ikeo K, Iwama A, Ishikawa T, Jakt M, Kanapin A, Katoh M, Kawasawa Y, Kelso J, Kitamura H, Kitano H, Kollias G, Krishnan SP, Kruger A, Kummerfeld SK, Kurochkin IV, Lareau LF, Lazarevic D, Lipovich L, Liu J, Liuni S, McWilliam S, Madan Babu M, Madera M, Marchionni L, Matsuda H, Matsuzawa S, Miki H, Mignone F, Miyake S, Morris K, Mottagui-Tabar S, Mulder N, Nakano N, Nakauchi H, Ng P, Nilsson R, Nishiguchi S, Nishikawa S, Nori F, Ohara O, Okazaki Y, Orlando V, Pang KC, Pavan WJ, Pavesi G, Pesole G, Petrovsky N, Piazza S, Reed J, Reid JF, Ring BZ, Ringwald M, Rost B, Ruan Y, Salzberg SL, Sandelin A, Schneider C, Schonbach C, Sekiguchi K, Semple CA, Seno S, Sessa L, Sheng Y, Shibata Y, Shimada H, Shimada K, Silva D, Sinclair B, Sperling S, Stupka E, Sugiura K, Sultana R, Takenaka Y, Taki K, Tammoja K, Tan SL, Tang S, Taylor MS, Tegner J, Teichmann SA, Ueda HR, van Nimwegen E, Verardo R, Wei CL, Yagi K, Yamanishi H, Zabarovsky E, Zhu S, Zimmer A, Hide W, Bult C, Grimmond SM, Teasdale RD, Liu ET, Brusic V, Quackenbush J, Wahlestedt C, Mattick JS, Hume DA, Kai C, Sasaki D, Tomaru Y, Fukuda S, Kanamori-Katayama M, Suzuki M, Aoki J, Arakawa T, Iida J, Imamura K, Itoh M, Kato T, Kawaji H, Kawagashira N, Kawashima T, Kojima M, Kondo S, Konno H, Nakano K, Ninomiya N, Nishio T, Okada M, Plessy C, Shibata K, Shiraki T, Suzuki S, Tagami M, Waki K, Watahiki A, Okamura-Oho Y,

Suzuki H, Kawai J, Hayashizaki Y (2005) The transcriptional landscape of the mammalian genome. *Science* 309(5740):1559-1563.

Cecconi D, Tessari M, Wille DR, Zoli M, Domenici E, Righetti PG, Carboni L (2008) Serum proteomic analysis during nicotine self-administration, extinction and relapse in rats. *Electrophoresis* 29(7):1525-1533.

Changeux JP (2010) Nicotine addiction and nicotinic receptors: lessons from genetically modified mice. Nat Rev Neurosci 11(6):389-401.

Choe LH, Lee KH (2003) Quantitative and qualitative measure of intralaboratory two-dimensional protein gel reproducibility and the effects of sample preparation, sample load, and image analysis. *Electrophoresis* 24(19-20):3500-3507.

Claverie JM (2005) Fewer genes, more noncoding RNA. *Science* 309(5740):1529-1530.

Colombo G, Rusconi F, Rubino T, Cattaneo A, Martegani E, Parolaro D, Bachi A, Zippel R (2009) Transcriptomic and proteomic analyses of mouse cerebellum reveals alterations in RasGRF1 expression following in vivo chronic treatment with delta 9-tetrahydrocannabinol. *J Mol Neurosci* 37(2):111-122.

Consortium IHGS (2004) Finishing the euchromatic sequence of the human genome. *Nature* 431(7011):931-945.

Contet C, Kieffer BL, Befort K (2004) Mu opioid receptor: a gateway to drug addiction. *Curr Opin Neurobiol* 14(3):370-378.

Cooper CA, Joshi HJ, Harrison MJ, Wilkins MR, Packer NH (2003) GlycoSuiteDB: a curated relational database of glycoprotein glycan structures and their biological sources. 2003 update. *Nucleic Acids Res* 31(1):511-513.

Damodaran S, Dlugos CA, Wood TD, Rabin RA (2006) Effects of chronic ethanol administration on brain protein levels: a proteomic investigation using 2-D DIGE system. *Eur J Pharmacol* 547(1-3):75-82.

del Castillo C, Morales L, Alguacil LF, Salas E, Garrido E, Alonso E, Perez-Garcia C (2009) Proteomic analysis of the nucleus accumbens of rats with different vulnerability to cocaine addiction. *Neuropharmacology* 57(1):41-48.

DeRisi J, Penland L, Brown PO, Bittner ML, Meltzer PS, Ray M, Chen Y, Su YA, Trent JM (1996) Use of a cDNA microarray to analyse gene expression patterns in human cancer. *Nat Genet* 14(4):457-460.

Dowsey AW, Dunn MJ, Yang GZ (2003) The role of bioinformatics in two-dimensional gel electrophoresis. *Proteomics* 3(8):1567-1596.

Dunckley T, Lukas RJ (2003) Nicotine modulates the expression of a diverse set of genes in the neuronal SH-SY5Y cell line. *J Biol Chem* 278(18):15633-15640.

Ekins R, Chu F, Biggart E (1990) Multispot, multianalyte, immunoassay. *Ann Biol Clin* (Paris) 48(9):655-666.

Faure JJ, Hattingh SM, Stein DJ, Daniels WM (2009) Proteomic analysis reveals differentially expressed proteins in the rat frontal cortex after methamphetamine treatment. *Metab Brain Dis* 24(4):685-700.

Fenselau C (1997) MALDI MS and strategies for protein analysis. *Anal Chem* 69(21):661A-665A.

Ferre S, Lluis C, Justinova Z, Quiroz C, Orru M, Navarro G, Canela EI, Franco R, Goldberg SR (2010) Adenosine-cannabinoid receptor interactions. Implications for striatal function. *Br J Pharmacol* 160(3):443-453.

Fountoulakis M (2004) Application of proteomics technologies in the investigation of the brain. *Mass Spectrom Rev* 23(4):231-258.

Fountoulakis M, Schuller E, Hardmeier R, Berndt P, Lubec G (1999) Rat brain proteins: two-dimensional protein database and variations in the expression level. *Electrophoresis* 20(18):3572-3579.

Franzen B, Yang Y, Sunnemark D, Wickman M, Ottervald J, Oppermann M, Sandberg K (2003) Dihydropyrimidinase related protein-2 as a biomarker for temperature and time dependent post mortem changes in the mouse brain proteome. *Proteomics* 3(10):1920-1929.

Freeman WM, Brebner K, Amara SG, Reed MS, Pohl J, Phillips AG (2005) Distinct proteomic profiles of amphetamine self-administration transitional states. *Pharmacogenomics* J 5(3):203-214.

Freeman WM, Hemby SE (2004) Proteomics for protein expression profiling in neuroscience. *Neurochem Res* 29(6):1065-1081.

Garavelli JS (2004) The RESID Database of Protein Modifications as a resource and annotation tool. *Proteomics* 4(6):1527-1533.

Garbis S, Lubec G, Fountoulakis M (2005) Limitations of current proteomics technologies. *J Chromatogr A* 1077(1):1-18.

Gasteiger E, Hoogland C, Gattiket A, Duvaud S, R. WM, D. AR, A. B (2005) Protein Identification and Analysis Tools on the ExPASy Server. In: Walker JM (ed) The Proteomics Protocols Handbook. Human Press Inc., Totowa, NJ, pp 571-607..

Gauss C, Kalkum M, Lowe M, Lehrach H, Klose J (1999) Analysis of the mouse proteome. (I) Brain proteins: separation by two-dimensional electrophoresis and identification by mass spectrometry and genetic variation. *Electrophoresis* 20(3):575-600.

Gerner C, Frohwein U, Gotzmann J, Bayer E, Gelbmann D, Bursch W, Schulte-Hermann R (2000) The Fas-induced apoptosis analyzed by high throughput proteome analysis. *J Biol Chem* 275(50):39018-39026.

Gibbs RA, Weinstock GM, Metzker ML, Muzny DM, Sodergren EJ, Scherer S, Scott G, Steffen D, Worley KC, Burch PE, Okwuonu G, Hines S, Lewis L, DeRamo C, Delgado O, Dugan-Rocha S, Miner G, Morgan M, Hawes A, Gill R, Celera, Holt RA, Adams MD, Amanatides PG, Baden-Tillson H, Barnstead M, Chin S, Evans CA, Ferriera S, Fosler C, Glodek A, Gu Z, Jennings D, Kraft CL, Nguyen T, Pfannkoch CM, Sitter C, Sutton GG, Venter JC, Woodage T, Smith D, Lee HM, Gustafson E, Cahill P, Kana A, Doucette-Stamm L, Weinstock K, Fechtel K, Weiss RB, Dunn DM, Green ED, Blakesley RW, Bouffard GG, De Jong PJ, Osoegawa K, Zhu B, Marra M, Schein J, Bosdet I, Fjell C, Jones S, Krzywinski M, Mathewson C, Siddiqui A, Wye N, McPherson J, Zhao S, Fraser CM, Shetty J, Shatsman S, Geer K, Chen Y, Abramzon S, Nierman WC, Havlak PH, Chen R, Durbin KJ, Egan A, Ren Y, Song XZ, Li B, Liu Y, Qin X, Cawley S, Cooney AJ, D'Souza LM, Martin K, Wu JQ, Gonzalez-Garay ML, Jackson AR, Kalafus KJ, McLeod MP, Milosavljevic A, Virk D, Volkov A, Wheeler DA, Zhang Z, Bailey JA, Eichler EE, Tuzun E, Birney E, Mongin E, Ureta-Vidal A, Woodwark C, Zdobnov E, Bork P, Suyama M, Torrents D, Alexandersson M, Trask BJ, Young JM, Huang H, Wang H, Xing H, Daniels S, Gietzen D, Schmidt J, Stevens K, Vitt U, Wingrove J, Camara F, Mar Alba M, Abril JF, Guigo R, Smit A, Dubchak I, Rubin EM, Couronne O, Poliakov A, Hubner N, Ganten D, Goesele C, Hummel O, Kreitler T, Lee YA, Monti J, Schulz H, Zimdahl H, Himmelbauer H, Lehrach H, Jacob HJ, Bromberg S, Gullings-Handley J,

Jensen-Seaman MI, Kwitek AE, Lazar J, Pasko D, Tonellato PJ, Twigger S, Ponting CP, Duarte JM, Rice S, Goodstadt L, Beatson SA, Emes RD, Winter EE, Webber C, Brandt P, Nyakatura G, Adetobi M, Chiaromonte F, Elnitski L, Eswara P, Hardison RC, Hou M, Kolbe D, Makova K, Miller W, Nekrutenko A, Riemer C, Schwartz S, Taylor J, Yang S, Zhang Y, Lindpaintner K, Andrews TD, Caccamo M, Clamp M, Clarke L, Curwen V, Durbin R, Eyras E, Searle SM, Cooper GM, Batzoglou S, Brudno M, Sidow A, Stone EA, Payseur BA, Bourque G, Lopez-Otin C, Puente XS, Chakrabarti K, Chatterji S, Dewey C, Pachter L, Bray N, Yap VB, Caspi A, Tesler G, Pevzner PA, Haussler D, Roskin KM, Baertsch R, Clawson H, Furey TS, Hinrichs AS, Karolchik D, Kent WJ, Rosenbloom KR, Trumbower H, Weirauch M, Cooper DN, Stenson PD, Ma B, Brent M, Arumugam M, Shteynberg D, Copley RR, Taylor MS, Riethman H, Mudunuri U, Peterson J, Guyer M, Felsenfeld A, Old S, Mockrin S, Collins F (2004) Genome sequence of the Brown Norway rat yields insights into mammalian evolution. *Nature* 428(6982):493-521.

Gorg A, Obermaier C, Boguth G, Harder A, Scheibe B, Wildgruber R, Weiss W (2000) The current state of two-dimensional electrophoresis with immobilized pH gradients. *Electrophoresis* 21(6):1037-1053.

Gorg A, Weiss W, Dunn MJ (2004) Current two-dimensional electrophoresis technology for proteomics. *Proteomics* 4(12):3665-3685.

Gras R, Muller M (2001) Computational aspects of protein identification by mass spectrometry. *Curr Opin Mol The*r 3(6):526-532.

Gygi SP, Corthals GL, Zhang Y, Rochon Y, Aebersold R (2000) Evaluation of two-dimensional gel electrophoresis-based proteome analysis technology. *Proc Natl Acad Sci U S A* 97(17):9390-9395.

Gygi SP, Rist B, Gerber SA, Turecek F, Gelb MH, Aebersold R (1999) Quantitative analysis of complex protein mixtures using isotope-coded affinity tags. *Nat Biotechnol* 17(10):994-999.

Haab BB, Dunham MJ, Brown PO (2001) Protein microarrays for highly parallel detection and quantitation of specific proteins and antibodies in complex solutions. *Genome Biol* 2(2):RESEARCH0004.

Habermann B, Oegema J, Sunyaev S, Shevchenko A (2004) The power and the limitations of cross-species protein identification by mass spectrometry-driven sequence similarity searches. *Mol Cell Proteomics* 3(3):238-249.

Hemby SE (2006) Assessment of genome and proteome profiles in cocaine abuse. P*rog Brain Res* 158173-195.

Hemby SE (2010) Cocainomics: new insights into the molecular basis of cocaine addiction. *J Neuroimmune Pharmacol* 5(1):70-82.

Herbert BR, Harry JL, Packer NH, Gooley AA, Pedersen SK, Williams KL (2001) What place for polyacrylamide in proteomics? *Trends Biotechnol* 19(10 Suppl):S3-9.

Hernandez P, Muller M, Appel RD (2006) Automated protein identification by tandem mass spectrometry: issues and strategies. *Mass Spectrom Rev* 25(2):235-254.

Huber LA (2003) Is proteomics heading in the wrong direction? *Nat Rev Mol Cell Biol* 4(1):74-80.

Hwang YY, Li MD (2006) Proteins differentially expressed in response to nicotine in five rat brain regions: identification using a 2-DE/MS-based proteomics approach. *Proteomics* 6(10):3138-3153.

Iwazaki T, McGregor IS, Matsumoto I (2006) Protein expression profile in the striatum of acute methamphetamine-treated rats. *Brain Res* 1097(1):19-25.

Iwazaki T, McGregor IS, Matsumoto I (2007) Protein expression profile in the striatum of rats with methamphetamine-induced behavioral sensitization. *Proteomics* 7(7):1131-1139.

Iwazaki T, McGregor IS, Matsumoto I (2008) Protein expression profile in the amygdala of rats with methamphetamine-induced behavioral sensitization. *Neurosci Lett* 435(2):113-119.

Jenison R, La H, Haeberli A, Ostroff R, Polisky B (2001) Silicon-based biosensors for rapid detection of protein or nucleic acid targets. *Clin Chem* 47(10):1894-1900.

Jiang L, He L, Fountoulakis M (2004) Comparison of protein precipitation methods for sample preparation prior to proteomic analysis. *J Chromatogr A* 1023(2):317-320.

Kanamoto T, Hellman U, Heldin CH, Souchelnytskyi S (2002) Functional proteomics of transforming growth factor-beta1-stimulated Mv1Lu epithelial cells: Rad51 as a target of TGFbeta1-dependent regulation of DNA repair. *Embo J* 21(5):1219-1230.

Kashem MA, Etages HD, Kopitar-Jerala N, McGregor IS, Matsumoto I (2009) Differential protein expression in the corpus callosum (body) of human alcoholic brain. *J Neurochem* 110(2):486-495.

Kashem MA, Harper C, Matsumoto I (2008) Differential protein expression in the corpus callosum (genu) of human alcoholics. *Neurochem Int* 53(1-2):1-11.

Kashem MA, James G, Harper C, Wilce P, Matsumoto I (2007) Differential protein expression in the corpus callosum (splenium) of human alcoholics: a proteomics study. *Neurochem Int* 50(2):450-459.

Kauer JA, Malenka RC (2007) Synaptic plasticity and addiction. *Nat Rev Neurosci* 8(11):844-858.

Kerns RT, Ravindranathan A, Hassan S, Cage MP, York T, Sikela JM, Williams RW, Miles MF (2005) Ethanol-responsive brain region expression networks: implications for behavioral responses to acute ethanol in DBA/2J versus C57BL/6J mice. *J Neurosci* 25(9):2255-2266.

Kim SI, Voshol H, van Oostrum J, Hastings TG, Cascio M, Glucksman MJ (2004) Neuroproteomics: expression profiling of the brain's proteomes in health and disease. *Neurochem Res* 29(6):1317-1331.

Kim SY, Chudapongse N, Lee SM, Levin MC, Oh JT, Park HJ, Ho IK (2005) Proteomic analysis of phosphotyrosyl proteins in morphine-dependent rat brains. *Brain Res Mol Brain Res* 133(1):58-70.

Kim SY, Chudapongse N, Lee SM, Levin MC, Oh JT, Park HJ, Ho IK (2004) Proteomic analysis of phosphotyrosyl proteins in the rat brain: effect of butorphanol dependence. *J Neurosci Res* 77(6):867-877.

Klose J, Nock C, Herrmann M, Stuhler K, Marcus K, Bluggel M, Krause E, Schalkwyk LC, Rastan S, Brown SD, Bussow K, Himmelbauer H, Lehrach H (2002) Genetic analysis of the mouse brain proteome. *Nat Genet* 30(4):385-393.

Kobeissy FH, Warren MW, Ottens AK, Sadasivan S, Zhang Z, Gold MS, Wang KK (2008) Psychoproteomic analysis of rat cortex following acute methamphetamine exposure. *J Proteome Res* 7(5):1971-1983.

Konu O, Kane JK, Barrett T, Vawter MP, Chang R, Ma JZ, Donovan DM, Sharp B, Becker KG, Li MD (2001) Region-specific transcriptional response to chronic nicotine in rat brain. *Brain Res* 909(1-2):194-203.

Konu O, Xu X, Ma JZ, Kane J, Wang J, Shi SJ, Li MD (2004) Application of a customized pathway-focused microarray for gene expression profiling of cellular homeostasis upon exposure to nicotine in PC12 cells. *Brain research Molecular brain research* 121(1-2):102-113.

Krapfenbauer K, Fountoulakis M, Lubec G (2003) A rat brain protein expression map including cytosolic and enriched mitochondrial and microsomal fractions. *Electrophoresis* 24(11):1847-1870.

Langen H, Berndt P, Roder D, Cairns N, Lubec G, Fountoulakis M (1999) Two-dimensional map of human brain proteins. *Electrophoresis* 20(4-5):907-916.

Larsson A, Engel JA (2004) Neurochemical and behavioral studies on ethanol and nicotine interactions. *Neurosci Biobehav Rev* 27(8):713-720.

Lee JR, Magee DM, Gaster RS, LaBaer J, Wang SX (2013) Emerging protein array technologies for proteomics. *Expert Rev Proteomics* 10(1):65-75.

Lehrmann E, Afanador ZR, Deep-Soboslay A, Gallegos G, Darwin WD, Lowe RH, Barnes AJ, Huestis MA, Cadet JL, Herman MM, Hyde TM, Kleinman JE, Freed WJ (2008) Postmortem diagnosis and toxicological validation of illicit substance use. *Addict Biol* 13(1):105-117.

Lewis TS, Hunt JB, Aveline LD, Jonscher KR, Louie DF, Yeh JM, Nahreini TS, Resing KA, Ahn NG (2000) Identification of novel MAP kinase pathway signaling targets by functional proteomics and mass spectrometry. *Mol Cell* 6(6):1343-1354.

Lewohl JM, Van Dyk DD, Craft GE, Innes DJ, Mayfield RD, Cobon G, Harris RA, Dodd PR (2004) The application of proteomics to the human alcoholic brain. *Ann N Y Acad Sci* 102514-26.

Li KW, Jimenez CR, van der Schors RC, Hornshaw MP, Schoffelmeer AN, Smit AB (2006) Intermittent administration of morphine alters protein expression in rat nucleus accumbens. *Proteomics* 6(6):2003-2008.

Li MD, Kane JK, Wang J, Ma JZ (2004) Time-dependent changes in transcriptional profiles within five rat brain regions in response to nicotine treatment. *Brain research Molecular brain research* 132(2):168-180.

Li MD, Konu O, Kane JK, Becker KG (2002) Microarray technology and its application on nicotine research. *Molecular neurobiology* 25(3):265-285.

Li MD, Wang J (2007) Neuroproteomics and its applications in research on nicotine and other drugs of abuse. *Proteomics Clin Appl* 1(11):1406-1427.

Li Q, Zhao X, Zhong LJ, Yang HY, Wang Q, Pu XP (2009) Effects of chronic morphine treatment on protein expression in rat dorsal root ganglia. *Eur J Pharmacol* 612(1-3):21-28.

Li X, Wang H, Qiu P, Luo H (2008) Proteomic profiling of proteins associated with methamphetamine-induced neurotoxicity in different regions of rat brain. *Neurochem Int* 52(1-2):256-264.

Liebler DC (2002) Proteomic approaches to characterize protein modifications: new tools to study the effects of environmental exposures. *Environ Health Perspect* 110 Suppl 13-9.

Lilley KS, Friedman DB (2004) All about DIGE: quantification technology for differential-display 2D-gel proteomics. *Expert Rev Proteomics* 1(4):401-409.

Lilley KS, Razzaq A, Dupree P (2002) Two-dimensional gel electrophoresis: recent advances in sample preparation, detection and quantitation. *Curr Opin Chem Biol* 6(1):46-50.

Link AJ, Eng J, Schieltz DM, Carmack E, Mize GJ, Morris DR, Garvik BM, Yates JR, 3rd (1999) Direct analysis of protein complexes using mass spectrometry. *Nat Biotechnol* 17(7):676-682.

Lubec G, Krapfenbauer K, Fountoulakis M (2003) Proteomics in brain research: potentials and limitations. *Prog Neurobiol* 69(3):193-211.

Lull ME, Freeman WM, VanGuilder HD, Vrana KE (2010) The use of neuroproteomics in drug abuse research. *Drug Alcohol Depend* 107(1):11-22.

Lull ME, Freeman WM, Vrana KE, Mash DC (2008) Correlating human and animal studies of cocaine abuse and gene expression. *Ann N Y Acad Sci* 114158-75.

Luscher C, Malenka RC (2011) Drug-evoked synaptic plasticity in addiction: from molecular changes to circuit remodeling. *Neuron* 69(4):650-663.

Manning G, Whyte DB, Martinez R, Hunter T, Sudarsanam S (2002) The protein kinase complement of the human genome. *Science* 298(5600):1912-1934.

Matsuda-Matsumoto H, Iwazaki T, Kashem MA, Harper C, Matsumoto I (2007) Differential protein expression profiles in the hippocampus of human alcoholics. *Neurochem Int* 51(6-7):370-376.

Matsumoto I (2009) Proteomics approach in the study of the pathophysiology of alcohol-related brain damage. *Alcohol Alcohol* 44(2):171-176.

McBride WJ, Schultz JA, Kimpel MW, McClintick JN, Wang M, You J, Rodd ZA (2009) Differential effects of ethanol in the nucleus accumbens shell of alcohol-preferring (P), alcohol-non-preferring (NP) and Wistar rats: a proteomics study. *Pharmacol Biochem Behav* 92(2):304-313.

McClung CA, Nestler EJ, Zachariou V (2005) Regulation of gene expression by chronic morphine and morphine withdrawal in the locus ceruleus and ventral tegmental area. *J Neurosci* 25(25):6005-6015.

Mehta AI, Ross S, Lowenthal MS, Fusaro V, Fishman DA, Petricoin EF, 3rd, Liotta LA (2003) Biomarker amplification by serum carrier protein binding. *Dis Markers* 19(1): 1-10.

Mendes Soares LM, Valcarcel J (2006) The expanding transcriptome: the genome as the 'Book of Sand'. *Embo J* 25(5):923-931.

Miller JC, Zhou H, Kwekel J, Cavallo R, Burke J, Butler EB, Teh BS, Haab BB (2003) Antibody microarray profiling of human prostate cancer sera: antibody screening and identification of potential biomarkers. *Proteomics* 3(1):56-63.

Millioni R, Miuzzo M, Sbrignadello S, Murphy E, Puricelli L, Tura A, Bertacco E, Rattazzi M, Iori E, Tessari P (2010) Delta2D and Proteomweaver: Performance evaluation of two different approaches for 2-DE analysis. *Electrophoresis* 31(8):1311-1317.

Molloy MP, Brzezinski EE, Hang J, McDowell MT, VanBogelen RA (2003) Overcoming technical variation and biological variation in quantitative proteomics. *Proteomics* 3(10):1912-1919.

Moron JA, Abul-Husn NS, Rozenfeld R, Dolios G, Wang R, Devi LA (2007) Morphine Administration Alters the Profile of Hippocampal Postsynaptic Density-associated Proteins: A Proteomics Study Focusing on Endocytic Proteins. *Mol Cell Proteomics* 6(1):29-42.

Morrison RS, Kinoshita Y, Johnson MD, Uo T, Ho JT, McBee JK, Conrads TP, Veenstra TD (2002) Proteomic analysis in the neurosciences. *Mol Cell Proteomics* 1(8):553-560.

Mukherjee S, Das SK, Vaidyanathan K, Vasudevan DM (2008) Consequences of alcohol consumption on neurotransmitters -an overview. *Curr Neurovasc Res* 5(4):266-272.

Neasta J, Uttenweiler-Joseph S, Chaoui K, Monsarrat B, Meunier JC, Mouledous L (2006) Effect of long-term exposure of SH-SY5Y cells to morphine: a whole cell proteomic analysis. *Proteome* Sci 423.

Nestler EJ (1997) Molecular mechanisms of opiate and cocaine addiction. *Curr Opin Neurobiol* 7(5):713-719.

Nestler EJ (2001) Molecular basis of long-term plasticity underlying addiction. *Nat Rev Neurosci* 2(2):119-128.

Nestler EJ (2004) Molecular mechanisms of drug addiction. *Neuropharmacology* 47 Suppl 124-32.

Nestler EJ (2005) Is there a common molecular pathway for addiction? Nat Neurosci 8(11):1445-1449.

Nguyen DN, Becker GW, Riggin RM (1995) Protein mass spectrometry: applications to analytical biotechnology. *J Chromatogr A* 705(1):21-45.

Nielsen UB, Cardone MH, Sinskey AJ, MacBeath G, Sorger PK (2003) Profiling receptor tyrosine kinase activation by using Ab microarrays. *Proc Natl Acad Sci U S A* 100(16):9330-9335.

O'Farrell PH (1975) High resolution two-dimensional electrophoresis of proteins. *J Biol Chem* 250(10):4007-4021.

Park B, Jeong SK, Lee WS, Seong JK, Paik YK (2004) A simple pattern classification method for alcohol-responsive proteins that are differentially expressed in mouse brain. *Proteomics* 4(11):3369-3375.

Piubelli C, Cecconi D, Astner H, Caldara F, Tessari M, Carboni L, Hamdan M, Righetti PG, Domenici E (2005) Proteomic changes in rat serum, polymorphonuclear and mononuclear leukocytes after chronic nicotine administration. *Proteomics* 5(5):1382-1394.

Pollock JD (2002) Gene expression profiling: methodological challenges, results, and prospects for addiction research. *Chem Phys Lipids* 121(1-2):241-256.

Potenza MN, Sofuoglu M, Carroll KM, Rounsaville BJ (2011) Neuroscience of behavioral and pharmacological treatments for addictions. *Neuron* 69(4):695-712.

Predic J, Soskic V, Bradley D, Godovac-Zimmermann J (2002) Monitoring of gene expression by functional proteomics: response of human lung fibroblast cells to stimulation by endothelin-1. *Biochemistry* 41(3):1070-1078.

Prokai L, Zharikova AD, Stevens SM, Jr. (2005) Effect of chronic morphine exposure on the synaptic plasma-membrane subproteome of rats: a quantitative protein profiling study based on isotope-coded affinity tags and liquid chromatography/mass spectrometry. *J Mass Spectrom* 40(2):169-175.

Quinn HR, Matsumoto I, Callaghan PD, Long LE, Arnold JC, Gunasekaran N, Thompson MR, Dawson B, Mallet PE, Kashem MA, Matsuda-Matsumoto H, Iwazaki T, McGregor IS (2008) Adolescent rats find repeated Delta(9)-THC less aversive than adult rats but display greater residual cognitive deficits and changes in hippocampal protein expression following exposure. *Neuropsychopharmacology* 33(5):1113-1126.

Raman B, Cheung A, Marten MR (2002) Quantitative comparison and evaluation of two commercially available, two-dimensional electrophoresis image analysis software packages, Z3 and Melanie. *Electrophoresis* 23(14):2194-2202.

Renthal W, Nestler EJ (2009) Chromatin regulation in drug addiction and depression. *Dialogues Clin Neurosci* 11(3):257-268.

Resing KA, Ahn NG (2005) Proteomics strategies for protein identification. *FEBS Lett* 579(4):885-889.

Reynolds JL, Mahajan SD, Bindukumar B, Sykes D, Schwartz SA, Nair MP (2006) Proteomic analysis of the effects of cocaine on the enhancement of HIV-1 replication in normal human astrocytes (NHA). *Brain Res* 1123(1):226-236.

Rhodes JS, Crabbe JC (2005) Gene expression induced by drugs of abuse. *Curr Opin Pharmacol* 5(1):26-33.

Roepstorff P (2012) Mass spectrometry based proteomics, background, status and future needs. *Protein Cell* 3(9):641-647.

Rogers M, Graham J, Tonge RP (2003) Using statistical image models for objective evaluation of spot detection in two-dimensional gels. *Proteomics* 3(6):879-886.

Rosengren AT, Salmi JM, Aittokallio T, Westerholm J, Lahesmaa R, Nyman TA, Nevalainen OS (2003) Comparison of PDQuest and Progenesis software packages in the analysis of two-dimensional electrophoresis gels. *Proteomics* 3(10):1936-1946.

Ross PL, Huang YN, Marchese JN, Williamson B, Parker K, Hattan S, Khainovski N, Pillai S, Dey S, Daniels S, Purkayastha S, Juhasz P, Martin S, Bartlet-Jones M, He F, Jacobson A, Pappin DJ (2004) Multiplexed protein quantitation in Saccharomyces cerevisiae using amine-reactive isobaric tagging reagents. *Mol Cell Proteomics* 3(12):1154-1169.

Rubino T, Realini N, Braida D, Alberio T, Capurro V, Vigano D, Guidali C, Sala M, Fasano M, Parolaro D (2009) The depressive phenotype induced in adult female rats by adolescent exposure to THC is associated with cognitive impairment and altered neuroplasticity in the prefrontal cortex. *Neurotox Res* 15(4):291-302.

Russo SJ, Dietz DM, Dumitriu D, Morrison JH, Malenka RC, Nestler EJ (2010) The addicted synapse: mechanisms of synaptic and structural plasticity in nucleus accumbens. *Trends Neurosci* 33(6):267-276.

Santoni V, Molloy M, Rabilloud T (2000) Membrane proteins and proteomics: un amour impossible? *Electrophoresis* 21(6):1054-1070.

Sari Y, Zhang M, Mechref Y (2010) Differential expression of proteins in fetal brains of alcohol-treated prenatally C57BL/6 mice: a proteomic investigation. *Electrophoresis* 31(3):483-496.

Schweitzer B, Roberts S, Grimwade B, Shao W, Wang M, Fu Q, Shu Q, Laroche I, Zhou Z, Tchernev VT, Christiansen J, Velleca M, Kingsmore SF (2002) Multiplexed protein profiling on microarrays by rolling-circle amplification. *Nat Biotechnol* 20(4):359-365.

Shui HA, Ho ST, Wang JJ, Wu CC, Lin CH, Tao YX, Liaw WJ (2007) Proteomic analysis of spinal protein expression in rats exposed to repeated intrathecal morphine injection. *Proteomics* 7(5):796-803.

Silzel JW, Cercek B, Dodson C, Tsay T, Obremski RJ (1998) Mass-sensing, multianalyte microarray immunoassay with imaging detection. *Clin Chem* 44(9):2036-2043.

Southern E, Mir K, Shchepinov M (1999) Molecular interactions on microarrays. *Nat Genet* 21(1 Suppl):5-9.

Sreekumar A, Nyati MK, Varambally S, Barrette TR, Ghosh D, Lawrence TS, Chinnaiyan AM (2001) Profiling of cancer cells using protein microarrays: discovery of novel radiation-regulated proteins. *Cancer Res* 61(20):7585-7593.

Stoevesandt O, Taussig MJ, He M (2009) Protein microarrays: high-throughput tools for proteomics. *Expert Rev Proteomics* 6(2):145-157.

Suder P, Bodzon-Kulakowska A, Mak P, Bierczynska-Krzysik A, Daszykowski M, Walczak B, Lubec G, Kotlinska JH, Silberring J (2009) The proteomic analysis of primary cortical astrocyte cell culture after morphine administration. *J Proteome Res* 8(10):4633-4640.

Tam SW, Wiese R, Lee S, Gilmore J, Kumble KD (2002) Simultaneous analysis of eight human Th1/Th2 cytokines using microarrays. *J Immunol Methods* 261(1-2):157-165.

Tan KR, Rudolph U, Luscher C (2011) Hooked on benzodiazepines: GABAA receptor subtypes and addiction. *Trends Neurosci* 34(4):188-197.

Tannu N, Mash DC, Hemby SE (2007) Cytosolic proteomic alterations in the nucleus accumbens of cocaine overdose victims. *Mol Psychiatry* 12(1):55-73.

Tannu NS, Howell LL, Hemby SE (2010) Integrative proteomic analysis of the nucleus accumbens in rhesus monkeys following cocaine self-administration. *Mol Psychiatry* 15(2):185-203.

Taoka M, Wakamiya A, Nakayama H, Isobe T (2000) Protein profiling of rat cerebella during development. *Electrophoresis* 21(9):1872-1879.

Thrift RN, Forte TM, Cahoon BE, Shore VG (1986) Characterization of lipoproteins produced by the human liver cell line, Hep G2, under defined conditions. *J Lipid Res* 27(3):236-250.

Tonge R, Shaw J, Middleton B, Rowlinson R, Rayner S, Young J, Pognan F, Hawkins E, Currie I, Davison M (2001) Validation and development of fluorescence two-dimensional differential gel electrophoresis proteomics technology. *Proteomics* 1(3):377-396.

Tsugita A, Kawakami T, Uchida T, Sakai T, Kamo M, Matsui T, Watanabe Y, Morimasa T, Hosokawa K, Toda T (2000) Proteome analysis of mouse brain: two-dimensional electrophoresis profiles of tissue proteins during the course of aging. *Electrophoresis* 21(9):1853-1871.

Ueyama J, Kitaichi K, Iwase M, Takagi K, Hasegawa T (2003) Application of ultrafiltration method to measurement of catecholamines in plasma of human and rodents by high-performance liquid chromatography. *J Chromatogr B Analyt Technol Biomed Life Sci* 798(1):35-41.

Van den Oever MC, Lubbers BR, Goriounova NA, Li KW, Van der Schors RC, Loos M, Riga D, Wiskerke J, Binnekade R, Stegeman M, Schoffelmeer AN, Mansvelder HD, Smit AB, De Vries TJ, Spijker S (2010) Extracellular matrix plasticity and GABAergic inhibition of prefrontal cortex pyramidal cells facilitates relapse to heroin seeking. *Neuropsychopharmacology* 35(10):2120-2133.

Wang J, Li MD (2010) Common and unique biological pathways associated with smoking initiation/progression, nicotine dependence, and smoking cessation. *Neuropsychopharmacology* 35(3):702-719.

Wang J, Yuan W, Li MD (2011) Genes and pathways co-associated with the exposure to multiple drugs of abuse, including alcohol, amphetamine/methamphetamine, cocaine, marijuana, morphine, and/or nicotine: a review of proteomics analyses. *Molecular neurobiology* 44(3):269-286.

Waterston RH, Lindblad-Toh K, Birney E, Rogers J, Abril JF, Agarwal P, Agarwala R, Ainscough R, Alexandersson M, An P, Antonarakis SE, Attwood J, Baertsch R, Bailey J, Barlow K, Beck S, Berry E, Birren B, Bloom T, Bork P, Botcherby M, Bray N, Brent MR, Brown DG, Brown SD, Bult C, Burton J, Butler J, Campbell RD, Carninci P, Cawley S, Chiaromonte F, Chinwalla AT, Church DM, Clamp M, Clee C, Collins FS, Cook LL, Copley RR, Coulson A, Couronne O, Cuff J, Curwen V, Cutts T, Daly M, David R, Davies J, Delehaunty KD, Deri J, Dermitzakis ET, Dewey C, Dickens NJ, Diekhans M, Dodge S, Dubchak I, Dunn DM, Eddy SR, Elnitski L, Emes RD, Eswara P, Eyras E, Felsenfeld A, Fewell GA, Flicek P, Foley K, Frankel WN, Fulton LA, Fulton RS, Furey TS, Gage D, Gibbs RA, Glusman G, Gnerre S, Goldman N, Goodstadt L, Grafham D, Graves TA, Green ED, Gregory S, Guigo R, Guyer M, Hardison RC, Haussler D, Hayashizaki Y, Hillier LW, Hinrichs A, Hlavina W, Holzer T, Hsu F, Hua A, Hubbard T, Hunt A, Jackson I, Jaffe DB, Johnson LS, Jones M, Jones TA, Joy A, Kamal M, Karlsson EK, Karolchik D, Kasprzyk A, Kawai J, Keibler E, Kells C, Kent WJ, Kirby A, Kolbe DL, Korf I, Kucherlapati RS, Kulbokas EJ, Kulp D, Landers T, Leger JP, Leonard S, Letunic I, Levine R, Li J, Li M, Lloyd C, Lucas S, Ma B, Maglott DR, Mardis ER, Matthews L, Mauceli E, Mayer JH, McCarthy M, McCombie WR, McLaren S, McLay K, McPherson JD, Meldrim J, Meredith B, Mesirov JP, Miller W, Miner TL, Mongin E, Montgomery KT, Morgan M, Mott R, Mullikin JC, Muzny DM, Nash WE, Nelson JO, Nhan MN, Nicol R, Ning Z, Nusbaum C, O'Connor MJ, Okazaki Y, Oliver K, Overton-Larty E, Pachter L, Parra G, Pepin KH, Peterson J, Pevzner P, Plumb R, Pohl CS, Poliakov A, Ponce TC, Ponting CP, Potter S, Quail M, Reymond A, Roe BA, Roskin KM, Rubin EM, Rust AG, Santos R, Sapojnikov V, Schultz B, Schultz J, Schwartz MS, Schwartz S, Scott C, Seaman S, Searle S, Sharpe T, Sheridan A, Shownkeen R, Sims S, Singer JB, Slater G, Smit A, Smith DR, Spencer B, Stabenau A, Stange-Thomann N, Sugnet C, Suyama M, Tesler G, Thompson J, Torrents D, Trevaskis E, Tromp J, Ucla C, Ureta-Vidal A, Vinson JP, Von Niederhausern AC, Wade CM, Wall M, Weber RJ, Weiss RB, Wendl MC, West AP, Wetterstrand K, Wheeler R, Whelan S, Wierzbowski J, Willey D, Williams S, Wilson RK, Winter E, Worley KC, Wyman D, Yang S, Yang SP, Zdobnov EM, Zody MC, Lander ES (2002) Initial sequencing and comparative analysis of the mouse genome. *Nature* 420(6915):520-562.

Westbrook JA, Yan JX, Wait R, Welson SY, Dunn MJ (2001) Zooming-in on the proteome: very narrow-range immobilised pH gradients reveal more protein species and isoforms. *Electrophoresis* 22(14):2865-2871.

Wheelock AM, Buckpitt AR (2005) Software-induced variance in two-dimensional gel electrophoresis image analysis. *Electrophoresis* 26(23):4508-4520.

Whitehouse PJ, Lynch D, Kuhar MJ (1984) Effects of postmortem delay and temperature on neurotransmitter receptor binding in a rat model of the human autopsy process. *J Neurochem* 43(2):553-559.

Wiese R, Belosludtsev Y, Powdrill T, Thompson P, Hogan M (2001) Simultaneous multianalyte ELISA performed on a microarray platform. *Clin Chem* 47(8):1451-1457.

Williams K, Wu T, Colangelo C, Nairn AC (2004) Recent advances in neuroproteomics and potential application to studies of drug addiction. *Neuropharmacology* 47 Suppl 1148-166.

Witzmann FA, Li J, Strother WN, McBride WJ, Hunter L, Crabb DW, Lumeng L, Li TK (2003) Innate differences in protein expression in the nucleus accumbens and hippocampus of inbred alcohol-preferring and -nonpreferring rats. *Proteomics* 3(7):1335-1344.

Witzmann FA, Strother WN (2004) Proteomics and alcoholism. *Int Rev Neurobiol* 61189-214.

Wong CC, Mill J, Fernandes C (2011) Drugs and addiction: an introduction to epigenetics. *Addiction* 106(3):480-489.

Wu CC, Yates JR, 3rd (2003) The application of mass spectrometry to membrane proteomics. *Nat Biotechnol* 21(3):262-267.

Yang JW, Czech T, Lubec G (2004) Proteomic profiling of human hippocampus. *Electrophoresis* 25(7-8):1169-1174.

Yang L, Sun ZS, Zhu YP (2007) Proteomic analysis of rat prefrontal cortex in three phases of morphine-induced conditioned place preference. *J Proteome Res* 6(6):2239-2247.

Yang MH, Kim S, Jung MS, Shim JH, Ryu NK, Yook YJ, Jang CG, Bahk YY, Kim KW, Park JH (2008) Proteomic analysis of methamphetamine-induced reinforcement processes within the mesolimbic dopamine system. *Addict Biol* 13(3-4):287-294.

Yeom M, Shim I, Lee HJ, Hahm DH (2005) Proteomic analysis of nicotine-associated protein expression in the striatum of repeated nicotine-treated rats. *Biochemical and biophysical research communications* 326(2):321-328.

Yuferov V, Nielsen D, Butelman E, Kreek MJ (2005) Microarray studies of psychostimulant-induced changes in gene expression. *Addict Biol* 10(1):101-118.

Zhang J, Goodlett DR, Peskind ER, Quinn JF, Zhou Y, Wang Q, Pan C, Yi E, Eng J, Aebersold RH, Montine TJ (2005) Quantitative proteomic analysis of age-related changes in human cerebrospinal fluid. *Neurobiol Aging* 26(2):207-227.

Zhang S, Day IN, Ye S (2001) Microarray analysis of nicotine-induced changes in gene expression in endothelial cells. *Physiol Genomics* 5(4):187-192.

In: Recent Advances in Proteomics Research
Editor: Ming D. Li

ISBN: 978-1-62948-218-7

Chapter 3

Current Proteomics Methods for Analysing the Human Microbiome

Luis Pinto[1, 2, 3, 4], Patrícia Poeta[3, 4], Concha Gil[5], and Gilberto Igrejas[1 2,*]

[1]Department of Genetics and Biotechnology, University of Trás-os-Montes and Alto Douro, Vila Real, Portugal

[2]Institute for Biotechnology and Bioengineering, Centre of Genomics and Biotechnology, University of Trás-os-Montes and Alto Douro, Vila Real, Portugal

[3]Centre of Studies of Animal and Veterinary Sciences, University of Trás-os-Montes and Alto Douro, Vila Real, Portugal

[4]Veterinary Science Department, University of Trás-os-Montes and Alto Douro, Vila Real, Portugal

[5]Department of Microbiology II, Faculty of Pharmacy, Complutense University of Madrid, Madrid, Spain

Abstract

Improving our knowledge of human or animal microbiomes will be important in coming to understand how biological environments function as a whole, and more particularly the significance of certain biochemical reactions like those involved in antibiotic resistance or stress responses. Large-scale programs such as the Human Microbiome Project have initiated extensive surveys of the range of microbial genes and genomes present in certain physical or biological environments. Microbiomes can also be characterized by several proteomics methods. Sodium dodecyl sulfate polyacrylamide gel electrophoresis (SDS-PAGE), high-performance liquid chromatography (HPLC) or capillary electrophoresis are commonly used to analyze biological samples by separating proteins according to their molecular weights. Proteins can also be separated according to their isoelectric point by isoelectric focusing. In two-dimensional electrophoresis (2-DE)

* Corresponding Author address: Email: gigrejas@utad.pt.

proteins are separated first by molecular weight then by isoelectric point thus maximizing the separation factor between proteins. Once separated, specific proteins can be identified by mass spectrometry (MS) followed by bioinformatics searches for signature sequences. Advanced mass spectrometry techniques such as matrix-assisted laser desorption-ionization time-of-flight mass spectrometry (MALDI-TOF MS), liquid chromatography mass spectrometry (LC-MS) and electrospray ionization mass spectrometry (ESI-MS) can be used to obtain peptide mass fingerprints, and are ways of obtaining accurate information on protein expression levels. Pathway analysis can be used to screen profiles of genes associated with specific biological processes and pinpoint possible functional interactions between genes.

Recent technological developments that make protein and peptide separation more rapid have improved the accuracy and level of detail possible in large-scale proteome analysis of microbiomes. Such approaches have been used to study the impact of chemical exposure, stress caused by an antimicrobial agent, or environmental conditions on the composition of the proteome. Although it is difficult to detect certain protein classes like low-abundance or hydrophobic proteins and overcome certain instrument limitations such as 2D electrophoresis reproducibility, new biological information has been discovered through proteomics and its applications in pure and applied research, like the discovery of useful biomarkers.

This chapter will consider how proteomic approaches can contribute to the study of human or animal microbiomes and will be illustrated by many case studies.

Keywords: Proteomics, Microbiome, Mass Spectrometry, Bioinformatics, Electrophoresis

Introduction

Proteomics is an important part of biological research nowadays enabling accurate protein identification, discovery of disease mechanisms and new drug targets, and biomarker development. Techniques for protein separation, mass spectrometry and bioinformatics are combined to produce complete and accurate proteome records. When compared to the genome, the proteome is highly dynamic, much more complex and extensive considering that for each gene one or more proteins can be expressed. Thus, proteomics faces greater technological challenges than genomics or transcriptomics. Some of these limitations are being overcome over time but there is also scope for better and more accurate usage of current proteomic tools.

The Human Microbiome Project (HMP) was created to characterize microbial communities populating niches at different sites of the human body, with the aim of defining a core human microbiome if such exists. This would give a reference from which to observe the global effects of drugs, diseases or stress on protein expression in a given microbiome. The use of proteomics for microbiome research has led to the creation of various databases. The biological and biochemical knowledge of the microbiome can be thus taken to a higher level, by interpreting, analyzing and cross-referencing the swathes of information available in these different systems.

Proteomics in Bacterial Research

Human Microbiome Project and Related Research

The concept of a 'microbiome' was first suggested by Joshua Lederberg to collectively describe the ecological community of microorganisms that inhabit higher organisms like *Homo sapiens* (Group et al. 2009). The human body and its microbiome have probably co-evolved convergently and divergently to establish communities made up of *mutualistic* (symbiotically beneficial), *commensal* (neither harmful nor beneficial) and *pathogenic* (detrimental to the host) microbe-host relationships in physiologically distinct niches at different body sites (Aagaard et al. 2013).

The Human Microbiome Project (HMP) was created by the National Institutes of Health, belonging to the United States Department of Health and Human Services, with the goal of characterizing microbial communities in different parts of the body, such as the gastrointestinal tract or oral cavity, which from the microbial point of view can be considered as different environmental niches. The task is to determine whether a core human microbiome exists, one that is necessary for normal physiological functions, and then investigate whether changes in the human microbiome are caused or correlated with disease (Figure 1) (Dave et al. 2012). The first step was to catalogue the microorganisms *in situ* by initially sequencing copies of the 16S ribosomal RNA gene (16S), a standard phylogenetic and taxonomic marker used to identify members of microbial communities (Gevers et al. 2012).

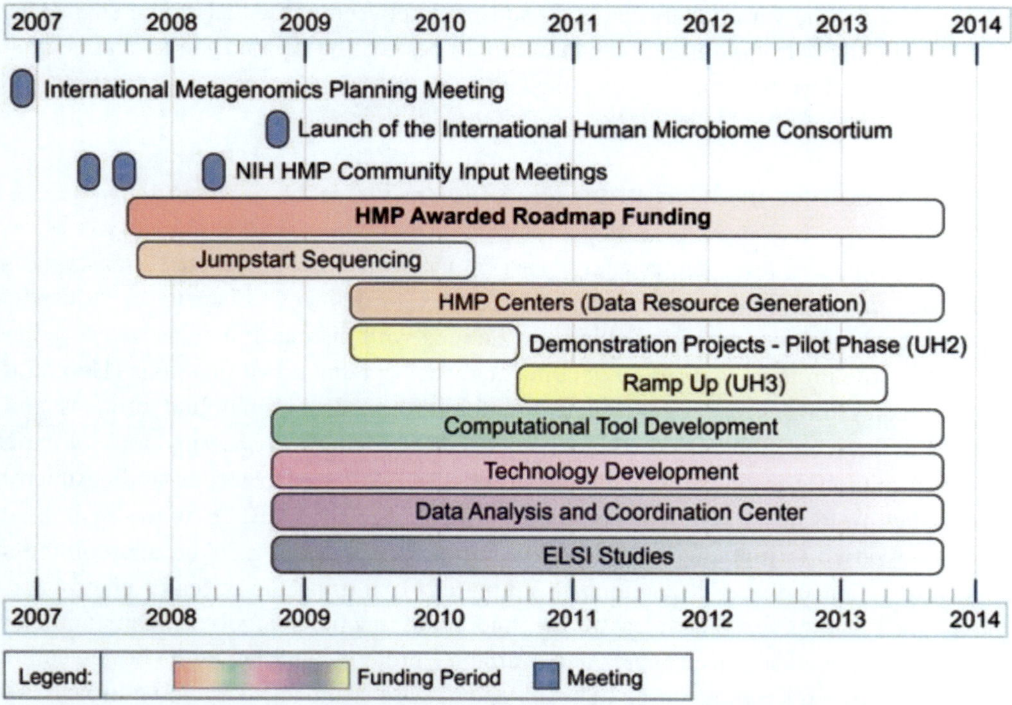

Reproduced by permission from the NIH Working Group (Group et al. 2009): The NIH Human Microbiome Project. Genome Research, 2009. 19 (12).

Figure 1. Human Microbiome Project Timeline.

The HMP then focused on generating reference genomes (viral, bacterial and eukaryotic), which provide a vital framework for subsequent metagenomic annotation and analysis, and a baseline description of microbial community structure and function from an adult cohort (Human Microbiome Project 2012). The HMP aims to establish an integrated "systems metagenomics" approach to describe the microbiota, the microbiome, the "meta-transcriptome", the "meta-proteome" and the "meta-metabolome" (Turnbaugh et al. 2007). Both conceptually and experimentally, the HMP is a logical extension of the Human Genome Project (HGP) in that it is an interdisciplinary effort consisting of multiple projects carried out worldwide that benefit from high-throughput data generation like highly parallel DNA sequences and precise and sensitive mass spectra (Turnbaugh et al. 2007). The first phase of the HMP was recently completed demonstrating that 1) the composition and function of the human microbiota contributes to metabolic and regulatory networks that are vital for the survival of the human body, and 2) the composition of the microbiota is unique in each individual while functions are conserved between individuals (Foxman and Rosenthal 2013).

One of the projects related to the HMP financed by the European Commission is Metagenomics of the Human Intestinal Tract (MetaHIT) which has the objective of establishing associations between genes of the human intestinal microbiota and human health and disease, with a major focus on inflammatory bowel disease and obesity (Koren et al. 2013). Large-scale surveys like MetaHIT and HMP show that we are at the threshold of a richer functional understanding of the human microbiome (Morgan et al. 2013). MetaHIT and HMP are often correlated and cross-linked through the use of the most recent sequencing technology using 16S rRNA or whole-genome shotgun metagenomic data, as this allows patterns of microbial variety between body sites and between individuals to be characterized (Koren et al. 2013). In the future, human microbiomes will be defined for several body sites and for different developmental stages (Aagaard et al. 2013).

Use of Proteomics in the Human Microbiome Project

At the single-cell level, omics analysis – genome sequences, epigenetics, transcriptomes, proteomes, metabolomes, interactomes and others – delineate discrete or "quantized" subpopulations of microbes representing populations with distinct functions. However more than 60% of the proteins defined by genome analysis have unknown functions (Hood 2012). Several studies have been conducted to identify thousands of microbial proteins and to understand their functions (Verberkmoes et al. 2009). Knowing the functions of human-associated microbes is of particular interest and the HMP has therefore studied microbial communities from many different body sites (Abubucker et al. 2012). When an individual human is exposed to immune, environmental, or dietary changes, a given microbiome may adapt to maintain the overall community structure. Monitoring the series of changes at the protein level will give some idea of the microbial functions involved. Determining the functions of uncharacterized core and variable protein families will be essential in understanding the microbiota's role in health and disease (Human Microbiome Project 2012). MALDI-TOF MS carried out by peptide mass fingerprinting is largely used, based on an accurate mass measure of a set of peptides derived from sequence-specific proteolysis. The determination of a protein's specific mass fingerprint is then a major value for proteomics and for the advancement of HMP (Marvin et al. 2003).

On the other hand, for high-performance comparative metagenomics open source software named Metagenomics Reports (METAREP) was created at the J. Craig Venter Institute (JCVI) for effective storage and dynamic querying of annotation data. METAREP allows the direct analysis of results produced by two pipelines used in the HMP. JCVI Prokaryotic Metagenomics Annotation Pipeline (JPMAP) is used for annotation of open reading frames from assemblies and provides an efficient system for identifying and functionally classifying the proteins encoded in shotgun metagenomics sequencing data from sampled communities of prokaryotic organisms (Goll et al. 2012). HMP Unified Metabolic Analysis Network (HUMAnN) is used for short reads annotations and is a scalable methodology to reconstruct the functional potential of microbial communities from metagenomic sequences and it can also be applied to metatranscriptomic or metaproteomic data using genes or pathways of interest (Figure 2) (Abubucker et al. 2012; Goll et al. 2012).

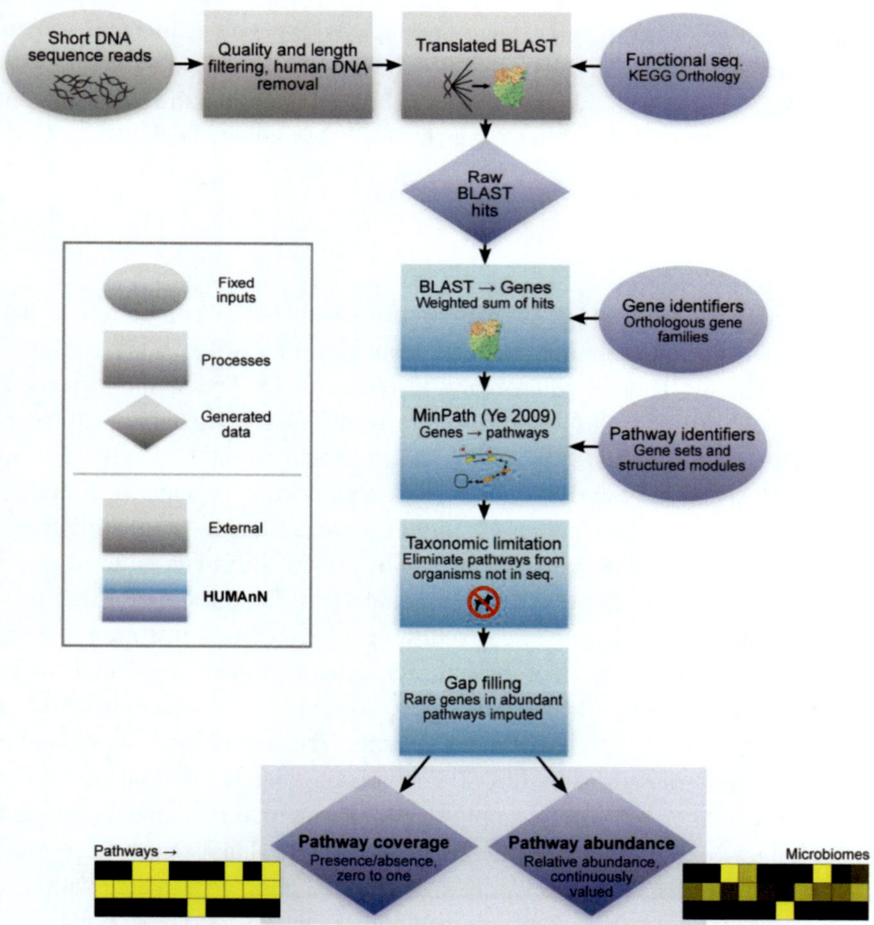

Abubucker S., Segata N., Goll J., Schubert A.M., et al. (2012) Metabolic reconstruction for metagenomic data and its application to the human microbiome. PLoS Comput Biol 8(6): e1002358. Doi: 10.1371/journal.pcbi.1002358. http://www.ploscompbiol.org/article/info:doi/10.1371/journal.pcbi.1002358 (Abubucker et al. 2012).

Figure 2. Overview of the HUMAnN method for metabolic and functional reconstruction from metagenomic data.

The first model-based taxonomic profiling approach, the Taxy-Pro method of Klingenberg and colleagues (2013), utilizes the frequencies of domain families designated by the Pfam database to define protein signatures of microbes. Protein regions that share a significant degree of sequence similarity are likely to have homologous functions. Taxy-Pro includes 92 phage genomes and 102 viral metagenomes in addition to 1,730 bacterial, 122 archaeal and 50 eukaryotic genomes giving 2,098 functional reference signatures (Punta et al. 2012; Klingenberg et al. 2013). Another example of a proteomic tool is the free software MultiAlign that has already been used to study the mouse spleen signaling response to anthrax, where data from LTQ-Orbitrap was analyzed and the peaks matched to phosphopeptide peaks identified by MASCOT (Matrix Science) analysis (see Proteomic Tools and Table 1 for more explanation of these methods) (Manes et al. 2011).

The Integrated Microbial Genomes and Metagenomes (IMG-M) system provides support for comparative analysis of metagenome sequence data generated by sequencing microbial communities, in an effort to integrate the continuously expanding universe of genome and metagenome datasets. IMG/M-HMP is an IMG/M Expert Review data mart focused on metagenome datasets produced by the US National Institute of Health HMP and is part of the HMP Data Analysis and Coordination Center (DACC) (Markowitz et al. 2012).

Case Studies

To characterize the ecology of human-associated microbial communities, the HMP is committed to analyzing microbiome samples from sets of distinct body habitats from the largest possible cohort (Human Microbiome Project 2012). Fodor and colleagues (2012) compared the HMP's 16S-based survey data from 18 different body habitats in more than 200 "healthy" volunteers to another 16S reference collection to first identify the operational taxonomic units (OTUs) and then draw up a list of high priority organisms to study further. If two 16S sequences show 97% sequence identity they are defined as originating from the same species, so imperfect 16S relatedness (in the 3% of non-identity) can be used as a proxy for taxonomic and genomic relatedness. Less common HMP OTUs are expected to show less identity to 16S data from other human derived samples, suggesting that the less frequent taxa are transitory organisms in the microbiome, not generally maintained in healthy humans (Figure 3) (Fodor et al. 2012). In another study, Aagaard and colleagues (2013) evaluated longitudinal changes in an individual's microbiome. The microbiome was sampled two or three times from various body sites of 300 individuals, yielding a total of 11,174 primary specimens, from which 12,479 DNA samples were submitted to four centers for metagenomic sequencing. Some of the differences found between individuals may be correlated with human phenotypic differences like sex or ethnicity with a minimal variance within subjects of a single sex (Aagaard et al. 2013). Obesity, metabolic syndrome and type 2 diabetes are major public health challenges and the human intestinal microbiota is believed to play a potential role in the prevalence of these disorders. To understand such disorders and their treatment, the HMP consortium has sequenced approximately 800 human-associated reference genomes creating a useful database of the human microbiome (Devaraj et al. 2013).

Unrelated to the HMP, there are also many other microbiome studies. Lu and colleagues (2012) demonstrated that *Helicobacter hepaticus* infection of mice to model inflammatory bowel disease largely changed many metabolite structures. Involvement of the gut microflora in altering the blood metabolites of infected animals was also demonstrated. Sensitive MS-based profiling of relevant animal models might suggest potential biomarkers for diagnosis and progression of inflammatory bowel disease and similar disorders (Lu et al. 2012).

To study microbial communities, metagenomic studies are used to determine the pool of genetic diversity. However, in order to understand the functional activities of the cells in various communities, metaproteomics can be used to assess the proteomes of two or more species in an environmental sample. Thus, through metaproteomics it is possible to determine the expressed proteins of a type of cell at specific and culture conditions (Seifert et al. 2012). Early metaproteome research dealt with low complexity microbial consortia from acid mine drainage biofilms or sludge water bioreactors, a good starting point for the study of more complex systems (Hettich et al. 2012).

Proteomic Tools

Methods for Protein Separation

Developed independently by O'Farrell, Klose and Scheele in 1975, two-dimensional polyacrylamide gel electrophoresis (2-DE) is a powerful technique that allows the separation of complex protein mixtures into discrete protein 'spots' in a gel, based on their isoelectric point (p*I*) and molecular weight (MW) (Figure 3). In addition to resolving a large number of proteins, one advantage of 2-DE as a separation technique is that the relative abundance of the proteins can be quantified when they are stained (Issaq and Veenstra 2008). In one of the first studies of 'bacterial proteomics', O'Farrell separated *Escherichia coli* cell extracts into thousands of protein components using the 2-DE approach (Curreem et al. 2012). Since then 2-DE has become the most well established technique used in proteomics, yielding the highest resolution for proteins besides MS (Fujii et al. 2006). In fact, nowadays 2-DE followed by enzymatic in-gel digestion of the peptides and subsequent identification by MS is the most frequently used methodology for analyzing complex protein mixtures (Ihling and Sinz 2005). The initial protocol of O'Farrell used chaotropes and detergents for protein solubilization and separation in both dimensions – Isoelectric focusing (IEF) and SDS-PAGE – for p*I* and MW, respectively), but these have been replaced by immobilized pH gradients (IPGs) (Gorg et al. 2009). IPGs are based on acrylamide derivatives with the general structure CH_2 = CH-CO-NH-R, where R contains either an amino or a carboxyl or group, which form a series of acrylamide buffers with different p*K* values between p*K* 1.0 and 13 (Figure 3) (Gorg et al. 2004).

The most common methods for protein visualization after 2-DE are colloidal Coomassie Brilliant Blue and silver staining, whose limits of detection are at picomole and femtomole levels, respectively. Both are compatible with downstream MS analysis. The later development of fluorescent dyes with sensitivities similar to silver stain and which are also compatible with MS analysis was another significant advance in proteome comparison (Rotilio et al. 2012). Certain protein classes have a limited dynamic range making analysis difficult, so the trend in proteomics is towards the development of non-gel based technologies, which allow a more comprehensive and less biased analysis (Sandra et al. 2008). More attention has been given to bottom-up proteomics in which proteins are first digested into peptides and then separated by various types of liquid chromatography (LC) (Tao et al. 2011).

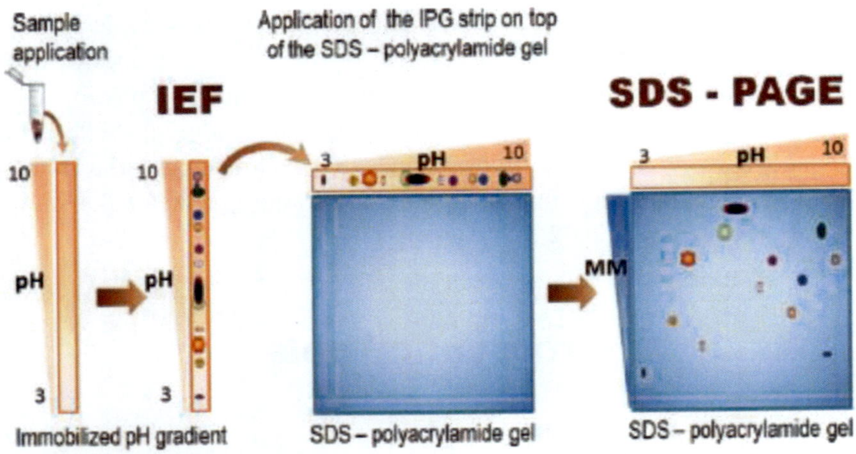

Reprinted by permission from Teixeira M.C. et al., Teaching expression proteomics: From the wet-lab to the laptop. Copyright © 2008 WILEY-VCH Verlag GmbH & Co. KGaA, Weinheim (Teixeira et al. 2009).

Figure 3. Schematic representation of the process of protein separation by two-dimensional electrophoresis. IEF, isoelectric focusing; SDS-PAGE, sodium dodecylsulphate-polyacrylamide gel electrophoresis.

LC is a liquid phase separation technique that has advanced significantly in the last decade. The physicochemical properties of peptides (like mass, charge and hydrophobicity) make them suitable for efficient LC separation. Reverse phase (RP) is the LC format used most often in proteomics because of its high-resolution in separating proteins and peptides on the basis of hydrophobicity and the compatibility of the LC mobile phases with subsequent MS (Xie et al. 2012). The principle of RP is based on partitioning of the analytes between a hydrophobic stationary phase and a polar hydrophilic mobile phase. Peptides are loaded onto an RP column under low-organic-solvent conditions, causing the peptides to partition into the stationary phase material with salts remaining in the low-organic-solvent. Separation is then achieved by gradually increasing the organic content of the mobile phase when the more hydrophobic peptides start to move out of the stationary phase as it moves down the column (Di Palma et al. 2012).

Proteome Comparison

In 1997, Ünlu and colleagues described a modification of 2-DE that allowed reproducible differences between two distinct protein samples to be visualized in a single gel. The principle of this technique is to differentially label samples with dyes that fluoresce at different wavelengths, samples are then combined before IEF and electrophoresed in the same 2-D gel (Arruda et al. 2011). This 2-DE technique was named difference gel electrophoresis (DIGE).

The most common fluorescent protein labeling dyes used are synthetic N-hydroxysuccinimidyl (NHS) ester derivatives of the cyanine dyes Cy3 and Cy5, although other NHS-fluorophores can be used to label target proteins through nucleophilic substitution reaction of N-terminal α-amino and lysine ε-amino groups (Figure 4) (Timms and Cramer 2008). With the development of a variety of fluorophores from cyanine dyes (Cy2, Cy3, Cy5), Alexa Fluor (488, 546, 594), infrared dyes (DY-680, DY-780) and DyLights (549, 649, 680, 8000), the DIGE technique became very easy to use with a wide range of options (Figure 5) (Riederer 2008).

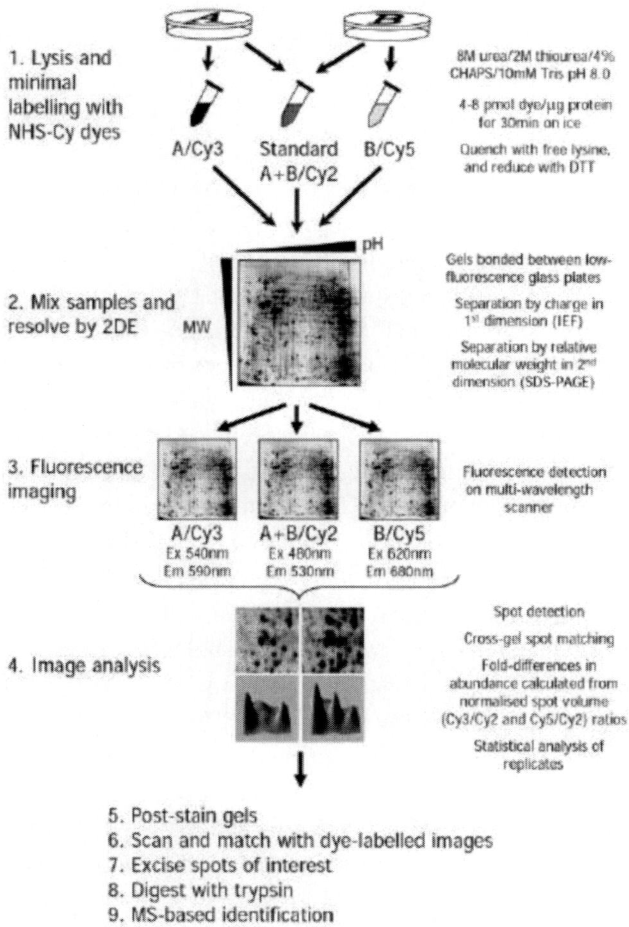

Reprinted by permission from Timms J.F. and Cramer R., Difference gel electrophoresis. Copyright © 2008 WILEY-VCH Verlag GmbH & Co. KGaA, Weinheim (Timms and Cramer 2008).

Figure 4. Schematic representation of the 2D DIGE protocol for minimal lysine labeling using an internal standard for normalization and cross-gel spot matching.

DIGE was developed to determine changes in protein expression in biological samples with higher accuracy and reproducibility than could be obtained in normal 2-DE while avoiding time-consuming gel staining and image analysis (Minden 2012; Becnel and McKenna 2012). In fact, it has been reported that the correlation between quantitation by DIGE and very precise metabolic stable isotope labeling is exceptionally good (Rotilio et al.

2012). DIGE has become widely accepted as a tool for separation, detection and quantitation, particularly in disease biomarker discovery (Issaq and Veenstra 2008).

Multiplexed proteomics (MP) is the detection of total proteins and specific proteins in the same gel or blot using reagents that have contrasting colors. MP makes it more straightforward to detect differences in protein composition and significantly reduces the total number of gels required for an experiment (Riederer 2008). The MP approach makes it possible to determine the phosphorylation, glycosylation and general protein expression patterns within a single gel electrophoresis experiment through a series of staining step using, for example, Pro-Q Diamond, Pro-Q Emerald and Sypro fluorescent staining, respectively (Rotilio et al. 2012).

Mass Spectrometry

One of the most popular high-throughput methods in proteomics is mass spectrometry (MS) which measures the mass-to-charge ratio (*m/z*) of charged particles to determine the mass and elemental composition in complex mixtures of individual proteins or the relative quantity of proteins of interest (Becnel and McKenna 2012). The use of MS in full-scale proteomics complements genomic techniques (Intelicato-Young and Fox 2013). The determination of accurate mass (AM) is used to confirm the elemental formulae of compounds. Since Aston (1927), magnetic sector mass analyzers have mostly been used for AM measurements, but nowadays these are being replaced by more accurate technologies with greater resolving power. Time-of-flight (TOF), hybrid quadrupole-TOF (Q-TOF), Fourier transform ion cyclotron resonance (FT-ICR) and more recent Orbitrap instruments are more commonly used now (Figure 6) (Cortes-Francisco et al. 2011).

The majority of MS assays involve proteins in the range of 500-3000Da because proteins are digested with proteases (usually trypsin) beforehand and it is the resulting peptides that are then analyzed and fragmented, in so called bottom-up MS. Larger molecular species (5-50kDa range) are analyzed in the gas phase as intact molecules and their multiple-charged ions are used as precursors in tandem mass analysis. This is top-down MS (Armirotti and Damonte 2010). After protein mixtures are separated by 2-DE and protein spots are excised from the gel and digested *in situ* by proteases, the resulting peptide mixtures must be vaporized and ionized before MS (Ihling and Sinz 2005). MALDI and ESI are the most commonly used ionization techniques in protein MS. MALDI is usually coupled with TOF technology to visualize biomolecules such as lipids smaller than 1kDa as well as peptides and proteins over 10kDa (Duncan et al. 2008; Saito et al. 2012). On the other hand, ESI coupled with MS allows the identification of distinct protein patterns among bacterial species, and coupled with tandem mass spectrometry (MS/MS) the identification of proteins by their peptide sequence (Intelicato-Young and Fox 2013). ESI produces gaseous ions from samples in solution so is also easily coupled to LC and capillary electrophoresis (Lane 2005). Generally MS/MS is routinely combined with MALDI and ESI in protein and peptide sequencing and identification of targeted microorganisms because of its flexibility, sensitivity, speed and accuracy (Intelicato-Young and Fox 2013; Dworzanski et al. 2006; Hesse et al. 2008).

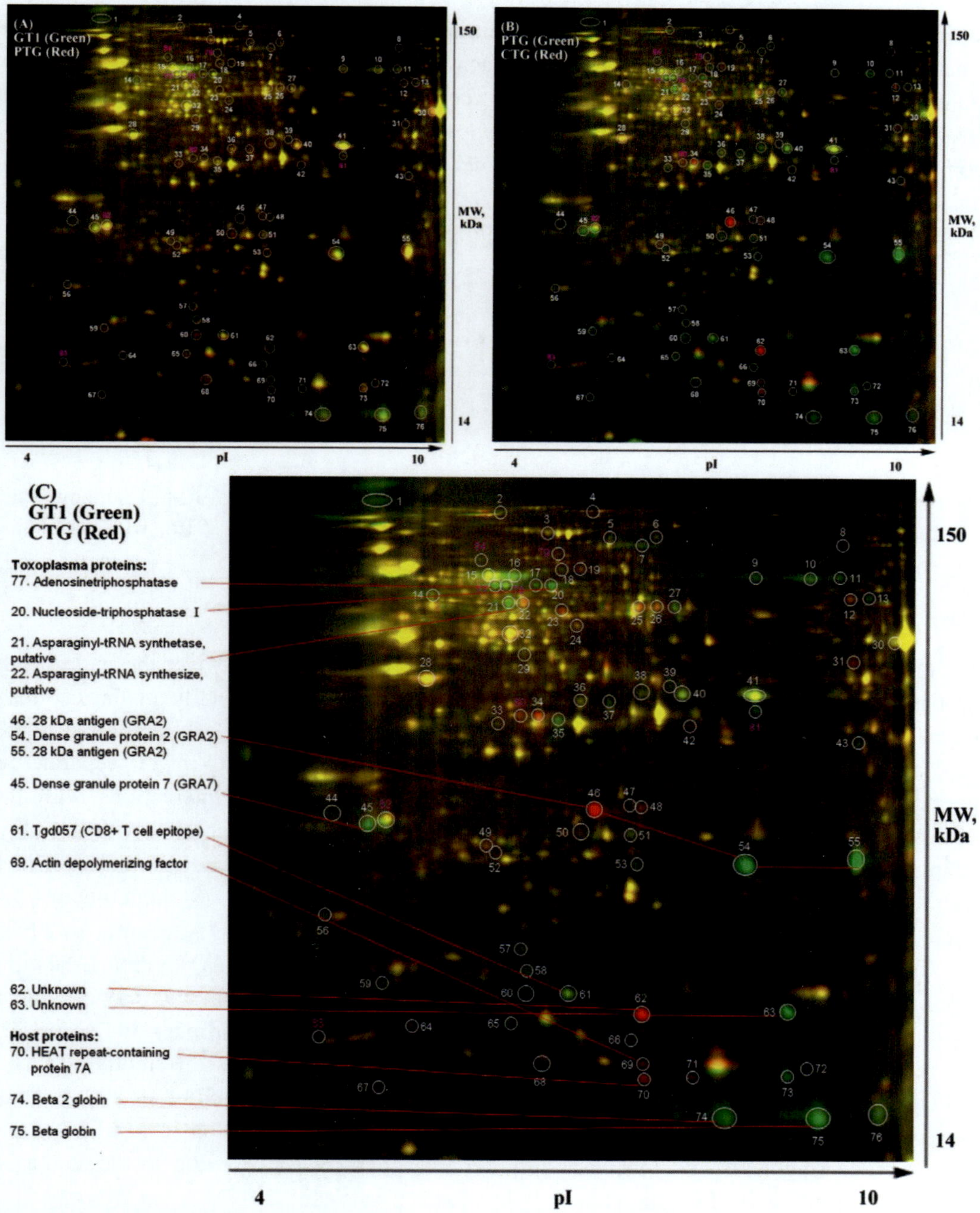

Reprinted from Experimental Parasitology, 133(4), Zhou H. et al., Differential proteomic profiles from distinct *Toxoplasma gondii* strains revealed by 2D-difference gel electrophoresis, 376-382. Copyright (2013) with permission from Elsevier (Zhou et al. 2013).

Figure 5. 2D-DIGE of *Toxoplasma gondii* proteins. Proteins were first separated in the first dimension at pH 4.0-9.0 and on a 12% SDS-PAGE gel in the second dimension.

When an Orbitrap mass analyzer is used, a significant population of ions is first trapped in an external injection device in a radio frequency (RF)-only gas-filled curved quadrupole (C-Trap). C-Trap injects ions into the Orbitrap analyzer in a short pulse so that each mass-to-charge (m/z) population forms a sub-microsecond pulse (Figure 6). Short ion packets then surround a curved central electrode of the Orbitrap analyzer, which are then submitted to an electrical field leading to axial oscillations. Each ion packet spreads rapidly forming a thin rotating ring that oscillates along the Orbitrap central electrode producing an image current. Broadband detection of such signal is then followed by a quick Fourier transform (FT) to convert the recorded time-domain signal into a *m/z* spectrum (Makarov and Scigelova 2010).

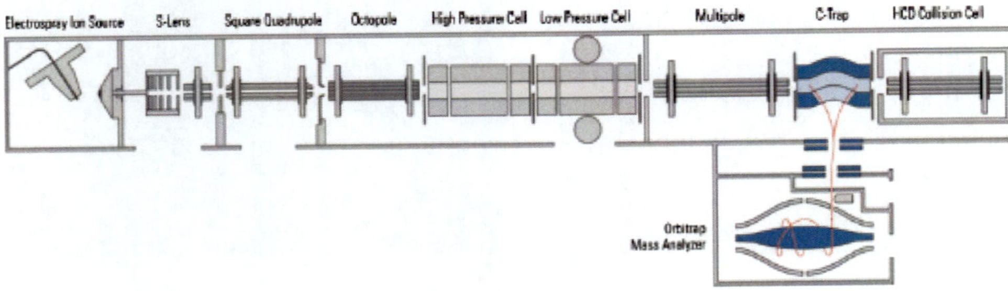

Reprinted from Journal of Chromatography A, 1217, Makarov A. and Scigelova M., Coupling liquid chromatography to Orbitrap mass spectrometry, 3938-3945, Copyright (2010), with permission from Elsevier (Makarov and Scigelova 2010).

Figure 6. Schematic layout of the LTQ Orbitrap Velos mass spectrometer.

Of all liquid phase separation methods, LC has advanced the most over the past decade probably because of its many applications in protein research, especially in the capillary format; the physicochemical properties of peptides make them particularly amenable to LC separation (Xie et al. 2012). HPLC is coupled to mass spectrometry (LC-MS) to separate the peptide mixture in the liquid phase on the basis of hydrophobic interactions with the stationary phase of the chromatography column. Column beads are decorated with C^{18} alkyl chains. After cell fractions are analyzed individually by LC-MS, peptides are selected then fragmented for MS/MS. MS and MS/MS spectra are able to serve as input queries for database search engines to identify the corresponding peptide sequences (Altelaar et al. 2013).

The isobaric tag for relative and absolute quantification (iTRAQ) peptide labeling approach has been successfully used in the discovery of biomarkers from a variety of clinical specimens. In this method the labels are isobaric so they react with primary and secondary amines of peptides. Each label has an identical mass but fractionates differentially during collision-induced fragmentation during the MS/MS procedure (Latterich et al. 2008). Thus, iTRAQ allows for simultaneous protein identification and quantification in MS/MS output from peptide fragments and low mass reporter ions in a robust way due to the stable N-hydroxysuccinimide (NHS) ester chemistry involved (Evans et al. 2012).

Shotgun proteomics is a way of using MS to detect and identify all proteins from a given lysed cell mixture without the need for gel-based separation or de novo sequencing. Peptides resulting from enzymatic digestion of the entire proteome are separated by LC and infused directly into rapidly scanning tandem mass spectrometers (2D-LC MS/MS) through ESI (Figure 7) (Verberkmoes et al. 2009).

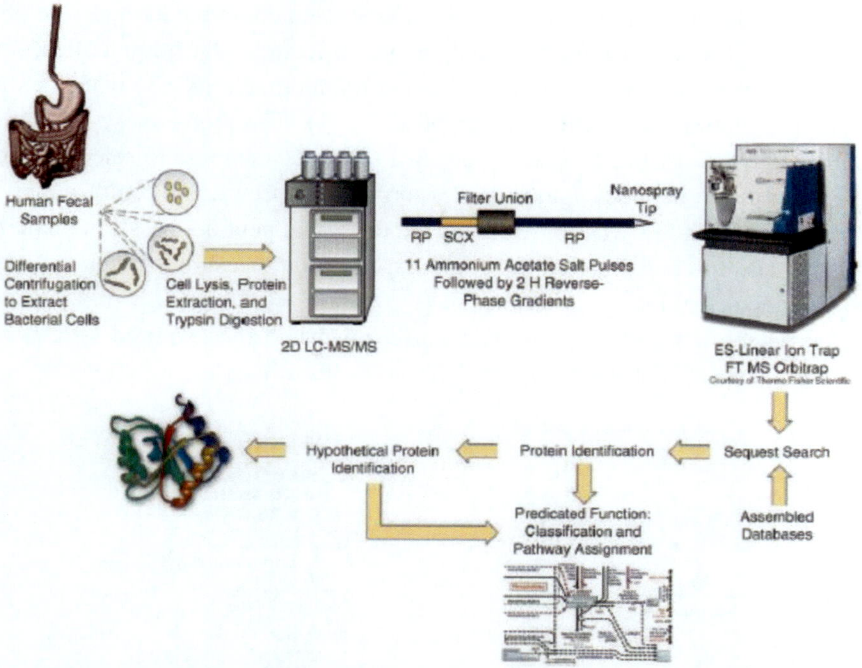

Reprinted by permission from Macmillan Publishers Ltd. The ISME Journal Shotgun metaproteomics of the human distal gut microbiota, 3, 179–189, copyright (2009) (Verberkmoes et al. 2009).

Figure 7. Shotgun metaproteomics approach used to identify microbial proteins in human faecal samples.

Imaging MS (IMS) is an emerging technique that is expected to overtake more conventional histological examinations like those using green fluorescent protein. In IMS mass spectra of proteins in small areas of a sample tissue section are collected systematically with reference to the spatial coordinates of the section. Simultaneous recording of mass spectra of various molecules can show their different expression patterns in tissues (Saito et al. 2012). Imaging can be performed either using a top-down approach identifying low-molecular-weight proteins by direct MS fragmentation (without prior digestion) or with the bottom-up approach where protein samples are digested by depositing trypsin solutions on the tissue sample before MS analysis (M et al. 2012).

Bioinformatics

The ultimate goal of proteomics is to integrate biology, analytical chemistry and bioinformatics to understand and then predict how cells respond to changes in their environment. For this bioinformatics tools need to integrate direct experimental evidence related to proteins into a model that describes the state of the cell (Panisko et al. 2002). Recent advances in MS and bioinformatics make it possible to estimate the absolute amount of proteins in terms of copy number of proteins per cell (Altelaar et al. 2013).

Computational analysis typically relies on two main components. First, a peptide identification algorithm assigns amino acid sequences to fragmentation spectra by searching an *in silico* digest database, a list of candidate peptides that could theoretically be obtained by digesting an organism's proteome. Second, a quantification algorithm collects peak data across mass spectra in the form of extracted ion chromatograms (XICs) over the duration of chromatographic elution of peptides (Khan et al. 2011). Therefore, most MS/MS database search tools read a set of MS/MS spectra, query a sequence database to select peptides of the right mass, score those peptides against the experimental spectra, and return putative peptide identifications. For each MS/MS spectrum, an experimental peptide mass can be derived from the precursor mass-to-charge ratio (m/z) and assumed or measured precursor charge state (Figure 8) (Eng et al. 2011).

Different database search algorithms are used to match the acquired spectra to peptide sequences from a protein database (Figure 8, Table 1) (Rohrbough et al. 2006).

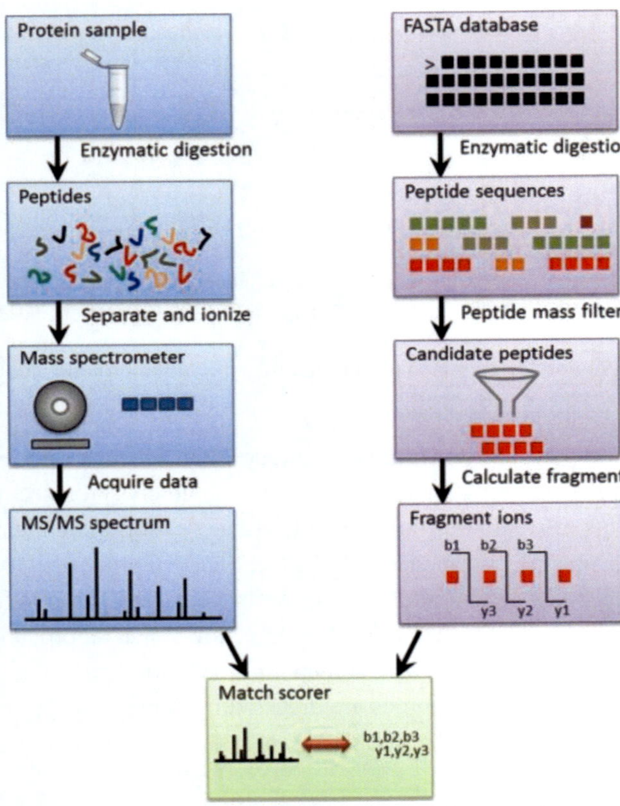

Reprinted by permission from Eng et al., A face in the crowd: recognizing peptides through database search, Mol Cell Proteomics, 2011. 10(11). Copyright © 2011, by the American Society for Biochemistry and Molecular Biology (Eng et al. 2011).

Figure 8. Data acquisition and database search. For a given experimental MS/MS spectrum, protein sequences from a database are *in silico* digested and peptides of the right mass are selected, scored against the experimental spectrum and the best matching peptides and their scores are reported.

Sequence databases, usually in FASTA format, are the source of protein and peptide sequences for database search algorithms. Rather than consider all possible sequences (the province of *de novo* algorithms), these tools limit their search to the sequences present in the database (Table 2) (Eng et al. 2011).

Table 1. Selected MS/MS spectral search engines

Name	Description
MASCOT	Identifies proteins via peptide mass fingerprint, sequence queries, and MS/MS ion searches from several public peptide/protein and contaminant databases
Open MS Search Algorithm (OMSSA)	Searches known sequence databases and assigns probability scores with a BLAST algorithm
Phenyx	Searches public databases and other search engines included here for validation as well as extra search functionalities
ProbID	Component of Trans-proteomic pipeline (TPP) that searches protein sequence databases
SEQUEST	Original and still popular engine for searching sequence databases now distributed by Thermo
X!Tandem	Searches against sequence databases and calculates statistical confidence to sequence assignments and maps these assignments to known protein sequences
MR-X!Tandem	Cloud-based (i.e. internet) version of X!Tandem that runs multiple searches in parallel on different servers, making it useful for very large peptide searches

Table 2. Overview of web accessible algorithms and databases

Name	Web address
dbEST	ftp://ftp.ncbi.nlm.nih.gov/blast/db/est.Z
NCBInr	ftp://ftp.ncbi.nlm.nih.gov/blast/db/nr.Z
Swiss-Prot	ftp://ftp.expasy.ch/databases/sp_tr_nrdb/sprot.fas.gz
Csm/Dictyostelium discoideum	ftp://ftp.csm.biol.tsukuba.ac.jp/pub/dna_csm/csm.fasta
OWL	ftp://ftp.ncbi.nlm.nih.gov/repository/OWL/owl.fasta.Z

Reprinted by permission from Reinders J. et al., Challenges in mass spectrometry-based proteomics. Copyright © 2008 WILEY-VCH Verlag GmbH & Co. KGaA, Weinheim (Reinders et al. 2004).

Protein Identification

Amino Acid Sequencing Using MS

Protein analysis is a tough challenge because amino acids, the "backbone" of proteins, have highly diverse physicochemical properties. The proteome includes many proteins with isoforms originating from alternative splicing or various post-translational modifications, and the complexity is further amplified by the interconnectivity of proteins in complexes and signaling networks (Altelaar et al. 2013). Thus, as already discussed, proteins need to be extracted, solubilized and separated by the processes of 2-DE or LC with further digestion in specific proteases. The resulting peptide mixture is mass-measurable and fragmentable by MS

(Krug et al. 2011). Through peptide fragmentation, MS can be used to identify sequences, deliver molecular mass profiles, and define post-translational modifications (PTMs) for both soluble and membrane proteins and even for *de novo* sequencing (Medzihradszky 2005; Souda et al. 2011).

MS is considered to be indisputably one of the best techniques for the analysis of the covalent structure of proteins, including those of therapeutic importance. It has also made a valuable impact in carbohydrate and nucleic acid structure analysis (Kaltashov et al. 2012). Protein quantitation by MS is accomplished through differential stable isotope labeling or label-free methods (Gygi et al. 1999; D'Imperio et al. 2010). If two or more samples are labeled differentially with light or heavy isotopes, the ratio of the heavy to light isotopes found in the resultant mass spectra can be used to calculate the relative amounts of protein present. In label-free methodologies the signal intensity of any peptide is compared directly between two or more samples with the advantage that data for each sample can be acquired independently and later compared *in silico* (Rotilio et al. 2012).

Considering the various possibilities and resources, MS has been used in several cases of peptide mass analysis, whether in studying biochemical pathways (Gorman et al. 2002; Upadhyay et al. 2013; Dziadosz et al. 2012), specific diseases or conditions (Clerico et al. 2007; Owens et al. 2011; Iagnocco and Naredo 2012; Grela et al. 2012), drug development (McDonald and Yates 2002; Wu and Vogt 2012; Meyer and Maurer 2012) or bacterial proteins (Boyer et al. 2011; Chen et al. 2012). Attempts have already been made to study so-called hypothetical proteins, peptides predicted to be encoded by nucleic acid sequences that have yet to be proven to exist experimentally. Such structures might represent up to half of the potential protein coding capacity of a genome (Lubec et al. 2005).

Effects of Antibiotics on Protein Expression of the Microbiome

Antibiotics are used to treat specific bacterial infections but they are also known to kill or inhibit other bacteria that are not the primary targets, such as resident commensal gut microbiota. The extent of the impact on the non-targeted microbial populations depends on the spectrum of action, mode of administration, dose and absorption rate (Ladirat et al. 2013). Antibiotics targeting protein biosynthesis have similar targets and, accordingly, induce a common set of genes that are not induced by any of the cell-wall targeting compounds, which induce more specific protein expression profiles. Compilation of comprehensive datasets of expression profiles have made it possible to identify marker genes for specific antibiotic classes (Wecke and Mascher 2011).

Bandow and colleagues (2003) established a database from 2-DE protein analysis of response of gram-positive bacteria *Bacillus subtilis* to treatment with antibiotics from all important established and emerging classes. This study concentrated on the p*I* range of 4 to 7 and used pulse-labeling with L-[^{35}S]methionine to detect the newly synthesized protein fraction (Buttner et al. 2001; Bandow et al. 2003). Changes in the protein synthesis pattern reflect the global response to damage caused in the cell and explains why each antimicrobial compound has a specific protein expression profile. For example, inhibition of Ile-tRNA synthetase caused a stringent response whereas protein synthesis inhibitors interfering with translation accuracy induced class I heat shock proteins known to be induced by misfolded proteins (Bandow et al. 2003). In another study, Roncada and colleagues (2009) characterized the protein expression involved in *Escherichia coli* virulence and antibiotic resistance using pathotypes with the same virulence factors but selecting different antibiotic profiles. Expression of proteins like peptidase, amidase, alkyl hydroperoxide reductase and stringent starvation protein A correlated with stress conditions like antibiotic resistance (Roncada et al. 2009).

Various 2-DE, MALDI-TOF and MS/MS proteomic assays have been used to study antibiotic-resistant extremophiles (AREs), bacteria which develop greater resistance to antibiotics by accumulating multiple genes or plasmids that harbor genes for multiple drug resistance (MDR). Piras and colleagues (2012) used MS/MS to identify differentially regulated proteins involved in metabolic pathways in a multidrug-resistant *E. coli* (Gabani et al. 2012). Membrane and transport proteins such as thiosulfate-binding protein, zinc transport protein B and protein Ox2 are all thought to be related to antibiotic resistance (Piras et al. 2012). Some of the most efficient antibiotics like vancomycin are cyclic peptides produced by non-ribosomal biosynthetic pathways, so for antibiotic resistance studies to be truly comprehensive MS methodologies need to be developed to recognize these kinds of structure (Mohimani et al. 2011).

Antibiotic resistance has also been found in wild animal microbiota, with the detection of genes related to antibiotic resistance or even pathogenicity or virulence. Specific protein expression from such genes was found in what were presumed to be commensal bacteria in the intestinal microbiota (Radhouani et al. 2010; Monteiro et al. 2012). Different concentrations of antibiotics may increase/decrease the expression of a given protein, so is an important issue to consider as more information on the microbiome is obtained with the use of more accurate technologies (Dengler et al. 2011).

Effects of Chemical Agents on the Proteome

Proteins are the most common type of drug target, so one of the major applications of charting the proteome is in target-based drug discovery by the pharmaceutical industry (Bantscheff and Drewes 2012). In fact, chemical proteomic techniques hold considerable potential as a tools in drug design (Miao et al. 2012). Chemoproteomics combines medicinal chemistry, biochemistry and cell biology to focus on the molecular mechanisms of the action of drugs and other bioactive molecules, by characterizing proteome-wide drug-induced changes in cellular protein expression and PTMs (Figure 9) (Bantscheff and Drewes 2012). There are two main approaches in this branch of proteomics. Global cell-wide profiling characterizes the response with respect to protein expression levels or defined PTMs while activity- or affinity-based target profiling employs engineered chemical probes to capture protein targets or even entire sub-proteomes (Figure 9) (Bantscheff and Drewes 2012).

Some case studies illustrate the power of chemoproteomics. Fleischer and colleagues (2010) used affinity-based proteomics to discover that the target of cytotoxic agent CB30865 is nicotinamide phosphoribosyltransferase, an enzyme of the NAD biosynthetic pathway vital for cancer cells to maintain elevated energy metabolism (Fleischer et al. 2010). Using a photoreactive GTP affinity probe, Cisar and colleagues (2013) were able to identify a variety of GTP- and ATP-binding proteins through MS, emphasizing the variability in the purine nucleotide selectivity of certain proteins (George Cisar et al. 2013).

In the near future, chemical proteomics could be seen in applications such as developing diagnostic kits, targeting biomarkers for clinical trials, or identifying novel therapeutic targets (Sutton 2012).

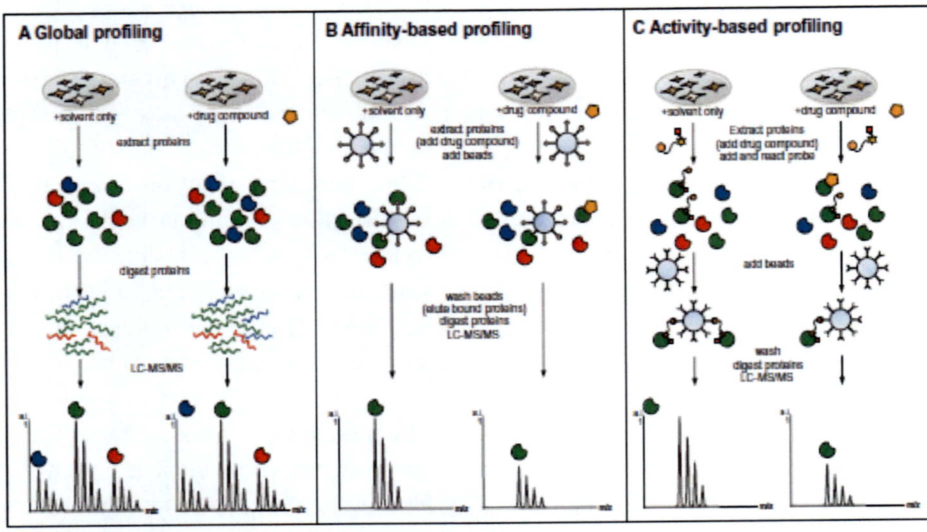

Reprinted from Bioorganic & Medicinal Chemistry, 20, Bantscheff M. and Drewes G. (2012) Chemoproteomic approaches to drug target identification and drug profiling. Copyright (2012) with permission from Elsevier (Bantscheff and Drewes 2012).

Figure 9. Experimental workflows in chemoproteomics. (a) Global proteomics approaches: cultured cells are treated with the compound of interest, followed by the preparation of a cell extract, digestion (typically with trypsin) to peptides, and analysis by liquid chromatography coupled to tandem mass spectrometry (LC–MS/MS). (b) Affinity-based profiling: The drug or compound under study, or more often, a functionalized derivative of the compound of interest, is immobilized covalently, or via a biotin tag, on a solid support. The immobilized drug is subsequently incubated with a cell extract to specifically enrich for target proteins that are subsequently identified by mass spectrometry. Competition with free excess inhibitor added to the cell extract reduces the abundance of captured target proteins, but not of nonspecific interacting proteins. (c) Activity-based profiling: a reactive probe is designed to specifically target the active site of an enzyme family. Following protein extraction, the cell extract is incubated with the probe to covalently attach to its targets. In the second step, probes and targets are purified using affinity chromatography before digestion and LC–MS/MS analysis. Pre-treatment of cells with a small molecule compound binding to the active site of the investigated enzyme family leads to reduced capturing of the target enzyme via the reactive probes.

Effects of Stress on the Proteome

One of the most common sources of stress in cells is heat shock which is efficiently countered by heat shock factors, transcription factors that bind upstream of heat shock protein (HSP) genes that ultimately modulate the heat shock-related gene expression (Xu et al. 2012).

Infection with a bacterial strain is also a cause of stress to the host microbiome, leading to changes in protein expression. In a study of inflammatory bowel disease, Lu and colleagues (2012) infected Rag2$^{-/-}$ mice with *Helicobacter hepaticus* for various periods of time in order to analyze global metabolite changes in healthy and differentially infected individuals. Serum samples were extracted and analyzed by ESI-TOF over a range of 80-1000 *m/z* and using a

C^{18} RP analytical column. The resulting ion chromatograms for samples from infected mice had several peaks of different intensities than corresponding ion chromatograms from samples from healthy mice (Figure 10) (Lu et al. 2012). Changes in metabolites were identified after database searches using exact masses coupled with MS/MS analysis. The conclusion was that the functioning of many metabolic pathways is altered by bacterial infection (Figure 11) (Lu et al. 2012).

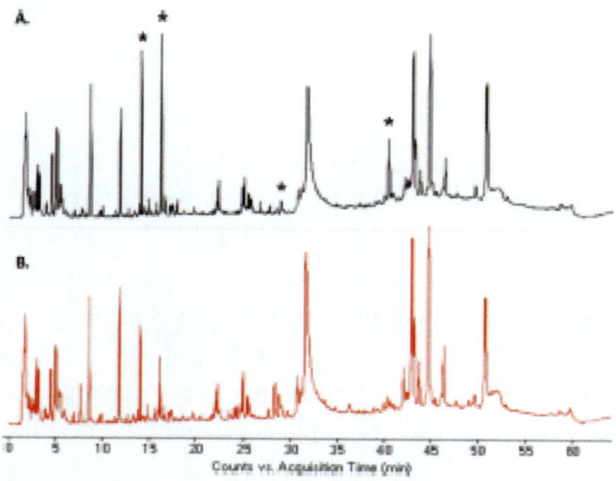

Figure 10. Typical LC-ESI-TOF total ion chromatograms of serum metabolites extracted from (A) uninfected and (B) infected animals (20 weeks postinfection). Data were acquired under positive ionization mode (80-1000 *m/z*).

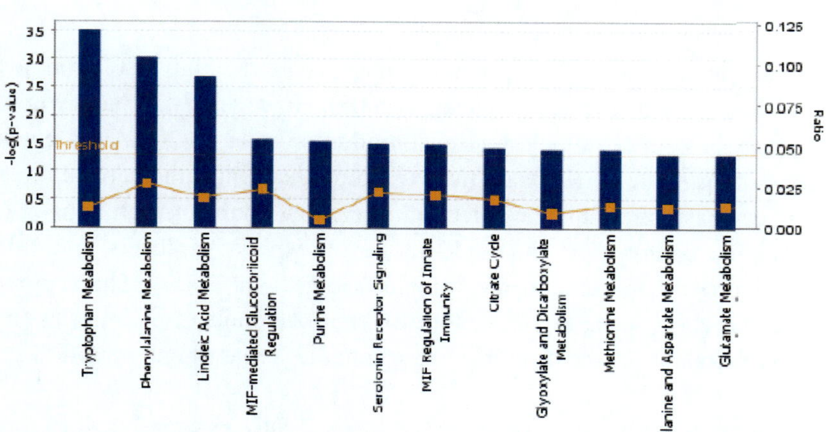

Figure 11. Perturbed metabolism pathways in $Rag2^{-/-}$ mice (20 weeks) infected by *H. hepaticus* revealed by Ingenuity Pathway software.

The Limits of Proteomics

The introduction of supplementary techniques including the use of fluorescent dyes, laser microdissection, highly sensitive MS and data-mining has resolved some of the limitations of 2-D PAGE such as poor reproducibility, the narrow dynamic range, difficulty with protein identification and laborious image analysis (Fujii et al. 2006).

Protein Abundance

Proteins are not expressed in equal amounts and there are large differences in protein levels in all proteomes (Garbis et al. 2005). Low-abundance proteins are a major problem encountered in almost all proteome analysis (Gorg et al. 2009; Renzone et al. 2005). The detection of low abundant proteins, often biologically the most significant (e.g., growth factors, receptors, membrane and signal transduction proteins) may be hampered by proteins of similar size and charge or simply not possible within the detection limits of the analytical techniques used (Rotilio et al. 2012). For a more comprehensive analysis of complex proteomes, the strategy is to divide the proteome by targeting a specific sub-proteome or pre-fractionating the sample (Gorg et al. 2009; Renzone et al. 2005). Low-abundance proteins can thus be enriched (Garbis et al. 2005). For mixtures containing fewer proteins, one-dimensional LC can be used to separate the digested peptides (Renzone et al. 2005).

The presence of high-abundance proteins limits the detection of low-abundance proteins, mainly in the search for biomarkers. Depletion of these proteins by chromatography and precipitation helps in the detection of important low-abundance proteins (Garbis et al. 2005).

2-DE Reproducibility

The main advantages of 2-DE in proteome analysis are the high resolution power and the good visualization of the results (Renzone et al. 2005). However, 2-DE has several limitations because not every protein is suitable for separation (Rotilio et al. 2012). Commonly cited 2-DE limitations often relate to (i) resolution, (ii) reproducibility, the separation of (iii) very acidic or (iv) very basic proteins, the detection of (v) low-abundance proteins or (vi) integral membrane proteins and/or (vii) the wide range of expression levels and differences in solubility of all proteins present in the sample (Gorg et al. 2009). The reproducibility of separation in 2-DE gels is usually satisfactory for small amounts of protein (less than 0.5 mg) but not for larger amounts (1 mg or more) because not all proteins enter the IPG strips (Figure 12) (Garbis et al. 2005).

Although reproducibility has been an issue in 2-DE, especially when profiling two protein mixtures, it has been greatly improved with the use of 2D-DIGE (Issaq and Veenstra 2008). Recent automation of protein identification and proteomic technologies has also increased the throughput and the reproducibility of 2-DE (Renzone et al. 2005).

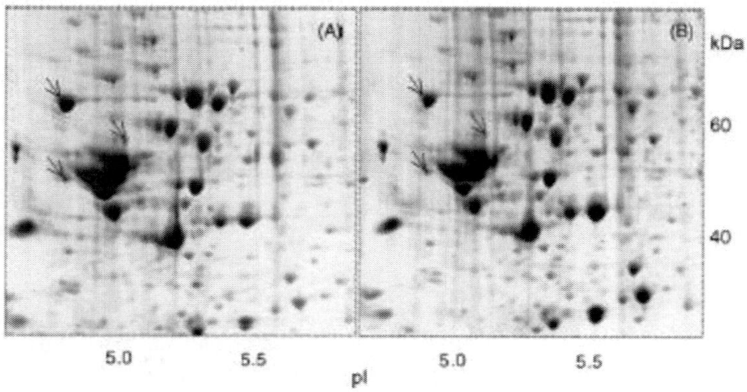

Reprinted from Journal of Chromatography A, 1077, Garbis S. et al., Limitations of current proteomics technologies, 1-18, Copyright (2005), with permission from Elsevier (Garbis et al. 2005).

Figure 12. Partial 2-DE gel images showing the reproducibility of separation of rat brain proteins (A, B). The proteins were separated on 3-10 non-linear (NL) IPG strips and the gels were stained with colloidal Coomassie Blue. (A, B) Separation of rat brain proteins, 1.0 and 1.5 mg, respectively. All spots are present in both gels and the spots in image B are larger in comparison with the corresponding spots of image A, where less protein was applied. However, there are differences between the two images, which should not exist (indicated by the arrowheads).

Protein Characteristics

Theoretical 2-DE maps of proteomes calculated from sequenced genomes indicate that nearly 30% of all proteins have alkaline p*I*s up to pH 12, but it is frequently reported that very alkaline proteins are not accessible to gel-based proteomics. By generating IPGs up to pH 12, e.g. IPGs 6-12 and 9-12 and modifying the standard protocol of IPG-Dalt, highly reproducible 2-DE patterns can be obtained. In particular, the pH gradient IPG 4-12 is extremely useful for the separation of very alkaline proteins as the gradient 'flattens out' between pH 9 and 12 (Gorg et al. 2009). On the other hand, very acidic IPGs can be generated by using acrylamide buffer pK 1.0. IPGs from 2.5 to 5.0 are straightforward to establish and run under standard settings without electroendosmotic flow or other problems. However, few acidic proteins with p*I*s below 3.0 are present in most samples (Gorg et al. 2009). The use of narrow pH range IPG strips, pre-fractionation of the protein mixtures and analysis by LC-MS are therefore recommended for detecting proteins with very acid or basic pHs (Garbis et al. 2005).

Proteins with a molecular weight lower than 10kDa or higher than 150kDa are hardly ever detected using conventional gels. Smaller proteins below the lower mass limit (about 8 kDa) are not efficiently stained by Coomassie blue and are difficult to identify in MALDI-TOF MS with typical matrices (Rotilio et al. 2012; Garbis et al. 2005). Tricine gels, preparative electrophoresis for enrichment of low-molecular-mass proteins and LC-MS are among the measures to improve the detection and identification of low-mass proteins (Garbis et al. 2005).

Hydrophobic proteins such as integral membrane proteins or cell surface-associated proteins generally have poor solubility under the chemical conditions used in the IEF step, because the buffer used is of low ionic strength without ionic detergents (Rotilio et al. 2012;

Garbis et al. 2005). Modifying the experimental conditions to favor protein solubilization often improves the resolution and detection of hydrophobic proteins (Renzone et al. 2005). As always when analyzing a protein mixture, the use of chromatography reduces the complexity of the sample by separating proteins by affinity, charge, hydrophobicity or size (Garbis et al. 2005).

MS Instrument Related Limitations

MS is used in proteomics in three main ways: to characterize recombinant proteins and other macromolecules; to identify proteins in biochemical or large-scale proteomic projects; and to detect and characterize PTMs (Mann et al. 2001). However, proteins have such diverse specific properties that MS may not be suited for every clinical or research aim. Insufficient and/or incorrect database entries hamper MS analysis, and the sensitivity of the mass spectrometer, the presence of isobaric masses and PTMs are also important limitations to consider (Intelicato-Young and Fox 2013; Garbis et al. 2005). Despite these limitations, the generation of stable gas-phase ions with significant chemical decomposition as in ESI and MALDI was a breakthrough, making MS more sensitive (Garbis et al. 2005). MALDI-TOF MS is the most efficient approach in high-throughput proteomics but it cannot identify multiple components of a given mixture, so must be preceded by 2-DE or HPLC for protein separation (Garbis et al. 2005). In ESI multiple-charged species are generated due to an increase in the mass of peptide ions (dilution effect of the monoisotropic mass), which makes the detection of higher mass ions less sensitive. If some peptides are very concentrated, the capacity for ESI to ionize all available peptides in a mixture may be limited due to depletion of protons and to upper mass density constraints to the dynamic range of the mass spectrometer (Garbis et al. 2005).

Even with well annotated protein databases using search algorithms, sequence specific peptide fragmentation or a peptide's side chain fragmentation may lead to false positive identifications (Reinders et al. 2004). There are well established databases for MALDI-TOF MS including the Bruker Biotyper protein profiling software and strain library and Shimadzu AXIMA-iD^{Plus} Microbial Identification System. It has been shown that adding certain species to the database significantly improves MALDI precision in identification (Intelicato-Young and Fox 2013). The MS/MS spectrum of each peptide present within a mixture gives a high degree of confidence in protein identification (Renzone et al. 2005).

Future Directions

Proteomics studies have identified important sets of proteins that contribute to or reflect the action of molecular mechanisms. However, proteomic methodologies are still not used systematically (Fujii et al. 2006). Three-dimensional homology analysis (*in silico* protein folding) may be an important next step which when combined with systems biology approaches could highlight gaps in biomolecular networks linked to physiological functions (Hood 2012). Integration of *in vivo* and *in vitro* proteomic data is also likely to be feasible using microdissection and 2-D DIGE with a proteome database (Fujii et al. 2006). Top-down

MS-based proteomics addressing to the identification of protein isoforms from amino acid modifications, transcript variations and PTMs is a promising vector for proteomics to evolve in the near future (Angel et al. 2012). A very important issue to be considered is the improvement of computational methods and algorithms in order to ensure the long-term storage of large amounts of data and biological information (Altelaar et al. 2013). Pathway analysis and metaproteomics are fields with some obvious applications in the short term and their scope will be broadened when they are coordinated with other technologies and methodologies

Proteins are the major type of drug target, so proteomics-based approaches present many possibilities (Bantscheff and Drewes 2012). In this context, targeted proteomics is playing an progressively important role in the accurate measurement of protein targets in biological samples in the hope of elucidating the molecular mechanisms of cellular function via the understanding of intricate protein networks and pathways (Boja and Rodriguez 2012). The goal of targeted proteomics is the monitoring of a small set of proteins with high sensitivity, reproducibility and quantitative accuracy (Doerr 2013).

Pathway Analysis

Omics technologies will enable researchers to look at the complete expression and regulation of genes, proteins and metabolites. However, at the present time, most statistical analyses are often based on fairly simplistic one-by-one comparisons of markers between treated and/or diseased groups. Recently, analytical approaches, like pathway analysis and structural equation modeling, are being used to integrate analyses of biological functions and responses to environmental factors (Vlaanderen et al. 2010). Pathway analysis has been used to the study gene ontology terms, physical interaction networks, kinetic simulation of pathways, steady-state pathway analysis and to infer pathways from expression and sequence data (although sometimes the use of this term is not strictly appropriate) (Khatri et al. 2012). In pathway analysis, the profile of genes associated with a specific biological process is tested in relation to the functional interactions between genes. There are still large gaps in our knowledge of biological pathways, but each new study adds to the information base necessary for these types of analyses (Vlaanderen et al. 2010). The amount of information on biochemical pathways is increasing very rapidly, making it possible to perform detailed analysis of metabolism for entire organisms. The size and complexity of many metabolic pathways have yet to be determined. Pathway analysis is therefore a challenge but with the development of the right tools has potential applications in antimicrobial research, drug design or human disease studies (Deville et al. 2003). There are several databases on metabolic pathways such as Kyoto Encyclopedia of Genes and Genomes (KEGG; for genes, enzymes and metabolic reactions), EMP (for enzymes and pathways), WIT (for metabolic pathway reconstruction), EcoCyc (for metabolic pathways in *Escherichia coli*), MetaCyc (for metabolic pathways of other organisms), aMAZE, CSNDB, PathDB, UM-BBD and SHARKdb. The BIND database contains information on signal transduction pathways (Deville et al. 2003). Regulatory pathways refer to transcriptional relationships or indirect relationships where one protein controls another protein's expression. BioCyc and WikiPathways are among the databases used to analyze these regulatory pathways (Goh et al. 2012).

Metaproteomics

The term "metaproteomics" was first proposed by Wilmes and Bond (2004) as "large-scale characterization of the entire protein complement of an environmental microbiota at a given point in time". As proteins are involved in biotransformation processes, metaproteome analysis is a way of characterizing the dynamics of microbial function (Maron et al. 2007; Wilmes and Bond 2004). In order to gather information about active functions, it is necessary to relate the expression profile (metatranscriptomics) to protein products (metaproteomics), both of which rely on technically demanding technologies that have only very recently been used to study microbial communities of the gut (Lamendella et al. 2012). Overall, metaproteomics essentially requires the availability of relevant genome and metagenome sequences as the mass spectra generated are used as input queries to search existing databases for protein identification (Siggins et al. 2012).

Early metaproteome research focused on low complexity microbial consortia from acid mine drainage or sludge water bioreactors, so provided a starting point in the development of proteomic methods and informatic resources for more complex systems (Figure 13) (Hettich et al. 2012). High-throughput proteomic analysis and a set of bioinformatics tools have since been used to profile the human mucosal metaproteome present in several anatomical regions of individual patients, for example (Li et al. 2011).

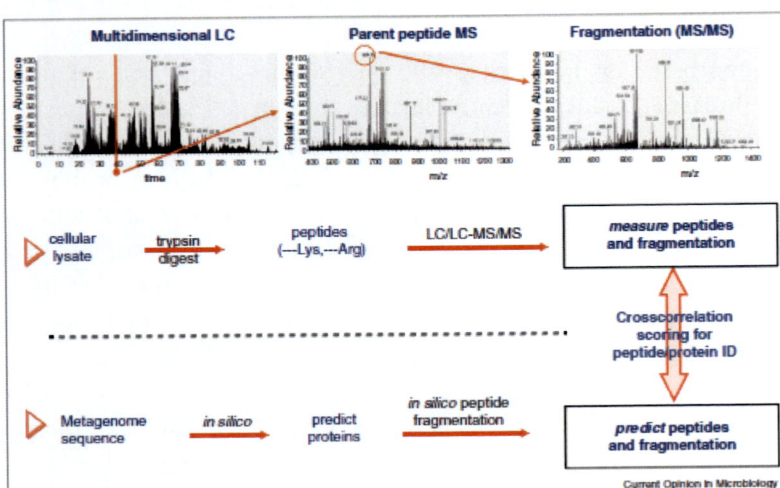

Reprinted from Current Opinion in Microbiology, 15, Hettich R. et al., Microbial metaproteomics: identifying the repertoire of proteins that microorganisms use to compete and cooperate in complex environmental communities, 373-380, Copyright (2012), with permission from Elsevier (Hettich et al. 2012).

Figure 13. Integrated experimental/computational approach for metaproteome measurements. A multidimensional liquid chromatography–tandem mass spectrometry approach is used to obtain high resolution parent ion and subsequent fragmentation information for proteolytic peptides in the environmental sample. To enable high-throughput peptide identifications, the metagenome sequence is used to predict all possible protein products, as well as their predicted proteolytic cleavage and tandem mass spectrometric fragmentation products. A cross-correlation scoring approach is used to compare the measured information on peptide parent masses and fragmentation with that predicted from the metagenome information. After peptides are scored, filtered, and identified, the predicted protein database is used to assemble the measured peptides computationally into respective protein families.

Large-scale metaproteomic studies are still needed to produce reference proteome databases of the host and microbial components of the mammalian gut. Such databases will be extremely valuable for studies of disease, vaccine development and diet/obesity studies (Lamendella et al. 2012).

Targeted Proteomics

Recent studies have shown that metabolic engineering can greatly benefit from sensitive and precise techniques for protein analysis. For this purpose, targeted proteomics can provide rapid detection and quantification of proteins (Singh et al. 2012). Unlike a typical discovery-based experiment, where peptide ions are automatically selected in the mass spectrometer for fragmentation, in a targeted workflow the mass spectrometer is programmed for detection of specific peptide ions derived from the proteins of interest on each assay (Doerr 2013).

Selected reaction monitoring (SRM) is being used in proteomics as a method for sensitive and selective quantification of targeted species of interest across a sample set. The emergence of SRM is due to the progress that has been made in instrumentation and software development. In fact, it has been used for the relative quantification of hundreds of peptides in a single multiplexed assay due to its high sensitivity, dynamic range, specificity and reproducibility (Bereman et al. 2012). Targeted proteomics technologies like multiplexed MS, protein arrays and enzyme-linked immunosorbent assays (ELISAs) provide quantitative measurement of selected proteins in a biological matrix of interest with higher accuracy and greater throughput when compared to regular MS (Boja and Rodriguez 2012).

Targeted proteomics studies supported by high-throughput technology, data integration and multidisciplinary approaches can therefore lead to the definition of new biological and chemical targets as well as the mechanisms of small-molecule action (Schenone et al. 2013).

References

Aagaard K, Petrosino J, Keitel W, Watson M, Katancik J, Garcia N, Patel S, Cutting M, Madden T, Hamilton H, Harris E, Gevers D, Simone G, McInnes P, Versalovic J (2013) The Human Microbiome Project strategy for comprehensive sampling of the human microbiome and why it matters. *The FASEB Journal* 27(3):1012-1022.

Abubucker S, Segata N, Goll J, Schubert AM, Izard J, Cantarel BL, Rodriguez-Mueller B, Zucker J, Thiagarajan M, Henrissat B, White O, Kelley ST, Methé B, Schloss PD, Gevers D, Mitreva M, Huttenhower C (2012) Metabolic Reconstruction for Metagenomic Data and Its Application to the Human Microbiome. *PLoS Comput Biol* 8(6):e1002358.

Altelaar AF, Munoz J, Heck AJ (2013) Next-generation proteomics: towards an integrative view of proteome dynamics. *Nature reviews Genetics* 14(1):35-48.

Angel TE, Aryal UK, Hengel SM, Baker ES, Kelly RT, Robinson EW, Smith RD (2012) Mass spectrometry-based proteomics: existing capabilities and future directions. *Chemical Society reviews* 41(10):3912-3928.

Armirotti A, Damonte G (2010) Achievements and perspectives of top-down proteomics. *Proteomics* 10(20):3566-3576.

Arruda SC, Barbosa Hde S, Azevedo RA, Arruda MA (2011) Two-dimensional difference gel electrophoresis applied for analytical proteomics: fundamentals and applications to the study of plant proteomics. *Analyst* 136(20):4119-4126.

Bandow JE, Brotz H, Leichert LI, Labischinski H, Hecker M (2003) Proteomic approach to understanding antibiotic action. *Antimicrob Agents Chemother* 47(3):948-955.

Bantscheff M, Drewes G (2012) Chemoproteomic approaches to drug target identification and drug profiling. *Bioorg Med Chem* 20(6):1973-1978.

Becnel LB, McKenna NJ (2012) Minireview: progress and challenges in proteomics data management, sharing, and integration. *Mol Endocrinol* 26(10):1660-1674.

Bereman MS, MacLean B, Tomazela DM, Liebler DC, MacCoss MJ (2012) The development of selected reaction monitoring methods for targeted proteomics via empirical refinement. *Proteomics* 12(8):1134-1141.

Boja ES, Rodriguez H (2012) Mass spectrometry-based targeted quantitative proteomics: achieving sensitive and reproducible detection of proteins. Proteomics 12(8):1093-1110.

Boyer AE, Gallegos-Candela M, Lins RC, Kuklenyik Z, Woolfitt A, Moura H, Kalb S, Quinn CP, Barr JR (2011) Quantitative mass spectrometry for bacterial protein toxins--a sensitive, specific, high-throughput tool for detection and diagnosis. *Molecules* 16(3):2391-2413.

Buttner K, Bernhardt J, Scharf C, Schmid R, Mader U, Eymann C, Antelmann H, Volker A, Volker U, Hecker M (2001) A comprehensive two-dimensional map of cytosolic proteins of *Bacillus subtilis*. *Electrophoresis* 22(14):2908-2935.

Chen LS, Saccone NL, Culverhouse RC, Bracci PM, Chen CH, Dueker N, Han Y, Huang H, Jin G, Kohno T, Ma JZ, Przybeck TR, Sanders AR, Smith JA, Sung YJ, Wenzlaff AS, Wu C, Yoon D, Chen YT, Cheng YC, Cho YS, David SP, Duan J, Eaton CB, Furberg H, Goate AM, Gu D, Hansen HM, Hartz S, Hu Z, Kim YJ, Kittner SJ, Levinson DF, Mosley TH, Payne TJ, Rao DC, Rice JP, Rice TK, Schwantes-An TH, Shete SS, Shi J, Spitz MR, Sun YV, Tsai FJ, Wang JC, Wrensch MR, Xian H, Gejman PV, He J, Hunt SC, Kardia SL, Li MD, Lin D, Mitchell BD, Park T, Schwartz AG, Shen H, Wiencke JK, Wu JY, Yokota J, Amos CI, Bierut LJ (2012) Smoking and genetic risk variation across populations of European, Asian, and African American ancestry--a meta-analysis of chromosome 15q25. *Genet Epidemiol* 36(4):340-351.

Clerico M, Contessa G, Durelli L (2007) Interferon-beta1a for the treatment of multiple sclerosis. *Expert Opin Biol Ther* 7(4):535-542.

Cortes-Francisco N, Flores C, Moyano E, Caixach J (2011) Accurate mass measurements and ultrahigh-resolution: evaluation of different mass spectrometers for daily routine analysis of small molecules in negative electrospray ionization mode. *Analytical and bioanalytical chemistry* 400(10):3595-3606.

Curreem SO, Watt RM, Lau SK, Woo PC (2012) Two-dimensional gel electrophoresis in bacterial proteomics. *Protein & cell* 3(5):346-363.

Dave M, Higgins PD, Middha S, Rioux KP (2012) The human gut microbiome: current knowledge, challenges, and future directions. Translational research : *the journal of laboratory and clinical medicine* 160(4):246-257.

Dengler V, Meier PS, Heusser R, Berger-Bachi B, McCallum N (2011) Induction kinetics of the *Staphylococcus aureus* cell wall stress stimulon in response to different cell wall active antibiotics. *BMC Microbiol* 1116.

Devaraj S, Hemarajata P, Versalovic J (2013) The human gut microbiome and body metabolism: implications for obesity and diabetes. *Clin Chem* 59(4):617-628.

Deville Y, Gilbert D, van Helden J, Wodak SJ (2003) An overview of data models for the analysis of biochemical pathways. *Briefings in bioinformatics* 4(3):246-259.

Di Palma S, Hennrich ML, Heck AJ, Mohammed S (2012) Recent advances in peptide separation by multidimensional liquid chromatography for proteome analysis. *J Proteomics* 75(13):3791-3813.

D'Imperio M, Della Corte A, Facchiano A, Di Michele M, Ferrandina G, Donati MB, Rotilio D (2010) Standardized sample preparation phases for a quantitative measurement of plasma peptidome profiling by MALDI-TOF. *J Proteomics* 73(7):1355-1367.

Doerr A (2013) Mass spectrometry-based targeted proteomics. *Nature methods* 10(1):23.

Duncan MW, Roder H, Hunsucker SW (2008) Quantitative matrix-assisted laser desorption/ionization mass spectrometry. *Briefings in functional genomics & proteomics 7(5):355-370.*

Dworzanski JP, Deshpande SV, Chen R, Jabbour RE, Snyder AP, Wick CH, Li L (2006) Mass spectrometry-based proteomics combined with bioinformatic tools for bacterial classification. *J Proteome Res* 5(1):76-87.

Dziadosz M, Lessig R, Bartels H (2012) HPLC-DAD protein kinase inhibitor analysis in human serum. *J Chromatogr B Analyt Technol Biomed Life Sci* 893-89477-81.

Eng JK, Searle BC, Clauser KR, Tabb DL (2011) A face in the crowd: recognizing peptides through database search. *Mol Cell Proteomics* 10(11):R111 009522.

Evans C, Noirel J, Ow SY, Salim M, Pereira-Medrano AG, Couto N, Pandhal J, Smith D, Pham TK, Karunakaran E, Zou X, Biggs CA, Wright PC (2012) An insight into iTRAQ: where do we stand now? *Analytical and bioanalytical chemistry* 404(4):1011-1027.

Fleischer TC, Murphy BR, Flick JS, Terry-Lorenzo RT, Gao ZH, Davis T, McKinnon R, Ostanin K, Willardsen JA, Boniface JJ (2010) Chemical proteomics identifies Nampt as the target of CB30865, an orphan cytotoxic compound. *Chem Biol* 17(6):659-664.

Fodor AA, DeSantis TZ, Wylie KM, Badger JH, Ye Y, Hepburn T, Hu P, Sodergren E, Liolios K, Huot-Creasy H, Birren BW, Earl AM (2012) The "most wanted" taxa from the human microbiome for whole genome sequencing. *PLoS One* 7(7):e41294.

Foxman B, Rosenthal M (2013) Implications of the Human Microbiome Project for Epidemiology. *Am J Epidemiol* 177(3):197-201.

Fujii K, Kondo T, Yamada M, Iwatsuki K, Hirohashi S (2006) Toward a comprehensive quantitative proteome database: protein expression map of lymphoid neoplasms by 2-D DIGE and MS. *Proteomics* 6(17):4856-4876.

Gabani P, Prakash D, Singh OV (2012) Emergence of antibiotic-resistant extremophiles (AREs). *Extremophiles* 16(5):697-713.

Garbis S, Lubec G, Fountoulakis M (2005) Limitations of current proteomics technologies. *J Chromatogr A* 1077(1):1-18.

George Cisar EA, Nguyen N, Rosen H (2013) A GTP affinity probe for proteomics highlights flexibility in purine nucleotide selectivity. *J Am Chem Soc* 135(12):4676-4679.

Gevers D, Knight R, Petrosino JF, Huang K, McGuire AL, Birren BW, Nelson KE, White O, Methé BA, Huttenhower C (2012) The Human Microbiome Project: A Community Resource for the Healthy Human Microbiome. *PLoS Biol* 10(8):e1001377.

Goh WW, Lee YH, Chung M, Wong L (2012) How advancement in biological network analysis methods empowers proteomics. *Proteomics* 12(4-5):550-563.

Goll J, Thiagarajan M, Abubucker S, Huttenhower C, Yooseph S, Methé BA (2012) A Case Study for Large-Scale Human Microbiome Analysis Using JCVI's Metagenomics Reports (METAREP). *PLoS One* 7(6):e29044.

Gorg A, Drews O, Luck C, Weiland F, Weiss W (2009) 2-DE with IPGs. *Electrophoresis* 30 Suppl 1S122-132.

Gorg A, Weiss W, Dunn MJ (2004) Current two-dimensional electrophoresis technology for proteomics. *Proteomics* 4(12):3665-3685.

Gorman JJ, Wallis TP, Pitt JJ (2002) Protein disulfide bond determination by mass spectrometry. *Mass Spectrom Rev* 21(3):183-216.

Grela A, Turek A, Piekoszewski W (2012) Application of matrix-assisted laser desorption/ionization time-of-flight mass spectrometry (MALDI-TOF MS) in Alzheimer's disease. *Clin Chem Lab Med* 50(8):1297-1304.

Group TNHW, Peterson J, Garges S, Giovanni M, McInnes P, Wang L, Schloss JA, Bonazzi V, McEwen JE, Wetterstrand KA, Deal C, Baker CC, Di Francesco V, Howcroft TK, Karp RW, Lunsford RD, Wellington CR, Belachew T, Wright M, Giblin C, David H, Mills M, Salomon R, Mullins C, Akolkar B, Begg L, Davis C, Grandison L, Humble M, Khalsa J, Little AR, Peavy H, Pontzer C, Portnoy M, Sayre MH, Starke-Reed P, Zakhari S, Read J, Watson B, Guyer M (2009) The NIH Human Microbiome Project. *Genome Res* 19(12):2317-2323.

Gygi SP, Rist B, Gerber SA, Turecek F, Gelb MH, Aebersold R (1999) Quantitative analysis of complex protein mixtures using isotope-coded affinity tags. *Nat Biotechnol* 17(10):994-999.

Hesse AM, Marcelo P, Rossier J, Vinh J (2008) Simple and universal tool to remove on-line impurities in mono- or two-dimensional liquid chromatography-mass spectrometry analysis. *J Chromatogr A* 1189(1-2):175-182.

Hettich RL, Sharma R, Chourey K, Giannone RJ (2012) Microbial metaproteomics: identifying the repertoire of proteins that microorganisms use to compete and cooperate in complex environmental communities. *Curr Opin Microbiol* 15(3):373-380.

Hood L (2012) Tackling the microbiome. *Science* 336(6086):1209.

Human Microbiome Project C (2012) A framework for human microbiome research. *Nature* 486(7402):215-221.

Human Microbiome Project C (2012) Structure, function and diversity of the healthy human microbiome. *Nature* 486(7402):207-214.

Iagnocco A, Naredo E (2012) Osteoarthritis: research update and clinical applications. *Rheumatology* (Oxford) 51 Suppl 7vii2-5.

Ihling C, Sinz A (2005) Proteome analysis of *Escherichia coli* using high-performance liquid chromatography and Fourier transform ion cyclotron resonance mass spectrometry. *Proteomics* 5(8):2029-2042.

Intelicato-Young J, Fox A (2013) Mass spectrometry and tandem mass spectrometry characterization of protein patterns, protein markers and whole proteomes for pathogenic bacteria. *J Microbiol Methods* 92(3):381-386.

Issaq H, Veenstra T (2008) Two-dimensional polyacrylamide gel electrophoresis (2D-PAGE): advances and perspectives. *Biotechniques* 44(5):697-698, 700.

Kaltashov IA, Bobst CE, Abzalimov RR, Wang G, Baykal B, Wang S (2012) Advances and challenges in analytical characterization of biotechnology products: mass spectrometry-based approaches to study properties and behavior of protein therapeutics. *Biotechnology advances* 30(1):210-222.

Khan Z, Amini S, Bloom JS, Ruse C, Caudy AA, Kruglyak L, Singh M, Perlman DH, Tavazoie S (2011) Accurate proteome-wide protein quantification from high-resolution 15N mass spectra. *Genome Biol* 12(12):R122.

Khatri P, Sirota M, Butte AJ (2012) Ten years of pathway analysis: current approaches and outstanding challenges. *PLoS Comput Biol* 8(2):e1002375.

Klingenberg H, Asshauer KP, Lingner T, Meinicke P (2013) Protein signature-based estimation of metagenomic abundances including all domains of life and viruses. *Bioinformatics* 29(8):973-980.

Koren O, Knights D, Gonzalez A, Waldron L, Segata N, Knight R, Huttenhower C, Ley RE (2013) A Guide to Enterotypes across the Human Body: Meta-Analysis of Microbial Community Structures in Human Microbiome Datasets. *PLoS Comput Biol* 9(1):e1002863.

Krug K, Nahnsen S, Macek B (2011) Mass spectrometry at the interface of proteomics and genomics. *Molecular bioSystems* 7(2):284-291.

Ladirat SE, Schols HA, Nauta A, Schoterman MH, Keijser BJ, Montijn RC, Gruppen H, Schuren FH (2013) High-throughput analysis of the impact of antibiotics on the human intestinal microbiota composition. *J Microbiol Methods* 92(3):387-397.

Lamendella R, VerBerkmoes N, Jansson JK (2012) 'Omics' of the mammalian gut--new insights into function. *Curr Opin Biotechnol* 23(3):491-500.

Lane CS (2005) Mass spectrometry-based proteomics in the life sciences. *Cell Mol Life Sci* 62(7-8):848-869.

Latterich M, Abramovitz M, Leyland-Jones B (2008) Proteomics: new technologies and clinical applications. *Eur J Cancer* 44(18):2737-2741.

Li X, LeBlanc J, Truong A, Vuthoori R, Chen SS, Lustgarten JL, Roth B, Allard J, Ippoliti A, Presley LL, Borneman J, Bigbee WL, Gopalakrishnan V, Graeber TG, Elashoff D, Braun J, Goodglick L (2011) A metaproteomic approach to study human-microbial ecosystems at the mucosal luminal interface. *PLoS One* 6(11):e26542.

Lu K, Knutson CG, Wishnok JS, Fox JG, Tannenbaum SR (2012) Serum metabolomics in a *Helicobacter hepaticus* mouse model of inflammatory bowel disease reveal important changes in the microbiome, serum peptides, and intermediary metabolism. *J Proteome Res* 11(10):4916-4926.

Lubec G, Afjehi-Sadat L, Yang JW, John JP (2005) Searching for hypothetical proteins: theory and practice based upon original data and literature. *Prog Neurobiol* 77(1-2):90-127.

M TM, Aydin B, Carlson RP, Hanley L (2012) Identification and imaging of peptides and proteins on *Enterococcus faecalis* biofilms by matrix assisted laser desorption ionization mass spectrometry. *Analyst* 137(21):5018-5025.

Makarov A, Scigelova M (2010) Coupling liquid chromatography to Orbitrap mass spectrometry. *J Chromatogr A* 1217(25):3938-3945.

Manes NP, Dong L, Zhou W, Du X, Reghu N, Kool AC, Choi D, Bailey CL, Petricoin EF, 3rd, Liotta LA, Popov SG (2011) Discovery of mouse spleen signaling responses to anthrax using label-free quantitative phosphoproteomics via mass spectrometry. *Mol Cell Proteomics* 10(3):M110 000927.

Mann M, Hendrickson RC, Pandey A (2001) Analysis of proteins and proteomes by mass spectrometry. *Annu Rev Biochem* 70437-473.

Markowitz VM, Chen IM, Chu K, Szeto E, Palaniappan K, Grechkin Y, Ratner A, Jacob B, Pati A, Huntemann M, Liolios K, Pagani I, Anderson I, Mavromatis K, Ivanova NN, Kyrpides NC (2012) IMG/M: the integrated metagenome data management and comparative analysis system. *Nucleic Acids Res* 40(Database issue):D123-129.

Maron PA, Ranjard L, Mougel C, Lemanceau P (2007) Metaproteomics: a new approach for studying functional microbial ecology. *Microb Ecol* 53(3):486-493.

Marvin LF, Roberts MA, Fay LB (2003) Matrix-assisted laser desorption/ionization time-of-flight mass spectrometry in clinical chemistry. *Clin Chim Acta* 337(1-2):11-21.

McDonald WH, Yates JR, 3rd (2002) Shotgun proteomics and biomarker discovery. *Dis Markers* 18(2):99-105.

Medzihradszky KF (2005) Peptide sequence analysis. *Methods Enzymol* 402209-244.

Meyer MR, Maurer HH (2012) Current applications of high-resolution mass spectrometry in drug metabolism studies. *Analytical and bioanalytical chemistry* 403(5):1221-1231.

Miao Q, Zhang CC, Kast J (2012) Chemical proteomics and its impact on the drug discovery process. *Expert review of proteomics* 9(3):281-291.

Minden JS (2012) DIGE: past and future. *Methods Mol Biol* 8543-8.

Mohimani H, Yang YL, Liu WT, Hsieh PW, Dorrestein PC, Pevzner PA (2011) Sequencing cyclic peptides by multistage mass spectrometry. *Proteomics* 11(18):3642-3650.

Monteiro R, Vitorino R, Domingues P, Radhouani H, Carvalho C, Poeta P, Torres C, Igrejas G (2012) Proteome of a methicillin-resistant *Staphylococcus aureus* clinical strain of sequence type ST398. *J Proteomics* 75(10):2892-2915.

Morgan XC, Segata N, Huttenhower C (2013) Biodiversity and functional genomics in the human microbiome. *Trends Genet* 29(1):51-58.

Owens GP, Gilden D, Burgoon MP, Yu X, Bennett JL (2011) Viruses and multiple sclerosis. *Neuroscientist* 17(6):659-676.

Panisko EA, Conrads TP, Goshe MB, Veenstra TD (2002) The postgenomic age: characterization of proteomes. *Exp Hematol* 30(2):97-107.

Piras C, Soggiu A, Bonizzi L, Gaviraghi A, Deriu F, De Martino L, Iovane G, Amoresano A, Roncada P (2012) Comparative proteomics to evaluate multi drug resistance in *Escherichia coli*. *Molecular bioSystems* 8(4):1060-1067.

Punta M, Coggill PC, Eberhardt RY, Mistry J, Tate J, Boursnell C, Pang N, Forslund K, Ceric G, Clements J, Heger A, Holm L, Sonnhammer ELL, Eddy SR, Bateman A, Finn RD (2012) The Pfam protein families database. *Nucleic Acids Res* 40(D1):D290-D301.

Radhouani H, Poeta P, Pinto L, Miranda J, Coelho C, Carvalho C, Rodrigues J, Lopez M, Torres C, Vitorino R, Domingues P, Igrejas G (2010) Proteomic characterization of *van*A-containing *Enterococcus* recovered from Seagulls at the Berlengas Natural Reserve, W Portugal. *Proteome science* 848.

Reinders J, Lewandrowski U, Moebius J, Wagner Y, Sickmann A (2004) Challenges in mass spectrometry-based proteomics. *Proteomics* 4(12):3686-3703.

Renzone G, D'Ambrosio C, Arena S, Rullo R, Ledda L, Ferrara L, Scaloni A (2005) Differential proteomic analysis in the study of prokaryotes stress resistance. *Ann Ist Super Sanita* 41(4):459-468.

Riederer BM (2008) Non-covalent and covalent protein labeling in two-dimensional gel electrophoresis. *J Proteomics* 71(2):231-244.

Rohrbough JG, Breci L, Merchant N, Miller S, Haynes PA (2006) Verification of single-peptide protein identifications by the application of complementary database search algorithms. Journal of biomolecular techniques : *JBT* 17(5):327-332.

Roncada P, Deriu F, Gaviraghi A, Martino PA, Bonizzi L (2009) Proteomic study of antibiotic resistance in *Escherichia coli* strains. *Vet Res Commun* 33 Suppl 1157-160.

Rotilio D, Della Corte A, D'Imperio M, Coletta W, Marcone S, Silvestri C, Giordano L, Di Michele M, Donati MB (2012) Proteomics: bases for protein complexity understanding. *Thromb Res* 129(3):257-262.

Saito Y, Waki M, Hameed S, Hayasaka T, Setou M (2012) Development of imaging mass spectrometry. *Biol Pharm Bull* 35(9):1417-1424.

Sandra K, Moshir M, D'Hondt F, Verleysen K, Kas K, Sandra P (2008) Highly efficient peptide separations in proteomics Part 1. Unidimensional high performance liquid chromatography. *J Chromatogr B Analyt Technol Biomed Life Sci* 866(1-2):48-63.

Schenone M, Dancik V, Wagner BK, Clemons PA (2013) Target identification and mechanism of action in chemical biology and drug discovery. *Nat Chem Biol* 9(4):232-240.

Seifert J, Taubert M, Jehmlich N, Schmidt F, Volker U, Vogt C, Richnow HH, von Bergen M (2012) Protein-based stable isotope probing (protein-SIP) in functional metaproteomics. *Mass Spectrom Rev* 31(6):683-697.

Siggins A, Gunnigle E, Abram F (2012) Exploring mixed microbial community functioning: recent advances in metaproteomics. *FEMS microbiology ecology* 80(2):265-280.

Singh P, Batth TS, Juminaga D, Dahl RH, Keasling JD, Adams PD, Petzold CJ (2012) Application of targeted proteomics to metabolically engineered *Escherichia coli*. *Proteomics* 12(8):1289-1299.

Souda P, Ryan CM, Cramer WA, Whitelegge J (2011) Profiling of integral membrane proteins and their post translational modifications using high-resolution mass spectrometry. *Methods* 55(4):330-336.

Sutton CW (2012) The role of targeted chemical proteomics in pharmacology. *Br J Pharmacol* 166(2):457-475.

Tao D, Zhang L, Shan Y, Liang Z, Zhang Y (2011) Recent advances in micro-scale and nano-scale high-performance liquid-phase chromatography for proteome research. *Analytical and bioanalytical chemistry* 399(1):229-241.

Teixeira MC, Santos PM, Rodrigues C, Sa-Correia I (2009) Teaching expression proteomics: From the wet-lab to the laptop. *Biochem Mol Biol Educ* 37(5):279-286.

Timms JF, Cramer R (2008) Difference gel electrophoresis. Proteomics 8(23-24):4886-4897.

Turnbaugh PJ, Ley RE, Hamady M, Fraser-Liggett CM, Knight R, Gordon JI (2007) The human microbiome project. *Nature* 449(7164):804-810.

Upadhyay RD, Balasinor NH, Kumar AV, Sachdeva G, Parte P, Dumasia K (2013) Proteomics in reproductive biology: beacon for unraveling the molecular complexities. *Biochim Biophys Acta* 1834(1):8-15.

Verberkmoes NC, Russell AL, Shah M, Godzik A, Rosenquist M, Halfvarson J, Lefsrud MG, Apajalahti J, Tysk C, Hettich RL, Jansson JK (2009) Shotgun metaproteomics of the human distal gut microbiota. *The ISME journal* 3(2):179-189.

Vlaanderen J, Moore LE, Smith MT, Lan Q, Zhang L, Skibola CF, Rothman N, Vermeulen R (2010) Application of OMICS technologies in occupational and environmental health research; current status and projections. *Occup Environ Med* 67(2):136-143.

Wecke T, Mascher T (2011) Antibiotic research in the age of omics: from expression profiles to interspecies communication. *J Antimicrob Chemother* 66(12):2689-2704.

Wilmes P, Bond PL (2004) The application of two-dimensional polyacrylamide gel electrophoresis and downstream analyses to a mixed community of prokaryotic microorganisms. *Environmental microbiology* 6(9):911-920.

Wu L, Vogt FG (2012) A review of recent advances in mass spectrometric methods for gas-phase chiral analysis of pharmaceutical and biological compounds. *J Pharm Biomed Anal* 69133-147.

Xie F, Smith RD, Shen Y (2012) Advanced proteomic liquid chromatography. *J Chromatogr A* 126178-90.

Xu YM, Huang DY, Chiu JF, Lau AT (2012) Post-translational modification of human heat shock factors and their functions: a recent update by proteomic approach. J Proteome Res 11(5):2625-2634.

Zhou H, Zhao Q, Singla LD, Min J, He S, Cong H, Li Y, Su C (2013) Differential proteomic profiles from distinct Toxoplasma gondii strains revealed by 2D-difference gel electrophoresis. *Exp Parasitol* 133(4):376-382.

In: Recent Advances in Proteomics Research
Editor: Ming D. Li

ISBN: 978-1-62948-218-7

Chapter 4

Analysis of Pleural Fluid: The Path Travelled with Proteomic Guidance

Suchit Kumar Chatterji* and Pao-Chi Liao
Department of Environmental and Occupational Health,
College of Medicine, National Cheng Kung University,
Tainan, Taiwan

Abstract

The study of body fluids is an important part of the quest for protein biomarkers of disease. Proteomics plays a very important role in this search. Body fluids close to a diseased organ have a much higher chance of containing disease specific proteins (and in a higher concentration) than plasma because of the dilution effect of blood that reduces the plasma concentration of potential biomarkers to levels below the detection (and quantification) limits of present proteomic analysis technology. It is because of this very reason that a pleural effusion can be regarded as a 'gold mine' for potential disease specific biomarkers. The various proteomic analysis guided efforts to study pleural fluid began in 2004 and continues to date. The various approaches to analyze the pleural proteome have their similarities and differences in goals, methodologies and study structure. It is thus not surprising that their results, also, have been varied. A number of disease specific potential biomarker candidates have been identified as a result of these studies and they open the door for exciting prospects in analysis of the pleural proteome in the near future.

Keywords: Biomarker; pleural fluid; pleural effusion; malignant pleural effusion; proteome; proteomic; shotgun proteomics; gel electrophoresis; lung cancer; lung adenocarcinoma; non-small cell lung cancer; exosomes; pigment epithelium derived factor;

* Corresponding Author address: Pao-Chi Liao, No.138, Shengli Rd., Tainan, 704, Taiwan; Email: liaopc@mail.ncku.edu.tw.

thrombospondin 2; tumor susceptibility gene 101; retinoblastoma binding protein 7; pigment epithelium derived factor; periostin; multimerin 2; activated leucocyte cell adhesion molecule; lysosome associated membrane glycoprotein 2; retinoblastoma associated binding protein 46; apolipoprotein C1; gelsolin; metalloproteinase inhibitor 2; alpha 2 HS glycoprotein; angiogenin; cystatin-c; insulin like growth factor binding protein 2; transthyretin; hemopexin; fibrinogen gamma; serum amyloid p; jumonji C-domain containing protein; apolipoprotein E

1. Introduction

Proteomic approaches to the investigation of disease, especially, cancer, were based on the ability of modern proteomic analysis to accurately identify and characterize proteins. The goal of this chapter is to identify and characterize specific and non-specific disease associated proteins. A majority of these investigations were in the search of cancer specific proteins, the potential cancer biomarkers. Proteomic study of pleural fluid was an important part of these investigative processes. In this chapter, we discuss the various approaches used in pleural fluid proteome analysis with a special emphasis on the quest for potential cancer biomarkers in the pleural fluid. We also discuss the successes and failures of various teams involved in this type of study and the road ahead.

2. Why Proteomics? - Limitations of Microscopic Analysis of Pleural Fluid

The study of body fluids of diseased people has traditionally been an important component of the diagnosis and follows up of diseases like cancer. Traditional pathology based approaches have relied principally on microscopic study of the centrifuged deposit of the fluid in two ways; the cytological analysis of cytospin preparations and the histological tissue processing (and subsequent microscopic analysis) of cell block preparations of body fluids. Like all microscopy dependent diagnostic approaches, these approaches suffered from the drawback of inter-observer variability. This drawback could lead to potential errors in the diagnosis of cancer (Figure 1).

Diagnostic analytical approaches relying on the presence of a cancer associated chemical (like a unique protein) that is measured and detected by standardized analytical methods seems to be a better option for diagnosis of cancer due to the absence of subjectivity/inter-observer variability in this type of diagnostic approach. Proteomic study of disease material has emerged as an important tool in the quest for unique cancer associated proteins. Modern proteomic analysis, augmented by the capability to search the vast online protein databases (from translated genomic databases) has the ability to accurately identify and characterize prospective cancer associated proteins; the so-called prospective biomarkers.

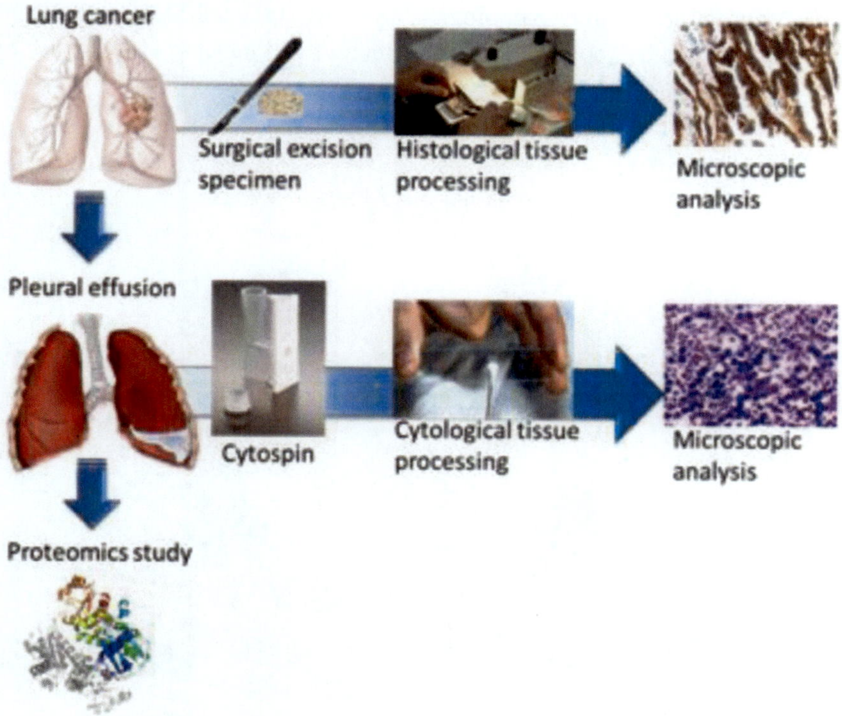

Figure 1. Flow Chart of Lung Cancer Tissue Analysis.

3. Pre-Analytic and Intra-Analytical Variables of the Pleural Fluid Proteome

Proteomic analysis of pathological material will yield prospective biomarkers only if both intra analytical and pre analytical processes are followed in an optimal way.

Intra analytical variables are easily dealt with by careful monitoring of analyzer performance (for example, mass spectrometer maintenance activity). The pre-analytical variables, however, are more difficult to standardize to optimal levels. Among them, optimal storage of the disease sample (for example, a sample of pleural fluid from a patient with lung cancer) plays a major role.

Prompt refrigeration at the correct temperature is very important here as it can prevent destruction of diagnostically important proteins in the sample; some of which may eventually turn out to be potential biomarkers (Figure 2). Because the goal here is to detect unique proteins that are often present in low concentrations, it is always preferred (though not always practically possible) to analyze a relatively fresh sample.

One should also bear in mind that levels of any particular protein in a body fluid (for example, pleural fluid) vary in a non-predictable way. This is because levels of a particular protein in a pathological fluid sample are directly related to the role played by that particular protein in the pathology of that particular disease (for example, lung cancer).

It is thus preferable to study pathological samples collected at various stages of cancer progression (if practically possible) to maximize the chances of discovering potential biomarkers (Figure 2).

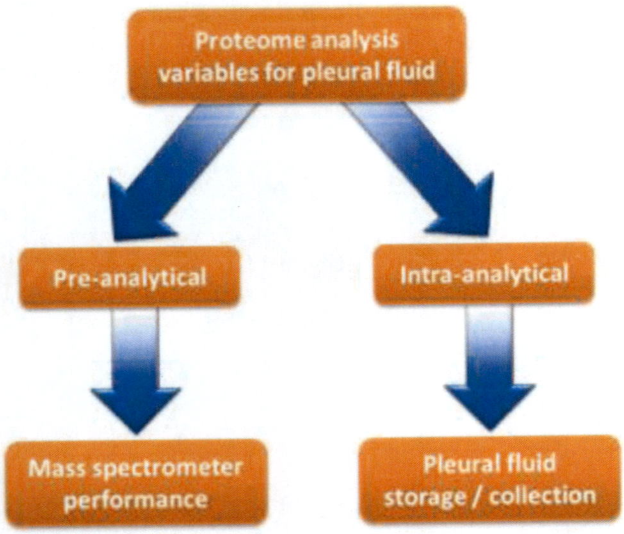

Figure 2. Variables in Proteomic Analysis.

These variables should always be borne in mind when comparison studies are undertaken of the relative successes or failures of various investigative teams carrying out proteomic studies of pathological samples with the aim of identifying potential biomarkers. A potential biomarker protein may be easily destroyed by suboptimal storage and may thus be present at concentrations that are virtually undetectable by present mass spectrometer technology.

4. The Pleural Fluid Proteome is a Great Place to Search for Cancer Biomarkers – Why?

The biological material studied for discovery efforts from 'gold standard' cases need not be plasma. Many biomarkers with clinical diagnostic value have concentrations reaching the limits of detection by MS. It is believed that if disease specific biomarkers arise locally from diseased tissue (e.g., a malignant tumor), a gradient of signal diminishes with distance from disease and steps off markedly with dilution and admixture into plasma. Also, there exists no analytical process similar to PCR in MS based Proteomics. Thus, the search for a biomarker by MS based proteomics is expected to be more fruitful if a biological fluid closer to or in direct contact with the disease site is analyzed (Rifai et al. 2006; Tyan et al. 2005; Hsieh et al. 2006; Rodriguez-Pineiro et al. 2010).

Pleural fluid is a plasma filtrate derived from capillaries of the parietal pleura. Normally, the pleural space contains a very small amount of this fluid. When the rate of production exceeds the rate of absorption, a pleural effusion occurs. Pleural fluid appears to be a good biological fluid candidate for biomarker discovery. Pleural effusions usually accumulate in the pleural cavity that also contains the diseased lung. A pleural effusion, also, usually occurs as a direct consequence of the primary lung disease (e.g., carcinoma) (Rifai et al. 2006; Tyan et al. 2005; Hsieh et al. 2006; Rodriguez-Pineiro et al. 2010; Soltermann et al. 2008; Pernemalm et al. 2009; Yu et al. 2011; Indovina et al. 2011; Wang et al. 2012).

Proximity to the affected organ and smaller dilution effect are believed to be the main reasons for considering pleural effusions as reservoirs of potential biomarkers. Many proteins have been identified in the pleural fluid. These proteins appear to be either sourced from the circulation or are locally released by inflammatory or malignant processes. Proteins present in pleural effusion appear to be secreted by many different types of human cells (e.g., inflammatory cells and malignant cells). It was found that each of these cell types contain at least 2000 to 6000 different primary proteins. Post-translational modifications like glycosylation, phosphorylation, ubiquination, acetylation and nitration significantly increase this number (Hsieh et al. 2006; Rodriguez-Pineiro et al. 2010; Yu et al. 2011; Wang et al. 2012; Tyan et al. 2005).

5. Early Efforts – Challenges and Promising Discoveries

In most cancerous and non-cancerous disease processes, proteins are subjected to numerous chemical modifications (including post-translational modification and proteolytic cleavage). This makes their proteomic detection more challenging. Difficulties in the proteomic analysis of pleural fluid (as a part of the quest for biomarkers) may arise due to low protein concentrations, high amounts of albumin and immunoglobulin and dynamic ranges of protein concentration of several orders of magnitude (Hsieh et al. 2006; Rodriguez-Pineiro et al. 2010; Yu et al. 2011; Wang et al. 2012; Tyan et al. 2005).

Efforts aimed at proteomic analysis began only in the 2000s with the first published reports of the same appearing only in 2004-2005. Bard et al. in 2004 conducted a proteomic analysis of exosomes isolated from malignant pleural effusions. Exosomes are endosomal membrane bound vesicles secreted by various cells, including malignant cells. Their study was inspired by the discovery of exosomes in malignant ascites effusions. They believed that exosomes could be used to design strategies for immunotherapy of cancer. Exosomes were purified by sucrose gradient ultracentrifugation (a method used with success to isolate exosomes from malignant ascitic fluid) and electron microscopy was used to verify their concentration and purity.

Proteins were separated by 1D SDS-PAGE and MALDI-TOF and Western Blotting identified protein bands. Immunoglobulins were found to be the most common proteins identified in the exosome containing fractions, showing the most intense bands on the SDS-PAGE gel. They thought their discovery of high immunoglobulin and complement levels in the exosomes of pleural fluid was a very significant finding because both these types of compounds play a major role in the anti-tumor immune response. Pleural IgG can originate by diffusion from the blood and from locally stimulated B-lymphocytes. All this can contribute to a strong anti-tumor response in the pleural space. Proteins not previously reported were identified (SNX 25, BTG-1, PEDF, Thrombospondin 2). PEDF had shown indirect tumor inhibitory activity in previous studies on neuroblastoma tumor cells.

Thrombospondin 2 had also been previously shown to have an inhibitory action on tumor growth via inhibition of angiogenesis; this essentially stops the tumor from getting the extra blood supply necessary to sustain its nutrient demand for rapid growth. Antigen presenting molecules (like MHC molecules), isolated from the pleural effusion are also believed to be important in the anti-cancer response in the pleural space. These proteins are important because of their ability to transfer antigens to antigen presenting cells and trigger an anti-cancer immune response locally [Figure 3; (Bard et al. 2004)].

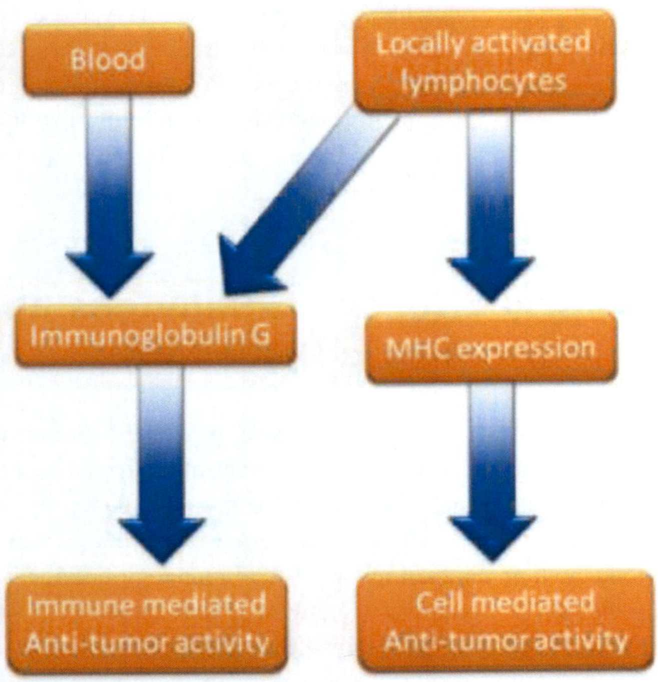

Figure 3. Anti-Tumor defenses of the pleural cavity.

We analyzed the global proteome of 43 human malignant pleural effusions by 2D Nano Liquid Chromatography Tandem Mass Spectrometry. Our report on the first global proteomics analysis of human pleural effusion confirmed our prior knowledge about the origins of pleural effusion. Large portions of pleural effusion proteins are common to those found in plasma. These proteins include complement factor B, immunoglobulin chain, albumin, transferrin, hemopexin, complement C3, complement C4, R-antitrypsin, vitamin D binding protein, fibrinogen, haptoglobin, apolipoprotein A-I, apolipoprotein A-II, plasminogen, transthyretin, hemoglobin, α2-microglobin, etc. We found 124 proteins among which were proteins that had not been reported in plasma and were unique to pleural effusion.

We found that the major proteins in pleural effusion are Albumin, Transferrin, Alpha 1 Antitrypsin, Apolipoprotein, Fibrinogen and Immunoglobulins. We also identified a number of proteins that may play a role in the development and progression of cancer like Tumor Susceptibility Gene 101 (TSG 101) and Retinoblastoma Binding Protein 7 (RBBP7). TSG101 is principally a negative growth regulator.

Mutations of the TSG101 gene have been identified in breast cancer and other neoplasms. In breast cancer, the mutations were related to aberrant splicing (Tyan et al. 2005). These findings suggest a possible role of mutant TSG 101 in the evolution of cancer (Figure 4). RBBP7 is a ubiquitously expressed nuclear protein. It has been found to possess the capability of interacting with the BRCA1 tumor suppressor gene. This ability of RBBP7 is believed to give it control of cell proliferation and differentiation, an important step in the promotion of carcinogenesis (Figure 5).

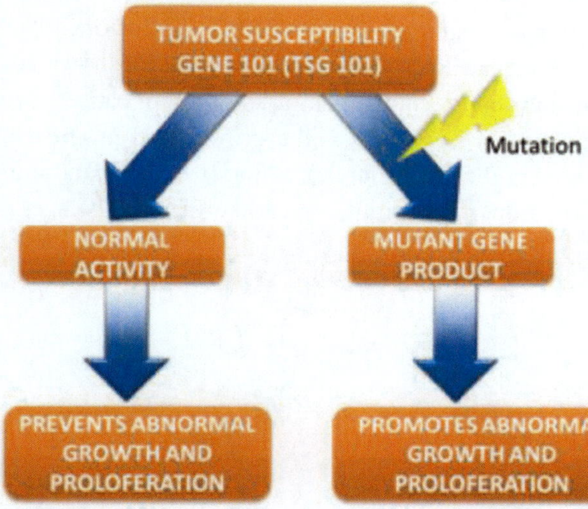

Figure 4. Possible Mechanism of TSG 101 mediated carcinogenesis (Tyan et al. 2005).

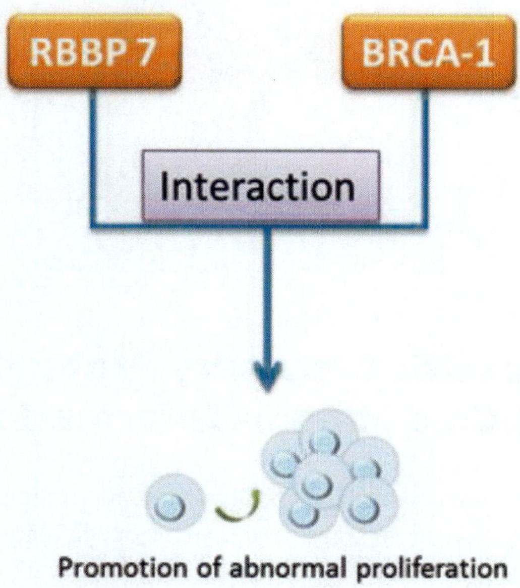

Figure 5. Suggested mechanism of RBBP 7 mediated carcinogenesis (Tyan et al. 2005).

In the following year, Hsieh et al. (2006) published a study where they studied the differentially expressed proteins between malignant and transudative pleural effusions. He used a more elaborate analytical technique involving initial 2D gel electrophoresis followed by identification of differentially expressed protein spots by MALDI, Q-TOF AND LC-MS/MS. Targeted proteins were further examined by ELISA and Western Immunoassays.

They found that, on 2D gel electrophoresis, there were seven spots whose expression was reduced in malignant pleural effusions. The abnormal spots were identified as Fibrinogen Gamma Chain Precursor, Fibrinogen Beta Chain Precursor and Pigment Epithelium Derived Factor. Subsequent Western Blot and ELISA showed that Pigment Epithelium Derived Factor levels were lower in malignant than in transudative pleural effusions.

Pigment Epithelium Derived Factor had previously been shown to have a potent anti-angiogenic activity. The anti-angiogenic action of any protein causes it to exert a strong anti-tumor effect since angiogenesis is central to the growth and progression of any malignant tumor, like, a carcinoma (Figure 6). Underexpression of this potent anti-cancer protein may thus be related to the formation of a malignant pleural effusion (Hsieh et al. 2006).

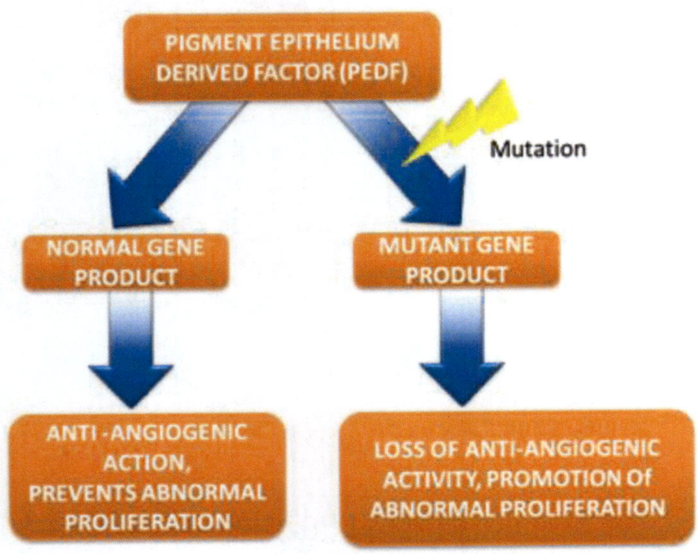

Figure 6. Possible mechanism of PEDF mutation mediated carcinogenesis.

6. Cancer Specific Discovery Proteomic Efforts – A Gold Mine of Information

In 2008, Soltermann et al conducted an N Glycoprotein profiling of lung adenocarcinoma pleural effusions by Shotgun Proteomics. The specific methodology to conduct this study involved samples of pleural effusions from both lung adenocarcinoma patients and non-malignant controls being used for triplicate N-Glycoprotein capture by solid phase extraction. After trypsin digest and PNGase F release, a LC-MS/MS analysis was carried out.

Several proteins associated with tumor progression or metastasis was identified including CA125, CD44, CD166, LAMP-2, multimerin2 and periostin. Periostin promotes oncogenesis by means of cell adhesion and spreading and is overexpressed in human cancers. It is also a marker for metastatic lung and breast cancers. It promotes invasion and metastatic growth. It is thought to be a promising potential biomarker. Multimerin 2 has been found to promote progression of melanoma metastases by promotion of neo angiogenesis around the metastatic tumor deposit. This ensures adequate nutrition to feed the cancer deposit and promote its proliferation at the metastatic site. CD166 (Activated Leucocyte Cell Adhesion Molecule or ALCAM) controls invasive tumor growth and has been found to be overexpressed in prostate, breast and colon cancers. Lysosome Associated Membrane Glycoprotein 2 (LAMP-2) has been implicated in pancreatic carcinoma metastasis. MS identification was correlated with the corresponding immunoreactivity in either effusion fluid or tumor tissue (Table 1). The authors concluded that reduction of sample complexity by N Glycoprotein capturing allows protein detection in the nanogram range to be achieved (Soltermann et al. 2008).

Table 1. Potential biomarkers identified by Soltermann et al (Soltermann et al. 2008)

NAME OF POTENTIAL BIOMARKER	POSSIBLE MECHANISM OF CARCINOGENESIS	CANCERS IMPLICATED
Periostin	Promotion of cell adhesion and spreading	Lung and Breast Cancer metastasis
CD 166	Controls tumor invasion and growth	Lung, Prostate, Breast and Colon cancers
Multimerin 2	Promotes neo angiogenesis	Lung, Malignant melanoma metastasis
LAMP-2	Unknown	Lung, Pancreatic Carcinoma metastasis

Hegmans et al. (2009) investigated the pleural fluid proteome of patients with mesothelioma. Using analytical techniques of pre-fractionation combined with SELDI-TOF and protein arrays, they analyzed 41 mesothelioma associated effusions and also 48 other effusions, 40 of them due to metastasis. They identified 5 differentially expressed proteins and validated one, Apolipoprotein C1 and concluded that Apolipoprotein C1 was a moderately accurate biomarker for pleural mesothelioma.

In 2009, Pernemalm et al applied narrow range peptide IEF to pleural effusion prior to LC-MS/MS. They used two methods for narrow range IEF, IPG strips and free-flow electrophoresis. Data from the study was compared with cell line data to evaluate the method performance in body fluids. To test the methods potential in quantitative biomarker discovery studies, plasma and pleural effusion from patients with lung adenocarcinoma were compared with inflammatory pleuritis using ITRAQ quantification. Using narrow range IEF on the peptide level, they were able to identify and quantify 282 proteins in plasma and 300 proteins in pleural effusion.

These body fluid proteomes demonstrated a high degree of overlap. Nine proteins were chosen for the initial validation with Western blot and one protein (NPC2) was found to be upregulated in patients with adenocarcinoma and was chosen for further validation with immunohistochemistry. Immunohistochemistry showed a strong expression of NPC2 in lung adenocarcinoma. It was noteworthy that the tumor specific up-regulation seen in pleural effusions and tumor tissue is not seen in plasma. The authors believe that the NPC2 in the pleural effusion is sourced directly from the tumor cells. This study shows the important role of study of pleural effusion in the quest for lung cancer biomarkers (Table 2). Overall, the quantitative results from IEF/LC/MS/MS showed good correlation with the results from

Western Blot and immunohistochemistry demonstrating the potential of this methodology in quantitative biomarker discovery studies.

The study also demonstrated that even after narrow range IEF, body fluids comprise a tough challenge for currently available MS technologies. Analyzing the identified proteins in depleted plasma and pleural effusion they found that plasma and pleural effusion were indeed very similar, with an overlap of 60–70% among the identified proteins. Based on these results they suggested the use of EDTA tubes for collection of pleural effusion for proteomics studies. Somewhat disappointing was the discovery that no higher content of tissue leakage (cellular) proteins could be seen in pleural effusion as compared with plasma (Pernemalm et al. 2009).

7. Secretome Analysis with Pleural Effusion Proteome Study – A Promising Path

In 2009, we conducted a Combined Analysis of Cancer Cell Secretome and Pleural Effusion Proteome. Here, we simultaneously analyzed the NSCLC cell secretome and proteomic profiles of pleural effusion from lung adenocarcinoma patients for NSCLC biomarker discovery. Retinoblastoma-associated binding protein 46 (RbAp46), one of the proteins detected both in NSCLC cell secretome and pleural effusion proteome, was eventually chosen for further evaluation. Both RbAp46 mRNA and protein levels were upregulated significantly in NSCLC cancer tissues. Serum levels of RbAp46 were found to be markedly higher in NSCLC patients than in healthy controls, and a combination of RbAp46 and CEA could potentially outperform CEA alone in discriminating NSCLC patients from healthy persons. Importantly, an elevated serum RbAp46 level was highly correlated with NSCLC distant metastasis. Knockdown of RbAp46 also inhibited the migration ability of lung cancer cells (Table 2). Our data suggested that RbAp46 may be a novel biomarker and prognosticator for NSCLC (since presence of metastasis indicates an advanced stage and poor prognosis) and appears to be involved in lung cancer cell migration. Previously, RbAp46 expression had been found to be associated with a migratory and invasive phenotype of breast cancer. Our study also indicated that the combination of secretome data obtained from specific cancer cell lines with tissue/organ secreted protein profiles will help us identify the body fluid biomarkers more efficiently (Wang et al. 2009).

Table 2. Potential biomarkers identified by Protein Arrays, Narrow range IEF and Combined (cancer cell secretome with effusion proteome) studies

NAME OF POTENTIAL BIOMARKER	DISCOVERY METHODS	CANCER TYPE
Apolipoprotein C1 (Hegmans et al. 2009)	SELDI with Protein Arrays	Pleural Mesothelioma
NPC2 (Pernemalm et al. 2009)	Narrow range IEF with ITRAQ	Lung Cancer
RbAp46 (Wang et al. 2009)	LC-MS/MS and real-time PCR with Immunohistochemistry	Breast Cancer, Non Small Cell Lung Cancer

8. Recent Studies with Mudpit and Other Methods – Many New Biomarker Candidates

In 2010, Rodriguez et al. did a comparative proteomic study of the serum and pleural effusion samples of lung cancer patients with those with benign diseases like pneumonia and tuberculosis. Samples were prefractionated through affinity chromatography prior to 2D-DIGE to detect proteins with altered expression in cancer patients. More potential biomarkers were identified in the pleural effusion than in the serum.

The biomarker candidates comprise both proteins, which were increased in malignant pleural effusions as well as proteins, which were decreased in malignant effusions. Gelsolin was increased in malignant effusions. This protein had, in previous studies, been shown to be associated with local invasion of the lymph nodes. Metalloproteinase Inhibitor 2 was also increased in malignancy. This protein was previously reported as associated with a variety of malignancy-associated processes. Among those decreased in malignant effusions were S-100-A8 and S-100-A9. These proteins are known for their ability to activate macrophages and modulate proliferation in an extracellular environment. Pigment Epithelium Derived Factor (PEDF) was overexpressed both in serum and pleural effusion of Non-Small Cell Lung Cancer Patients (Table 3). In fact, PEDF was the only protein found with significantly different levels both in the pleural effusion and the serum from NSCLC patients, when compared against tuberculosis and pneumonia (Rodriguez-Pineiro et al. 2010).

In 2011, Yu et al. published a study of comprehensive proteome analysis of malignant pleural effusion with the help of MUDPIT technology. Their analytical technique involved a removal system for high abundance proteins followed by one-dimensional SDS PAGE combined with Nano LC-MS/MS. They generated a malignant pleural effusion proteome data set with 482 non-redundant proteins. Next, they integrated the proteome data set with secretome data sets from three adenocarcinoma cell lines to identify potential pleural effusion biomarkers from malignant cells.

They found four potential candidates, Alpha 2 HS Glycoprotein, Angiogenin, Cystatin-C and Insulin like growth factor binding protein 2 which were subsequently selected for preclinical validation using ELISA. Both Alpha 2 HS Glycoprotein (AHSG) and Insulin like Growth Factor Binding Protein 2 levels were higher in malignant pleural effusions when compared with benign pleural effusions, more so the latter suggesting its possible potential as a biomarker. Alpha 2 HS Glycoprotein had been shown in recent studies to regulate immune response, control cell adhesion and inhibit tumor progression. Insulin like Growth Factor Binding Protein 2 had been previously found to be over expressed in many cancers, including breast cancer, lung cancer, prostate cancer, ovarian cancer and gastric cancer (Table 3). Over expression was found to be associated with a more aggressive tumor phenotype (Yu et al. 2011).

In 2012, Wang et al. published a study on the differential proteome profiling of pleural effusions of lung cancer and benign inflammatory disease patients. They conducted the study using 2D- DIGE followed by MALDI-TOF/TOF MS. 28 protein spots displayed significantly different expression levels. They were subsequently positively identified by MALDI-TOF-MS as representing 16 unique proteins.

Five identified protein candidates were further validated and analyzed in effusions, sera or tissues. Among them, Transthyretin, hemopexin and fibrinogen gamma were up-regulated in cancer samples. Hemopexin had previously been linked to invasion and metastasis in many cancer model systems; this may be linked to its ability to activate matrix metalloproteinase. Fibrinogen and its degradation products have been suggested to affect the progression of tumor growth and metastasis. Interestingly, fibrinogen deficiency was found to strongly diminish the development of metastasis in an animal model (Table 3). The role of transthyretin in malignancy is not well understood. The effusion concentration of Serum Amyloid P component was found to be significantly lower in lung cancer cases than benign effusions although these differences were not found on analysis of sera. Also, a Jumonji C-domain containing protein was found to be down-regulated in malignant effusions, lung cancer tissues and cancer cells (Wang et al. 2012).

In 2013, Wang et al. conducted a study to evaluate the diagnostic value of pleural effusion apoE in Non-Small Cell Lung Cancer (NSCLC). Their study was inspired by the fact that previous research had shown apoE to be elevated in pleural effusions of patients with NSCLC. In their study, they analyzed pleural effusion samples, both benign and malignant, of 200 patients (160 patients with histologically proven NSCLC and 40 patients with benign lung pathology such as tuberculosis and pneumonia) for apoE (by ELISA) and other markers.

Their study found apoE levels to be significantly increased in malignant pleural effusions. It is also interesting to note that they also found that serum apoE was not significantly elevated in malignancy. This finding agrees with our view that the pleural fluid proteome study has a higher chance of discovering prospective biomarkers for lung cancer than analysis of the serum proteome of lung cancer patients.

Table 3. Recently discovered promising biomarker candidates for lung cancer

NAME	REFERECNE
Gelsolin	(Rodriguez-Pineiro et al. 2010)
Metalloproteinase Inhibitor 2	(Rodriguez-Pineiro et al. 2010)
Pigment Epithelium Derived Factor	(Rodriguez-Pineiro et al. 2010)
Alpha 2 HS Glycoprotein	(Yu et al. 2011)
Insulin like Growth Factor Binding Protein 2	(Yu et al. 2011)
Transthyretin	(Wang et al. 2012)
Hemopexin	(Wang et al. 2012)
Fibrinogen Gamma	(Wang et al. 2012)
Apolipoprotein E	(Wang et al. 2013)

Conclusion

Proteomics has brought to the table of diagnostic cancer research a novel protein analysis methodology based on identification of unique cancer associated proteins by various forms of proteomic analysis (like MALDI or ESI).

The study of the pleural effusion proteome is a major step in the ever-continuing efforts of science to unlock the mysteries of cancer, especially lung cancer. A number of promising potential biomarker candidates has already been identified by proteomic analytical efforts from multiple countries. The importance of these findings lie in the fact that they help us gain a better understanding of the initiation and progression of lung cancer, especially, mechanisms involving the tissue response to a cancerous growth.

These findings may help in developing specific targeted anti-cancer drugs in the future. Proteomic study of pleural fluid has its own analytical challenges. Modern analysis methodologies employed by recent researchers have come a long way in solving the analytical difficulties faced by early researchers of the pleural fluid proteome. The greatest challenge, however, still remains in interpreting the functional significance of the large number of cancer associated proteins identified by modern proteomic analysis.

It seems that the presently identified potential biomarkers represent only a small part of circulating potential biomarkers in the pleural fluid proteome. There exists a very real possibility that many more potential biomarker candidates will be discovered by the research of pleural fluid proteome in the coming years.

References

Bard MP, Hegmans JP, Hemmes A, Luider TM, Willemsen R, Severijnen LA, van Meerbeeck JP, Burgers SA, Hoogsteden HC, Lambrecht BN (2004) Proteomic analysis of exosomes isolated from human malignant pleural effusions. *American journal of respiratory cell and molecular biology* 31(1):114-121.

Hegmans JP, Veltman JD, Fung ET, Verch T, Glover C, Zhang F, Allard WJ, T'JAMPENS D, Hoogsteden HC, Lambrecht BN (2009) Protein profiling of pleural effusions to identify malignant pleural mesothelioma using SELDI-TOF MS. *Technology in cancer research & treatment* 8(5):323-332.

Hsieh WY, Chen MW, Ho HT, You TM, Lu YT (2006) Identification of differentially expressed proteins in human malignant pleural effusions. *The European respiratory journal* 28(6):1178-1185.

Indovina P, Marcelli E, Maranta P, Tarro G (2011) Lung cancer proteomics: recent advances in biomarker discovery. *Int J Proteomics* 2011726869.

Pernemalm M, De Petris L, Eriksson H, Branden E, Koyi H, Kanter L, Lewensohn R, Lehtio J (2009) Use of narrow-range peptide IEF to improve detection of lung adenocarcinoma markers in plasma and pleural effusion. *Proteomics* 9(13):3414-3424.

Rifai N, Gillette MA, Carr SA (2006) Protein biomarker discovery and validation: the long and uncertain path to clinical utility. *Nat Biotechnol* 24(8):971-983.

Rodriguez-Pineiro AM, Blanco-Prieto S, Sanchez-Otero N, Rodriguez-Berrocal FJ, de la Cadena MP (2010) On the identification of biomarkers for non-small cell lung cancer in serum and pleural effusion. *J Proteomics* 73(8):1511-1522.

Soltermann A, Ossola R, Kilgus-Hawelski S, von Eckardstein A, Suter T, Aebersold R, Moch H (2008) N-glycoprotein profiling of lung adenocarcinoma pleural effusions by shotgun proteomics. *Cancer* 114(2):124-133.

Tyan Y-C, Wu H-Y, Lai W-W, Su W-C, Liao P-C (2005) Proteomic profiling of human pleural effusion using two-dimensional nano liquid chromatography tandem mass spectrometry. *Journal of proteome research* 4(4):1274-1286.

Tyan YC, Wu HY, Lai WW, Su WC, Liao PC (2005) Proteomic profiling of human pleural effusion using two-dimensional nano liquid chromatography tandem mass spectrometry. *J Proteome Res* 4(4):1274-1286.

Wang CL, Wang CI, Liao PC, Chen CD, Liang Y, Chuang WY, Tsai YH, Chen HC, Chang YS, Yu JS, Wu CC, Yu CJ (2009) Discovery of retinoblastoma-associated binding protein 46 as a novel prognostic marker for distant metastasis in nonsmall cell lung cancer by combined analysis of cancer cell secretome and pleural effusion proteome. *J Proteome Res* 8(10):4428-4440.

Wang Y, Chen Z, Chen J, Pan J, Zhang W, Pan Q, Ding H, Lin X, Wen X, Li Y, Meng QH (2013) The diagnostic value of apolipoprotein E in malignant pleural effusion associated with non-small cell lung cancer. *Clin Chim Acta* 421230-235.

Wang Z, Wang C, Huang X, Shen Y, Shen J, Ying K (2012) Differential proteome profiling of pleural effusions from lung cancer and benign inflammatory disease patients. *Biochim Biophys Acta* 1824(4):692-700.

Yu CJ, Wang CL, Wang CI, Chen CD, Dan YM, Wu CC, Wu YC, Lee IN, Tsai YH, Chang YS, Yu JS (2011) Comprehensive proteome analysis of malignant pleural effusion for lung cancer biomarker discovery by using multidimensional protein identification technology. *J Proteome Res* 10(10):4671-4682.

In: Recent Advances in Proteomics Research
Editor: Ming D. Li

ISBN: 978-1-62948-218-7

Chapter 5

Application of Quantitative Proteomics to Disease Diagnosis and Prognosis Using Stable Isotope Dimethyl Labeling Coupled with Shotgun Proteomics Strategy and Multiple Reaction Monitoring

Thomas Shyh-Horng Chiou*[1,2,3,*]*, Min-Yi Lee*[1]*,
***Chao-Jen Kuo*[1,2]* and Shih-Shin Liang*[2,4]**
[1]Graduate Institute of Medicine, College of Medicine,
Kaohsiung Medical University, Taiwan
[2]Quantitative Proteomics Center, Kaohsiung Medical University, Taiwan
[3]Institute of Biological Chemistry, Academia Sinica, Taipei, Taiwan
[4]Department of Biotechnology, College of Life Science,
Kaohsiung Medical University, Taiwan

Abstract

Mainly attributable to the advent of emerging proteomics, the analysis and identification of complex protein mixtures in biological tissues have recently become less tedious and more amendable to routine analysis. The study of proteins at the molecular and cellular levels by means of fast-evolving and state-of-the-art proteomics methodologies has provided a firm basis for understanding the complex proteome profiles of total protein mixtures from complex whole tissues or cells of various

* Corresponding author: Thomas S.-H. Chiou, Graduate Institute of Medicine, Kaohsiung Medical University, Kaohsiung 807, Taiwan; or Institute of Biological Chemistry, Academia Sinica, Taipei 115, Taiwan. Tel: 886-7-3220377; Fax: 886-7-3133434; E-mail: shchiou@kmu.edu.tw.

biological sources. As a result, proteomics has been increasingly valuable to examine dynamic changes in protein expression of various tissues or body fluids, resulting in the identification of clinically useful biomarkers for diagnosis, prognosis and disease staging. Comparative proteomic analysis of normal and diseased tissue samples from patients with a defined disease diagnosis is particularly powerful for analysis and correlation of proteins expressed differentially at certain disease stages. In contrast to the conventional gel-based proteomics, especially concerning some limitation and shortcomings of 2-D gel approach, we have made an endeavor based on gel-free shotgun proteomics to examine the manifestations and significance of biomarkers in various tissues and body fluids including urine with some rewarding success. In this chapter, we have focused on potential disease biomarkers for (1) coronary artery disease (CAD), (2) renal cell carcinoma (RCC), and (3) varicose vein. We have utilized a proteomic approach using stable isotope dimethyl labeling and shotgun proteomics strategy in combination with nano-liquid chromatography and tandem mass spectrometry (nanoLC-MS/MS). We aim to establish a high-throughput proteomics platform based on multiple reaction monitoring (MRM) for the detection and analysis of potential biomarkers at low concentrations in various normal and diseased tissues. The analysis platform can be uniquely applied to disease-oriented translational medicine for diagnosis and prognosis of diverse types of diseases.

Keywords: Disease proteomics; coronary artery disease; renal cell carcinoma; varicose vein; shotgun proteomics; multiple reaction monitoring (MRM); quantitative proteomics

1. Introduction

This chapter will give an overview and evaluation of proteomic strategies commonly-adopted in current mass spectrometry (MS)-based proteomic studies of various biomedical samples. A general review and comments on some drawbacks or pitfalls associated with the conventional gel-based proteomic methodologies for biomarker discovery and verification are presented. We then introduce the label-based quantitative proteomics to study samples of three clinical cases based on a stable isotope dimethyl labeling and shotgun proteomics strategy in combination with sensitive nano-liquid chromatography and tandem mass spectrometry (nanoLC-MS/MS).

The newer proteomic approach for these three clinically-relevant cases exemplified the potential application of shotgun quantitative proteomics to translational medicine. It is intended to provide some standard practice and guidelines to those clinical researchers first exposed to proteomics studies by means of advanced MS instrumentation.

Modern clinical research is increasingly dependent on evidence-based approach to biomedical research issues related to human diseases with the advent of postgenomic era and the completion of human genome sequences. There comes next the functional genomics with its major emphasis on interactions between various genome-encoded proteins, i.e. the proteomes of varied and diversified tissues or cells.

Thus proteomics in the context of the link between genes and proteins arose and evolved over time to become a major theme of central topics in modern biomedical research. Proteomics can reveal the global protein expression profiles of particular cells or tissues at defined time points.

Through the assistance of constantly increasing databases of complete genomes from various model organisms, including human and various submammalian species (Ideker et al. 2001; Fehniger and Marko-Varga 2010), proteomes of any samples from some exotic biological sources can now be analyzed at protein expression levels by various proteomic strategies coupled with mass spectrometry (MS).

The intricate and diversified protein identities (protein ID) in complex protein mixture of biological tissues and cells can then be compared with *in silico* peptide maps derived from enzyme-digested proteins or genomic databases deposited at various databanks. To date, proteomics through the fast-evolving MS instrumentation has developed to a critical stage of replacing the conventional protein chemistry in global protein characterization. Recently, clinician scientists who have gradually witnessed and appreciated the epoch-making potential and impact of proteomics to medicine began to use the innovative proteomic technology to clinical research especially on its application to translational medicine. They are especially interested in coupling the proteomics technology with bioinformatics mining of human genome to unravel the related molecular mechanisms of diseases. This has engendered a new ground for the development of so-called disease or clinical proteomics (Fehniger and Marko-Varga 2010; Alaiya et al. 2005; Anderson 2010; Colantonio and Chan 2005; Dominguez et al. 2007; Hendriks et al. 2007; Marko-Varga and Laurell 2010; Paik et al. 2008; Sinha et al. 2007; Tao and Lazarev 2007; Wu and Zhang 2007).

The newly developed discipline would focus on effective clinical monitoring and diagnosis of patients before and after medical treatment and the prognosis and assessment during the progress of disease treatment. This usually involves several phases of proteomics study, that is, (1) biomarker discovery; (2) biomarker verification; and finally (3) biomarker validation and clinical trial.

In this review, we have made an endeavor to compare and evaluate the following widely used methodological strategies and their corresponding workflows (Figure 1) for a holistic proteomic approach to characterize protein compositions or expression profiles of various disease-related tissues or body fluids with the aim of providing some application guidelines and strategies to clinical researchers interested in knowledge-based translational medicine.

2. Proteomic Methodologies and Platforms

2.1. 1-D or 2-D Gel-Based Proteomics

1-D and 2-D gel electrophoresis referring to one- or two-dimensional polyacrylamide gel electrophoresis (1-D or 2-D PAGE) have been used for the analysis of complex protein mixtures for more than three decades. The most common protocols for 1-D and 2-D gel electrophoreses are those described by Laemmli (Laemmli 1970) and O'Farrell (O'Farrell 1975) respectively long before the advent of proteomic era. The simplicity and high-resolution separation patterns obtained from the Laemmli's method and the fact that dilute samples with microgram amounts of sample proteins can routinely be analyzed, have made the report of Laemmli's SDS-PAGE (with the addition of strong anionic detergent sodium dodecyl sulfate, SDS, in the electrophoretic buffer) system one of the most cited methodological papers in life sciences. Actually 1-D SDS-PAGE methodology including its buffer containing SDS has been incorporated into the second-dimensional protein separation based on molecular size in O'Farrell's 2-D gel system.

They are both also the most widely used procedures for the determination of the complexity and molecular masses of constituent polypeptides in a protein sample on 1-D and 2-D slab gels.

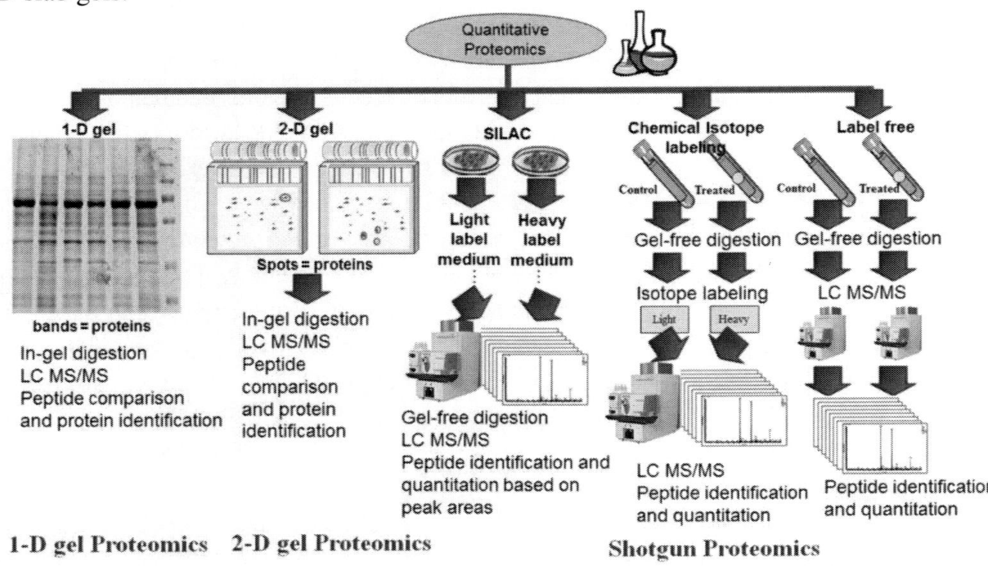

Figure 1. Scheme of the workflow for each of the proteomics strategies applied to translational medicine.

Combining the features of two different separation parameters of molecular size and charge, 2-D gel can resolve more than 1000 different proteins in the form of a 2-D polypeptide map from the protein extract of *E. coli* (O'Farrell 1975; Dunn 1983). There has been an increasing emphasis on the separation and characterization of polypeptides by 2-D gel for the pre-sample treatment of gel-based proteomics since the first 2-D gel proteomics report of Henzel et al. (1993), which has been claimed to be the most sensitive and high-resolution method for the separation of complex protein composition in any cell extract or protein mixture before MS-based proteomics analysis. Although up to 2000 individual polypeptide chains may be resolved on a single 2-D gel, enough to account for most of the abundant proteins in a bacterium, it is still far from a complete and satisfactory global proteomic analysis on a human cellular proteome containing more than 25000 protein species.

In the 2-D gel proteomics, the protein spots on polyacrylamide gels were treated with cleavage-specific proteases such as trypsin to generate peptides by *in situ* tryptic digestion of each of the separated protein sports on 2-D gels. The mass spectral patterns of protein fragments generated with trypsin digestion were then used for comparison with those fragments of previously known proteins (treated with trypsin under the same conditions) deposited in databanks to confirm the identified peptides, which can be used further for identification of intact proteins (protein ID). The 2-D gel proteomics were used so successfully in the initial stage of proteomics development such that 2-D gel were mistakenly taken as the necessary step for all proteomic analysis of any biological samples. With the development of state-of-the-art MS instruments both in terms of sensitivity and high-throughput operations, the limitation and pitfalls inherently associated with 2-D gel proteomics need to be addressed when the proteome maps derived from 2-D gel strategy shows some discrepancies with those obtained from newer gel-free shotgun proteomics (see below).

Although the theory and principle of 2-D gel electrophoresis is relatively simple and straightforward, methodology of 2-D gel is generally a technically demanding, time-consuming plus skill-intensive procedure in practice. The art of running good and reproducible 2-D gels can only be learned from experience gained in the laboratory. The advantages for using 2-D gel in proteomics lie in the separation and visualization capabilities of 2-D gel to reveal and enable the protein expression profiles of paired samples, *i.e.* normal and diseased ones to be readily compared with ease. However, comprehensive 2-D gel databases for samples of health and disease are still lacking. There is always an appraisal of the current status and advantages/shortcomings associated with 2-D gels that challenges its central position in conventional proteomics analysis. (1) First, protein spots resolved on 2-D gels and visualized using a variety of methods, including organic dye staining, silver nitrate, isotope labeling and immunoblotting are all sources of contaminated chemicals which may not be compatible with the MS instruments of high cost. (2) The serious drawback of low sensitivity can be attributed to the fact that the maximal numbers of several thousands of spots resolved and analyzed by 2-D gels is still relatively small as compared with the whole genome encoded functional proteins of about 20,000-30,000 proteins, especially being under-representative of some special classes of proteins, such as the extremely basic or acidic groups of proteins, some low-abundance transcription factors, and membrane proteins (Han et al. 2008; Leth-Larsen et al. 2010) because of the low solubility of these classes of proteins in the first dimensional isoelectric focusing gel of 2-D gels.

Currently, 1-D gel instead of 2-D gel is more often used for proteomic analysis of samples of poor solubility because the most effective denaturing agent of SDS is included in the 1-D gel and electrophoresis buffer in contrast to that of 2-D gel which contains SDS only in the second-dimensional separation. Figure 2 shows the representative 1-D gel resolution of complex protein components of normal and cataractous human lenses (Huang et al. 2011) under dissociating conditions in the presence of SDS. The separated protein bands or zones were then subjected to digestion with trypsin and the digested protein mixture was subsequently fed into a liquid chromatograph MS (LC-MS or LC-MS/MS) for protein ID determination. It can be seen that 1-D gel being less tedious and more time-saving than 2-D gel can still afford respectable and sometimes more protein identification than that of 2-D gel (based on shotgun proteomics analysis on protein bands in 1-D gel). The complete analysis of all proteins expressed within a cell is becoming possible by 1-D gel coupled with shotgun-proteomics analysis as described below.

2.2. Gel-Free Shotgun Proteomics

Currently, there are two fundamental strategies for proteomics study. One is bottom-up and the other is top-down. In bottom-up approach, purified proteins or complex protein mixtures are subjected to chemical or enzymatic cleavage and the peptide products are usually separated by chromatography followed by mass spectrometric analysis (Chait and 2006). In top-down proteomics, intact protein ions or large protein fragments are subjected to gas-phase fragmentation for MS analysis directly (Bogdanov and Smith 2005; Kelleher 2004;

Sze et al. 2002). Shotgun proteomics belongs to the category of bottom-up approach and it detects protease-digested peptide fragments throughout the whole cellular proteome in a random statistical sampling method similar to that of shotgun genomics.

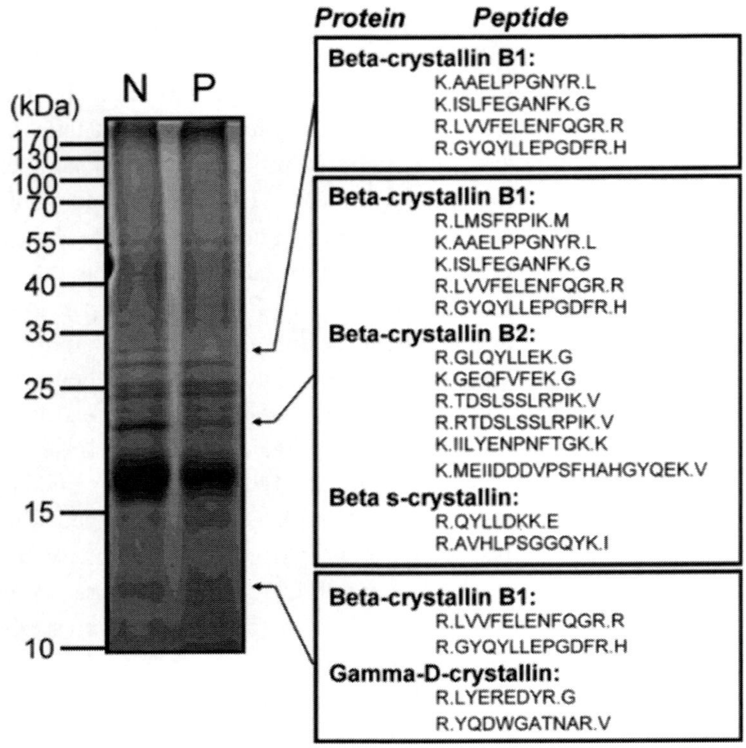

Figure 2. Comparative analysis of normal and cataractous human lens proteins by 1-D gel proteomics. As shown in the left panel, lens proteins extracted from normal (N) and cataractous (P) eye lenses. In the right panel, protein and peptide bands with different expression levels identified by nanoLC-ESI-MS/MS were indicated by arrows. The protein bands or zones separated by 1-D gel coupled with trypsin digestion can be directly injected to nanoLC-ESI-MS/MS for an extensive and comparative proteomic analysis of normal and cataractous human lens proteins. It is worth noting that the blurred and broad bands or zones shown in 1-D gel is due to overloading of protein samples in order to have protein identification for some low-abundance proteins.

The recent explosion in available genomic sequence information is providing a useful sequence infrastructure for clinical proteomics. A major aspect of various proteomics strategies is the validation of protein ID using analytical "fingerprints" or peptide mass fingerprinting (PMF) generated by digestion of proteins with specific enzymes, such as trypsin, followed by tandem mass (MS/MS) spectrometry, from which MS/MS spectra of peptide fragments can then be used for comparison search in available sequence databanks. The mixture of enzyme-digested proteins or peptides can usually be separated by multidimensional LC (MDLC) and analyzed by a tandem MS. Various MS/MS spectra can be algorithmically compared with predicted peptide spectra from sequence databases to identify the respective proteins. The digestion of intact protein or peptide mixtures with various enzymes of different specificities followed by the direct analysis of the resulting peptides by high-resolution LCs coupled with tandem MS/MS has facilitated the shotgun strategy for the identification of protein mixtures without the need for prior sample fractionation based on 1-

D or 2-D gels (*i.e.* gel-free proteomics). By combining with the recent development of capillary MDLC, this approach is now capable of characterizing proteins directly from entire cell lysates (MacCoss et al. 2002; McCormack et al. 1997; Washburn et al. 2001; Lescuyer et al. 2010).

In shotgun proteomics, MDLC is a necessity to reduce sample complexity and increase dynamic range of protein identification. For the past few years gel-free shotgun proteomics in conjunction with LC-tandem MS (LC-MS/MS) has been the method of choice for analysis of complex samples because of its ease of automation, high throughput, and sensitivity. Shotgun proteomics approach involving nanoLC-quadrupole-TOF MS has also been especially attractive for efficient proteomic analysis of limited and scarce samples. Recently mass spectrometric methods are being developed along the line that not only identifies proteins in a mixture but also compares the relative levels of protein expression between two different samples, *i.e.* gel-free shotgun proteomics has paved a way to quantitative proteomics circumventing the conventional semi-quantitative gel-based proteomic analysis based on image analysis of protein spots on 2-D gels.

2.3. Quantitative Proteomics

To date, the current bottom-up shotgun proteomic approach to find protein ID in various biological samples gradually becomes a routine practice. Now the major challenge lies in quantitative measurements for different proteins in unknown samples of interest. Biomarker discovery is especially dependent on the accurate quantitative difference between normal and diseased samples. In the search for disease associated protein determinants, reproducible and quantitative proteomic analysis remains a vital analytical part in the study of the translational effects of various protein biomarkers in regulating cellular activity and function. Therefore, quantitative determination of each individual protein deserves more efforts. On the other hand, the extension of dynamic range to measure ultra-low quantity of proteins inside a proteomic sample remains a high priority and a great demand. At present, there are two different approaches to quantitative proteomics. One is strictly based on mass spectra from MS/MS without labeling and the other relies on the labeling methodology.

2.3.1. Label-Free Quantitative Proteomics

Label-free approach for quantitative proteomics is more attractive because it can reduce the contaminant species during the labeling process before proteomic analysis with high sensitivity. Nevertheless, the entire process for proteomic analysis without a quantitative labeling tag is quite complex. It is also very difficult to assure quality control on every purification and analytical step in different laboratories or different personnel from the same laboratory. At the present time, there are still controversies over the reliability of label-free quantitative analysis. It is noteworthy that quantitative determination between two samples cannot be just based on qualitative protein ID analysis without a dedicated proteomic analysis platform to do data analysis. Heavy participation of bioinformatics software analysis is warranted for the future success of label-free quantitative analysis. MOEL

2.3.2. Label-Based Quantitative Proteomics

2.3.2.1. Isotopic-Coded Affinity Tag

The basic principle of Isotopic-coded affinity tag (ICAT) (Gygi et al. 1999) relies on the protein thiol-specific reagents.

It consists of three major steps: (1) a special affinity tag is used to react with cysteine residues in a protein; (2) labeled peptides are obtained from the enzymatic digestion of labeled proteins; and (3) labeled peptides are separated by LC or MDLC followed by MS analysis for relative quantitative measurements. ICAT is often used to measure the relative protein abundance between two samples, such as tissues from a healthy person and a patient. However, the analysis is usually long and expensive because of high costs of isotopic labeling reagents.

2.3.2.2. Isobaric Tags for Relative and Absolute Quantitation

The quantitation strategy for isobaric tags for relative and absolute quantitation (iTRAQ) labeling (Ross et al. 2004) of multiplexed sets of experiments is commonly based on a set of four isobaric reagents, each of which comprises three groups: reporter (mass from 114 to 117 Da), balance (mass from 28 to 31 Da), and reactive groups. After cell lysis, reduction/ alkylation, and protein digestion, the peptides in four states are separately labeled on N-terminal and internal lysine residues by the reactive group of iTRAQ. Each labeled peptide appears at the same mass in a MS scan. During the fragmentation in the MS, the label dissociates and releases the reporter group as a singly charged ion of masses 114, 115, 116, or 117 Da, respectively. Relative intensity of these peaks indicates the contribution of each sample to total peptide intensity and can provide the information on the relative abundance. The labeled peptides are then combined and analyzed by MDLC and MS. The multiplexed protein quantitation capability of the iTRAQ strategy permits simultaneous comparison of samples of different cell states and disease specimens. This advantage can be applied to a comparative study on the temporal and/or spatial distribution of proteins in cells. iTRAQ can be used to simultaneously label and analyze up to four different samples. Absolute quantitative measurement of a specific protein can also be achieved by iTRAQ. This involves comparing peptides from a target protein to a known amount of labeled standard peptide added to the sample.

2.3.2.3. Stable Isotope Labeling with Amino Acids in Cell Culture

Stable isotope labeling with amino acids in cell culture (SILAC) was developed by Ong et al. (Ong et al. 2002) to detect proteomic differences between two cell samples. One of the cultured cell populations is fed with normal amino acids. In contrast, second cell population is fed with amino acids labeled with stable (nonradioactive) isotopes. For example, the medium can contain arginine labeled with six ^{13}C instead of the normal ^{12}C. During the cell growing period, they incorporate arginine into all of their proteins. Therefore, all heavy arginine-containing peptides are heavier than their normal counterparts by 6 Da. Pairs of chemically identical peptides of different stable isotope composition can be clearly distinguished because of their mass difference. The ratio of peak intensities for such peptide pairs can accurately reflect the population ratio for the two proteins. SILAC is becoming a powerful tool to study cell signaling (Ong et al. 2003; Ong and Mann 2006; Ong and Mann 2007).

2.3.2.4. Stable Isotope Dimethyl Labeling

Among the various N-terminal labeling reagents, stable isotope dimethyl labeling method developed by Chen's group (Hsu et al. 2003) affords a simple and fast chemical modification on peptides or proteins with relatively high specificity.

As shown in the following equation all primary amino group of peptides such as unblocked N-terminus and lysine side chain can be labeled by H_2-formaldehyde (normal control sample) or D_2-formaldehyde (treated sample) to form a Schiff base that is reduced by sodium cyanoborohydride to form a secondary amine, which is relatively more reactive than the primary amine. Subsequently, the more-reactive species reacts with another formaldehyde unit and is then reduced to form a dimethylamino group. In this labeling reaction, reductive amination is used to modify the N-terminal group of proteins or peptides, which can be easily applied to quantitative proteomics.

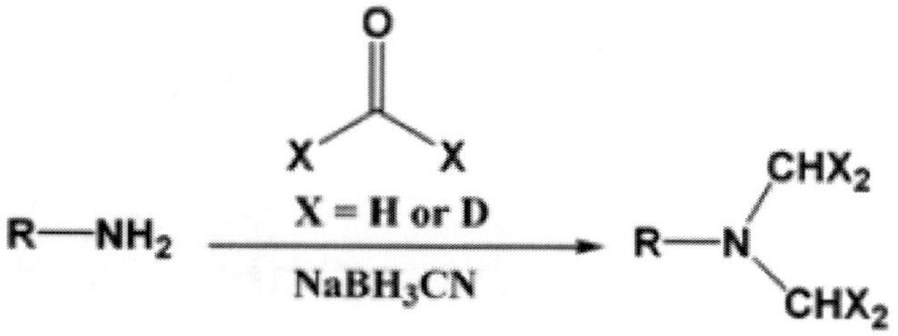

The applicability of this method for quantitative protein profiling is exemplified in the proteomic analysis for the following clinical cases (See section 3.1., 3.2. and 3.3.).

2.3.3. Multiple Reaction Monitoring Coupled Tandem MS Quantitative Proteomics

Quantitative targeted proteomics focusing on multiple reaction monitoring-MS (MRM-MS) methodologies has taken front stage in the proteomics research recently, especially being frequently applied to the verification of global proteomics data, the discovery of lower abundance proteins, protein PTMs, and biomarker discovery (Yocum and Chinnaiyan 2009). As an older methodology used for small molecule analysis, MRM-MS has been adapted as the next methodological advance to address previously challenging issues in global proteomics experimentation, namely dynamic range, identification of PTMs, and sensitivity/ selectivity of measurements. These developments have shown that targeted MRM analysis on non-redundant peptides of conventional MS/MS data of potential biomarkers can provide more exact quantification of homologous proteins. The methodology has facilitated the transition from routine biomarker-discovery proteomics to quantitative targeted-proteomics analysis, further allowing us to identify and verify more low-abundance potential biomarkers.

MRM is the extended version of Selected Reaction Monitoring (SRM), which arose as a technology to complement the discovery capabilities of shotgun proteomic strategies by its unique capability for reliable quantification of biomolecules (or termed analytes) present at low abundance in complex mixtures (Lange et al. 2008). SRM exploits the unique capabilities of triple quadrupole MS for quantitative analysis, which is ideally suited to address a major proteomic challenge in systems biology for the accurate quantification of some specific sets of proteins spanning the whole range of cellular proteomes from different samples of evolutionarily related organisms. In SRM, the first and the third quadrupoles act as filters to specifically select predefined m/z values corresponding to the peptide ion and a specific fragment ion of the peptide, whereas the second quadrupole serves as collision cell. Several such transitions from precursor peptide to their fragment ion pairs are monitored over time, yielding a set of chromatographic traces with the retention time and signal intensity for a specific transition as coordinates.

The two levels of mass selection by two quadrupoles with narrow mass windows result in a high selectivity, as coeluting background ions are filtered out very effectively. It is also important to note that SRM and MRM have been successfully applied to a variety of biological applications, including robust quantitative proteomic analysis of cellular signaling networks (Wolf-Yadlin et al. 2007), quantification of DNA adducts purified from tissue (Koc and Swenberg 2002), and detection of doping substances in human urine and plasma (Guan et al. 2005; Herrin et al. 2005; Ho et al. 2006; Thevis et al. 2001). To date, the use of SRM and MRM in proteomics is less widespread, although MRM has been used to quantify protein expression (Kirkpatrick et al. 2005), to find protein biomarkers for disease severity in rheumatoid arthritis (Liao et al. 2004), and to detect and quantify protein phosphorylation profiles (Wolf-Yadlin et al. 2007; Cox et al. 2005; Unwin et al. 2005; Zappacosta et al. 2006).

3. Proteomics Study on Three Exemplar Clinical Cases

3.1. Coronary Artery Disease (CAD)

We performed 1-D SDS gel followed by liquid chromatography coupled with tandem mass spectrometry (LC-MS/MS) to identify differentially expressed proteins present in urine samples of normal and CAD patients. We used this case to provide a platform for efficient separation of complex tissue extracts or body fluids (urine was used in this case). The protein bands or zones were treated with trypsin and protease-digested bands were then directly used for analysis using shotgun proteomics in combination with nano-liquid chromatography and tandem mass spectrometry (nanoLC-MS/MS). Shotgun analysis provided more than 100 identified proteins with different expression labels based on stable isotope (hydrogen/ deuterium) labeling.

This 1-D gel coupled with shotgun strategy for the proteomic comparison of paired samples between normal and CAD patients (Figure 3) revealed new insights in pinpointing the close association of urinary sCD14 and coronary atherosclerosis because of its consistent and conspicuous presence in CAD patients as compared to other protein components identified in pair-wise urine samples (unpublished results). We have analyzed the concentration of sCD14 in urine and serum of 73 patients with angiographically proven CAD and 35 control subjects using enzyme-linked immunosorbent assay (ELISA). The sCD14 in the urine and serum samples were also confirmed and verified by Western blot analysis. The unusual findings that release of sCD14 in urine coupled with higher CD14 mRNA expression in CAD patients was strongly correlated with severity of CAD point to the potential application of urinary sCD14 as a novel biomarker for screening and diagnosis of patients manifested with different extents of coronary atherosclerosis.

3.2. Renal Cell Carcinoma (RCC)

A previous report showed the up-regulation of αB-crystallin expression in normal renal and clear cell renal cell carcinoma (ccRCC) tissues (Ho et al. 2012). Renal cell carcinoma (RCC) is the most common type of kidney cancer in adults, and mainly arises from the proximal renal tubules (Motzer et al. 1996; Tomita 2006).

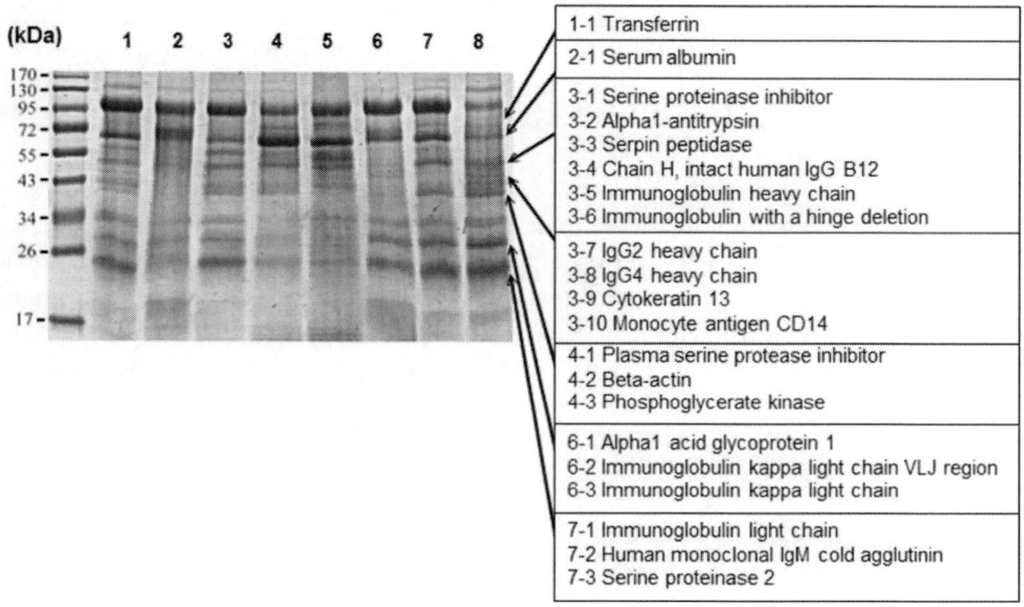

Figure 3. 1-D gel analysis of urinary proteins from normal and CAD patients. A total of 10 μg of urinary proteins derived from normal controls and patients with different severity of coronary artery disease manifested with SYNTAX scores were used for 1-D gel comparative analysis of urinary proteins. Proteomic bands shown with different intensities were excised and trypsin-digested and applied to LC-MS/MS for shotgun proteomic analysis. The identified proteins/peptides were shown at the right panel.

Histopathologically, about 80% of RCC are clear-cell type (ccRCC), about 15% of RCC are papillary type, and the remaining 5% are other types (Reuter and Presti 2000). Currently, there are only limited amounts of reliable diagnostic, prognostic, and therapeutic markers for ccRCC even though numerous investigations underlying the abnormalities of RCC based on biochemical, cellular and morphological characterization have been conducted.

Herein, 18 up-regulated and 48 down-regulated proteins were identified, respectively based on quantitative proteomics analysis by application of shutgun approach coupled with stable isotope dimethyl labeling. Many of these identified proteins in the expression profile have been previously shown for RCC patients, including down-regulation of serine hydroxymethyltransferase, betaine--homocysteine S-methyltransferase 1, glutamate dehydrogenase 1, delta-1-pyrroline-5-carboxylate dehydrogenase and aromatic-L-amino-acid decarboxylase involved in amino acid metabolism, fructose-bisphosphate aldolase B, fructose-1,6-bisphosphatase 1, and phosphoenolpyruvate carboxykinase involved in gluconeogenesis pathway, up-regulation of keratin type II cytoskeletal 8, vimentin, tropomyosin alpha-1 chain and annexin family, ATP synthase subunit beta, electron transfer flavoprotein subunit beta and alpha involved in ATP generation, signifying that this up- or down-regulated effect may be universal among patients. Additionally, several novel changes in differentially expressed proteins were found, including down-regulation of L-xylulose reductase, transthyretin, ribonuclease UK114, and up-regulation of 39S ribosomal protein L43. It has been observed that αB-crystallin is overexpressed in gliomas (Sorlie et al. 2001), breast carcinomas (Clark and Muchowski 2000; Sotiriou et al. 2003), and ductal carcinoma *in situ* compared with the matched normal breast tissues (Mehlen et al. 1995).

αB-Crystallin, a small heat shock protein (sHSP), is a major component of vertebrate lens, and is also expressed in nonlenticular tissues, such as renal tubular epithelium (Takashi et al. 1998; Lowe et al. 1992). This chapter showed the up-regulation of αB-crystallin expression in RCC tissues. Overall survival of RCC patients was significantly correlated with αB-crystallin expression in tumor tissues. Therefore, we would employ αB-crystallin as biomarker candidate for verification and evaluation of ccRCC by using ion scanning of peptides measured and quantified in multiple reaction monitoring (MRM) shown in Table 1. Precursor ion scanning is employed to confirm the identity of proteins with specific product ions. For precursor ion scanning, the second mass analyzer (MS2) is set to monitor product ions of a specific m/z ratio. The first mass analyzer (MS1) is set to scan the overall mass range including all the precursor ions which will be further bombarded and fragmented into the selected product ions. As the ions emit from the source and travel through an analyzer, they will be filtered based on mass to charge (m/z) ratio.

3.3. Varicose Vein

This chapter deals with a comparative proteomics analysis of varicose veins by shotgun proteomic approach. A scheme of sample processing, separation and the subsequent workflow concerning trypsin digestion, dimethyl labeling and shotgun analysis is presented in Figure 4. Respective tryptic peptides with dimethyl labeling were mixed in a 1:1 (w/w) ratio and then enriched by the reverse-phase C_{18} column. Owing to the fact that the enriched peptide population was too complex to be fully detected and characterized by a single LC-MS/MS run, the enriched peptides were fractionated by HILIC based on polarity difference and harvested into 10 fractions. Each fraction was analyzed by LC-LTQ-Orbitrap and the parameter used as searching for identified peptides was set to allow for no missed cleavage. Most of the peptides were separated from a single or two adjacent HILIC fractions, and peptides identified by in-house software were accepted if their individual ion score was higher than 20, which had been a cutoff point used for the lower-quality MS/MS spectra (Hou et al. 2007).

Once the differentially expressed proteins with confident score were identified, the peptide quantification ratio (D/H) was obtained by Mascot Distiller program using the average area of the first 3 isotopic peaks across each elution profile (Wu et al. 2011; Santos et al. 2012). Data from the publicly available Mascot searching engine as well as quantification results from each fraction were also merged by Mascot Distiller program into one file that combined peptides with more than one peptide matching the same sequence, which were harvested from different fractions or at different retention time and charge state. Herein, 12 proteins with unambiguous consistency of down-regulation in the protein expression levels were identified, signifying that this down-regulated effect may be universal among patients.

Table 1. m/z values of precursor and product ions scanning employed for verification and evaluation of αB-crystallin biomarker candidate in clear cell renal cancer carcinoma (RCC) verified by multiple reaction monitoring (MRM)

Mass	Position	Peptide sequence	Precursor ion m/z ratio	Product ion m/z ratio			
1496.68	57~69	APSWFDTGLSEMR	748.85	374.93	1328.59	1055.48	793.39
1388.73	1~11	MDIAIHHPWIR	694.87	347.94	1257.69	1029.57	845.45
1374.71	12~22	RPFFPFHSPSR	687.86	344.43	1218.61	974.48	583.29
1165.66	93~103	VLGDVIEVHGK	583.33	292.17	1066.59	781.46	569.30
1140.63	164~174	EEKPAVTAAPK	570.82	285.91	1011.58	657.39	386.24
1088.51	108~116	QDEHGFISR	544.76	272.88	960.45	716.38	522.30
986.49	83~90	HFSPEELK	493.75	247.38	849.44	615.33	389.24
921.50	75~82	FSVNLDVK	461.26	231.13	774.44	687.40	474.29
772.39	151~157	QVSGPER	386.70	193.85	644.34	458.24	304.16
700.44	158~163	TIPITR	350.72	175.86	599.39	389.25	276.17

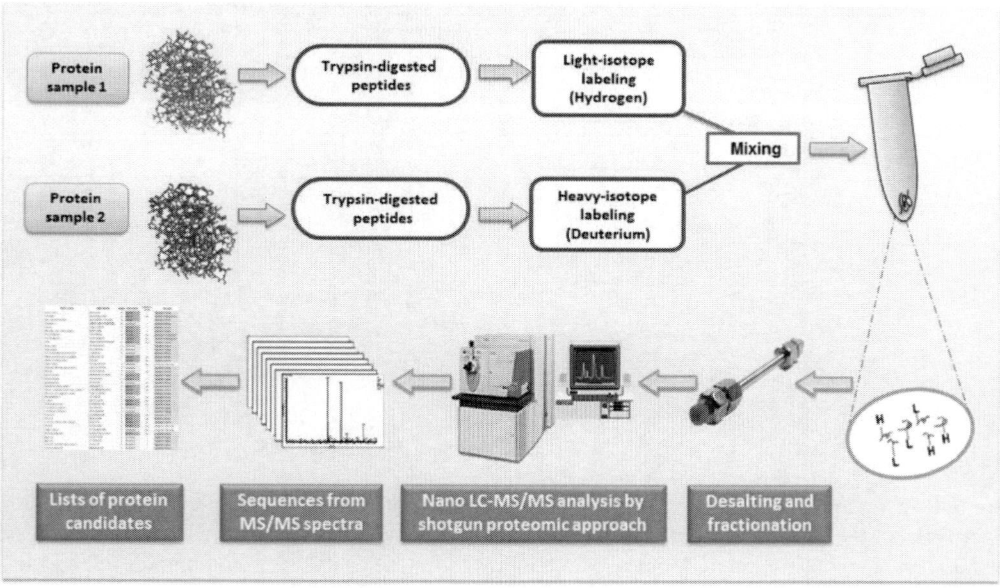

Figure 4. Schematic representation of the procedures used for the quantitative proteomics analysis. The quantitative shotgun analysis of proteome changes from clinical samples of varicose vein tissues was carried out by using HILIC-C18 peptide separation and nanoLC-MS/MS coupled with stable isotope dimethyl labeling sample treatment.

All these 12 proteins were directly or indirectly involved in the energy generation and utilization, eight of which were glycolysis-related proteins, which included glucose-6-phosphate isomerase, phosphoglycerate mutase 1, fructose-bisphosphate aldolase A, phosphoglycerate kinase 1, pyruvate kinase isozymes M1/M2, triosephosphate isomerase, alpha-enolase and anaerobic glycolysis enzyme identified as lactate dehydrogenase B chain. Two of which, calmodulin and MYLK, were found to correlate with ATP utilization and regulation. The other two proteins, identified to be mitochondrial ATP synthase β subunit and creatine kinase B type, were known to be involved in ATP generation and act as energy reservoir.

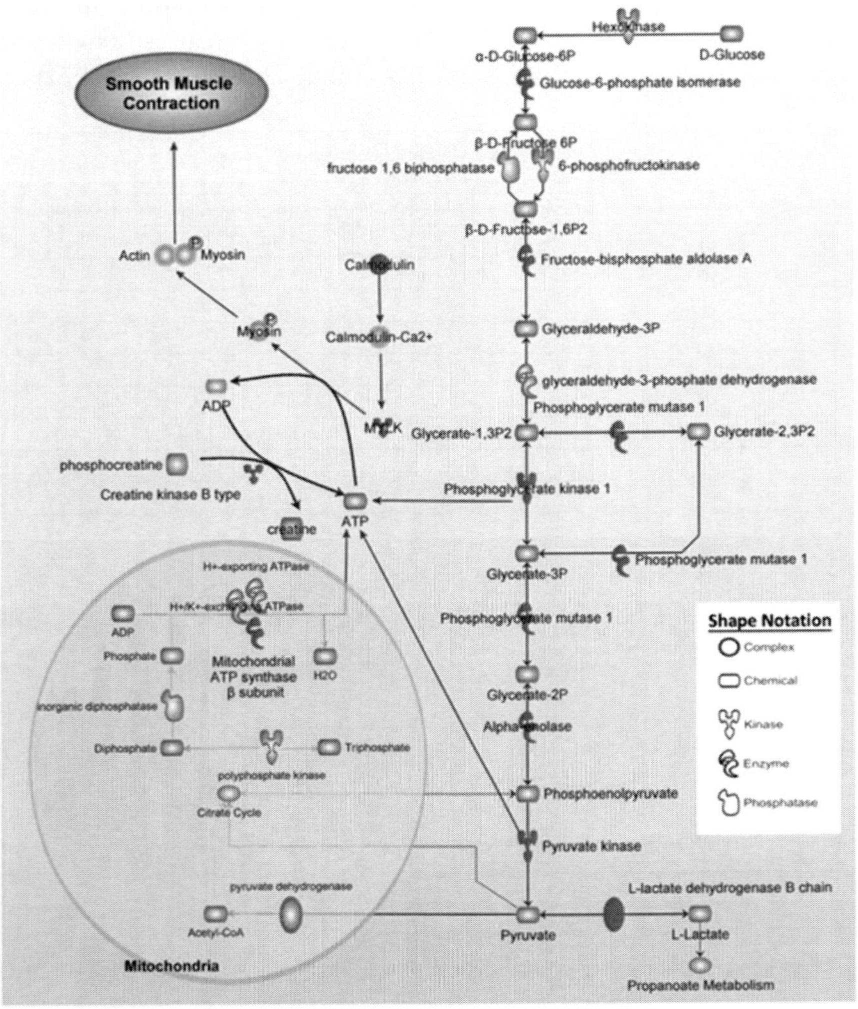

Figure 5. Schematic representation of derived pathways associated with ATP synthesis and utilization. The networks of these identified proteins mapping to the canonical pathways from Ingenuity Pathways Analysis (IPA, Ingenuity Systems) library were employed for the analysis of protein molecules with significant down-regulation between normal and diseased tissues. Identified proteins shown in deep green color were displayed with different shapes to indicate different functions. The biological interrelationships between molecules were represented as arrows in blue color. All drawn arrows were cited and supported by at least one reference from the literature, textbooks, or canonical information stored in the Ingenuity Knowledge Base. The systematic decrease of these enzymes may be responsible for the dysfunction of vascular smooth muscle, followed by morphological aberrations upon progressive development of varicose veins.

Taken together, a continuous supply of large quantities of ATP from metabolic processes of muscle is essential for maintaining the contraction machinery, energy storage and calcium transport. The inhibition of glycolysis pathway and down-regulation of ATP utilization attenuate not only a variety of physiological processes but also energy supply for physical activity.

In this chapter we only focus on 12 protein factors related to ATP generation and utilization because other identified protein species associated with cytoskeleton assembly or cellular morphology show no clear up- or down-regulation among these patients. The availability of low levels of both ATP-generating and ATP-utilizing enzymes may thus partly account for the dysfunction of smooth muscle lining in the vein vessel, followed by a progressive development with morphological aberrations of varicose veins (Figure 5) (Kuo et al. 2013).

Conclusion

All the above methods including ICAT, iTRAQ, SILAC, 1-D and 2-D gels in proteomics research are in general designed for relative quantitative proteomic analysis and comparison of selected proteins among different pair-wise samples under some specified conditions. In contrast, methods for absolute quantitative protein determination are more desirable than the relative methods from the quantitative aspects of proteomic approaches.

Proteomics based on absolute quantitative analysis such as MRM is aimed to measure the absolute level of protein biomarkers using a characteristic peptide unique to the studied protein. For absolute quantification, peptides generated from specific enzymatic digestions are synthesized with heavy isotopes.

A known quantity of the synthesized peptides with known molecular masses was added to proteomic samples as internal standards for intensity comparison to determine the quantities of the studied proteins.

This chapter did not intend to include an in-depth description on various types of strategies used in current proteomics analysis and its application to biomedical research. However some comparison between the old gel-based proteomics and newer sensitive and high-throughput gel-free shotgun quantitative strategy is given for the comparison and assessment of suitability of available proteomic strategies. Stable isotope labeling especially provides a compelling strategy for comparative proteomic studies. As regards the consideration for the selection of *in vivo* or *in vitro* isotope labeling methodology, the latter approach appears to be always preferable when clinical specimens are used for proteomic analysis.

For proteomic analysis it is also most desirable that the smaller amounts of sample proteins required and a larger capacity of high-throughput operation for a wider dynamic range of protein-size detection when we implement MS-based quantitative proteomics. We have applied the gel-free and isotope-labeled shotgun strategy to clinically obtained samples of tissues and body fluids including urine for three translational-relevant diseases with some rewarding success.

Of course the goal of proteomics is not just to make simple protein identification (protein ID); characterization of post-translational modifications (PTM) of cellular proteins remains a great challenge for current and future proteomics research. Along with the rapid improvement of algorithms and software for shotgun proteomic data analysis, the gel-free shotgun approach of quantitative proteomics similar to that of shotgun genomics should facilitate unraveling the complexity of human proteome in the foreseeable future.

Acknowledgments

This work was supported in part by the Quantitative Proteomics Center Grant (NSC99-2745-B-037-005) from the National Science Council, Taipei, Taiwan. We wish to thank the technical and instrumental support of the core-facility at the Center for Research Resources and Development, Kaohsiung Medical University.

References

Alaiya, A., Al-Mohanna, M., Linder, S. (2005) Clinical cancer proteomics: promises and pitfalls. *J. Proteome. Res.* 4(4):1213-1222.

Anderson, N. G. (2010) Adventures in clinical chemistry and proteomics: a personal account. *Clin. Chem.* 56(2):154-160.

Bogdanov, B., Smith, R. D. (2005) Proteomics by FTICR mass spectrometry: top down and bottom up. *Mass. Spectrom. Rev.* 24(2):168-200.

Chait, B. T., (2006) Mass Spectrometry: Bottom-Up or Top-Down? *Science* 31465-66.

Clark, J. I., Muchowski, P. J. (2000) Small heat-shock proteins and their potential role in human disease. *Curr. Opin. Struct. Biol.* 10(1):52-59.

Colantonio, D. A., Chan, D. W. (2005) The clinical application of proteomics. *Clin. Chim. Acta* 357(2):151-158.

Cox, D. M., Zhong, F., Du, M., Duchoslav, E., Sakuma, T., McDermott, J. C. (2005) Multiple reaction monitoring as a method for identifying protein posttranslational modifications. *J. Biomol. Tech.* 16(2):83-90.

Dominguez, D. C., Lopes, R., Torres, M. L. (2007) Proteomics: clinical applications. *Clin. Lab. Sci.* 20(4):245-248.

Dunn, M., Burghes, A. H. M. (1983) High resolution two-dimensional polyacrylamide gel electrophoresis. I. *Methodological procedures Electrophoresis* 4(2):97-116.

Fehniger, T. E., Marko-Varga, G. (2010) Proteomics and disease revisited: the challenge of providing proteomic tools into clinical practice. *J. Proteome. Res.* 9(3):1191-1192.

Guan, F., Uboh, C. E., Soma, L. R., Luo, Y., Rudy, J., Tobin, T. (2005) Detection, quantification and confirmation of anabolic steroids in equine plasma by liquid chromatography and tandem mass spectrometry. *J. Chromatogr. B Analyt. Technol. Biomed. Life Sci.* 829(1-2):56-68

Gygi, S. P., Rist, B., Gerber, S. A., Turecek, F., Gelb, M. H., Aebersold, R. (1999) Quantitative analysis of complex protein mixtures using isotope-coded affinity tags. *Nat. Biotechnol.* 17(10):994-999.

Han, C. L., Chien, C. W., Chen, W. C., Chen, Y. R., Wu, C. P., Li, H., Chen, Y. J. (2008) A multiplexed quantitative strategy for membrane proteomics: opportunities for mining therapeutic targets for autosomal dominant polycystic kidney disease. *Mol. Cell. Proteomics* 7(10):1983-1997.

Hendriks, M. M., Smit, S., Akkermans, W. L., Reijmers, T. H., Eilers, P. H., Hoefsloot, H. C., Rubingh, C. M., de Koster, C. G., Aerts, J. M., Smilde, A. K. (2007) How to distinguish healthy from diseased? Classification strategy for mass spectrometry-based clinical proteomics. *Proteomics* 7(20):3672-3680.

Henzel, W. J., Billeci, T. M., Stults, J. T., Wong, S. C., Grimley, C., Watanabe, C. (1993) Identifying proteins from two-dimensional gels by molecular mass searching of peptide fragments in protein sequence databases. *Proc. Natl. Acad. Sci. US* 90(11):5011-5015.

Herrin, G. L., McCurdy, H. H., Wall, W. H. (2005) Investigation of an LC-MS-MS (QTrap) method for the rapid screening and identification of drugs in postmortem toxicology whole blood samples. *J. Anal. Toxicol.* 29(7):599-606.

Ho, E. N., Leung, D. K., Wan, T. S., Yu, N. H. (2006) Comprehensive screening of anabolic steroids, corticosteroids, and acidic drugs in horse urine by solid-phase extraction and liquid chromatography-mass spectrometry. *J. Chromatogr. A* 1120(1-2):38-53.

Ho, P. Y., Chueh, S. C., Chiou, S. H., Wang, S. M., Lin, W. C., Lee, I. L., Yang, H. Y., Peng, H. C., Lai, M. K. (2012) alphaB-Crystallin in clear cell renal cell carcinoma: Tumor progression and prognostic significance. *Urol. Oncol.*

Hou, W., Ethier, M., Smith, J. C., Sheng, Y., Figeys, D. (2007) Multiplexed proteomic reactor for the processing of proteomic samples. *Anal. Chem.* 79(1):39-44.

Hsu, J. L., Huang, S. Y., Chow, N. H., Chen, S. H. (2003) Stable-isotope dimethyl labeling for quantitative proteomics. *Anal. Chem.* 75(24):6843-6852.

Huang, C. H., Wang, Y. T., Tsai, C. F., Chen, Y. J., Lee, J. S., Chiou, S. H. (2011) Phosphoproteomics characterization of novel phosphorylated sites of lens proteins from normal and cataractous human eye lenses. *Mol. Vis.* 17186-198.

Ideker, T., Thorsson, V., Ranish, J. A., Christmas, R., Buhler, J., Eng, J. K., Bumgarner, R., Goodlett, D. R., Aebersold, R., Hood, L. (2001) Integrated genomic and proteomic analyses of a systematically perturbed metabolic network. *Science* 292(5518):929-934.

Kelleher, N. L. (2004) Top-down proteomics. *Anal. Chem.* 76(11):197A-203A.

Kirkpatrick, D. S., Gerber, S. A., Gygi, S. P. (2005) The absolute quantification strategy: a general procedure for the quantification of proteins and post-translational modifications. *Methods* 35(3):265-273.

Koc, H., Swenberg, J. A. (2002) Applications of mass spectrometry for quantitation of DNA adducts. *J. Chromatogr. B Analyt. Technol. Biomed. Life Sci.* 778(1-2):323-343.

Kuo, C. J., Liang, S. S., His, E., Lin, S. D., Chiou, S. H. (2013) Quantitative proteomics analysis of varicose veins by shotgun proteomic approach identifies a set of differentially expressed proteins related to ATP generation and utilization. *Kaohsiung J. Med. Sci. In Press.*

Laemmli, U. K. (1970) Cleavage of structural proteins during the assembly of the head of bacteriophage T4. *Nature* 227(5259):680-685.

Lange, V., Picotti, P., Domon, B., Aebersold, R. (2008) Selected reaction monitoring for quantitative proteomics: a tutorial. *Mol. Syst. Biol.* 4222.

Lescuyer, P., Farina, A., Hochstrasser, D. F. (2010) Proteomics in clinical chemistry: will it be long? *Trends Biotechnol.* 28(5):225-229.

Leth-Larsen, R., Lund, R. R., Ditzel, H. J. (2010) Plasma membrane proteomics and its application in clinical cancer biomarker discovery. *Mol. Cell Proteomics* 9(7):1369-1382.

Liao, H., Wu, J., Kuhn, E., Chin, W., Chang, B., Jones, M. D., O'Neil, S., Clauser, K. R., Karl, J., Hasler, F., Roubenoff, R., Zolg, W., Guild, B. C. (2004) Use of mass spectrometry to identify protein biomarkers of disease severity in the synovial fluid and serum of patients with rheumatoid arthritis. *Arthritis Rheum.* 50(12):3792-3803.

Lowe, J., McDermott, H., Pike, I., Spendlove, I., Landon, M., Mayer, R. J. (1992) alpha B crystallin expression in non-lenticular tissues and selective presence in ubiquitinated inclusion bodies in human disease. *J. Pathol.* 166(1):61-68.

MacCoss, M. J., McDonald, W. H., Saraf, A., Sadygov, R., Clark, J. M., Tasto, J. J., Gould, K. L., Wolters, D., Washburn, M., Weiss, A., Clark, J. I., Yates, J. R., 3rd (2002) Shotgun identification of protein modifications from protein complexes and lens tissue. *Proc. Natl. Acad. Sci. US* 99(12):7900-7905.

McCormack, A. L., Schieltz, D. M., Goode, B., Yang, S., Barnes, G., Drubin, D., Yates, J. R., 3rd (1997) Direct analysis and identification of proteins in mixtures by LC/MS/MS and database searching at the low-femtomole level. *Anal. Chem.* 69(4):767-776.

Marko-Varga, G., Laurell, T. (2010) Clinical Proteomics; protein expression research within disease--A challenging task. *J. Proteomics* 73(6):1043-1044.

Mehlen, P., Preville, X., Chareyron, P., Briolay, J., Klemenz, R., Arrigo, A. P. (1995) Constitutive expression of human hsp27, Drosophila hsp27, or human alpha B-crystallin confers resistance to TNF- and oxidative stress-induced cytotoxicity in stably transfected murine L929 fibroblasts. *J. Immunol.* 154(1):363-374.

Motzer, R. J., Bander, N. H., Nanus, D. M. (1996) Renal-cell carcinoma. *N Engl. J. Med.* 335 (12):865-875.

O'Farrell, P. H. (1975) High resolution two-dimensional electrophoresis of proteins. *J. Biol. Chem.* 250(10):4007-4021.

Ong, S. E., Blagoev, B., Kratchmarova, I., Kristensen, D. B., Steen, H., Pandey, A., Mann, M. (2002) Stable isotope labeling by amino acids in cell culture, SILAC, as a simple and accurate approach to expression proteomics. *Mol. Cell. Proteomics* 1(5):376-386.

Ong, S. E., Kratchmarova, I., Mann, M. (2003) Properties of 13C-substituted arginine in stable isotope labeling by amino acids in cell culture (SILAC). *J. Proteome Res.* 2(2): 173-181.

Ong, S. E., Mann, M. (2006) A practical recipe for stable isotope labeling by amino acids in cell culture (SILAC). *Nat. Protoc.* 1(6):2650-2660.

Ong, S. E., Mann, M. (2007) Stable isotope labeling by amino acids in cell culture for quantitative proteomics. *Methods Mol. Biol.* 35937-52.

Paik, Y. K., Kim, H., Lee, E. Y., Kwon, M. S., Cho, S. Y. (2008) Overview and introduction to clinical proteomics. *Methods Mol. Biol.* 4281-31.

Reuter, V. E., Presti, J. C., Jr. (2000) Contemporary approach to the classification of renal epithelial tumors. *Semin. Oncol.* 27(2):124-137.

Ross, P. L., Huang, Y. N., Marchese, J. N., Williamson, B., Parker, K., Hattan, S., Khainovski, N., Pillai, S., Dey, S., Daniels, S., Purkayastha, S., Juhasz, P., Martin, S., Bartlet-Jones, M., He, F., Jacobson, A., Pappin, D. J. (2004) Multiplexed protein quantitation in Saccharomyces cerevisiae using amine-reactive isobaric tagging reagents. *Mol. Cell Proteomics* 3(12):1154-1169.

Santos, H. M., Kouvonen, P., Capelo, J. L., Corthals, G. L. (2012) Isotopic labelling of peptides in tissues enhances mass spectrometric profiling. *Rapid Commun. Mass Spectrom* 26(3):254-262.

Sinha, A., Singh, C., Parmar, D., Singh, M. P. (2007) Proteomics in clinical interventions: achievements and limitations in biomarker development. *Life Sci.* 80(15):1345-1354.

Sorlie, T., Perou, C. M., Tibshirani, R., Aas, T., Geisler, S., Johnsen, H., Hastie, T., Eisen, M. B., van de Rijn, M., Jeffrey, S. S., Thorsen, T., Quist, H., Matese, J. C., Brown, P. O.,

Botstein, D., Lonning, P. E., Borresen-Dale, A. L. (2001) Gene expression patterns of breast carcinomas distinguish tumor subclasses with clinical implications. *Proc. Natl. Acad. Sci. US* 98(19):10869-10874.

Sotiriou, C., Neo, S. Y., McShane, L. M., Korn, E. L., Long, P. M., Jazaeri, A., Martiat, P., Fox, S. B., Harris, A. L., Liu, E. T. (2003) Breast cancer classification and prognosis based on gene expression profiles from a population-based study. *Proc. Natl. Acad. Sci. US* 100(18):10393-10398.

Sze, S. K., Ge, Y., Oh, H., McLafferty, F. W. (2002) Top-down mass spectrometry of a 29-kDa protein for characterization of any posttranslational modification to within one residue. *Proc. Natl. Acad. Sci. US* 99(4):1774-1779.

Takashi, M., Katsuno, S., Sakata, T., Ohshima, S., Kato, K. (1998) Different concentrations of two small stress proteins, alphaB crystallin and HSP27 in human urological tumor tissues. *Urol. Res.* 26(6):395-399.

Thevis, M., Opfermann, G., Schanzer, W. (2001) High speed determination of beta-receptor blocking agents in human urine by liquid chromatography/tandem mass spectrometry. *Biomed. Chromatogr.* 15(6):393-402.

Tao, F., Lazarev, A. (2007) Clinical proteomics: opportunities for diagnostics, pharmaceuticals and the clinical laboratory. *Expert Rev. Proteomics* 4(1):9-11.

Tomita, Y. (2006) Early renal cell cancer. *Int. J. Clin. Oncol.* 11(1):22-27.

Unwin, R. D., Griffiths, J. R., Leverentz, M. K., Grallert, A., Hagan, I. M., Whetton, A. D. (2005) Multiple reaction monitoring to identify sites of protein phosphorylation with high sensitivity. *Mol. Cell Proteomics* 4(8):1134-1144.

Washburn, M. P., Wolters, D., Yates, J. R., 3rd (2001) Large-scale analysis of the yeast proteome by multidimensional protein identification technology. *Nat. Biotechnol.* 19(3): 242-247.

Wolf-Yadlin, A., Hautaniemi, S., Lauffenburger, D. A., White, F. M. (2007) Multiple reaction monitoring for robust quantitative proteomic analysis of cellular signaling networks. *Proc. Natl. Acad. Sci. US* 104(14):5860-5865.

Wu, C. J., Chen, Y. W., Tai, J. H., Chen, S. H. (2011) Quantitative phosphoproteomics studies using stable isotope dimethyl labeling coupled with IMAC-HILIC-nanoLC-MS/MS for estrogen-induced transcriptional regulation. *J. Proteome. Res.* 10(3):1088-1097.

Wu, K., Zhang, Y. (2007) Clinical application of tear proteomics: Present and future prospects. *Proteomics Clin. Appl.* 1(9):972-982.

Yocum, A. K., Chinnaiyan, A. M. (2009) Current affairs in quantitative targeted proteomics: multiple reaction monitoring-mass spectrometry. *Brief Funct. Genomic Proteomic.* 8(2): 145-157.

Zappacosta, F., Collingwood, T. S., Huddleston, M. J., Annan, R. S. (2006) A quantitative results-driven approach to analyzing multisite protein phosphorylation: the phosphate-dependent phosphorylation profile of the transcription factor Pho4. *Mol. Cell. Proteomics* 5(11):2019-2030.

In: Recent Advances in Proteomics Research
Editor: Ming D. Li
ISBN: 978-1-62948-218-7

Chapter 6

Mitochondria Protein Profiling in Striated Muscle: Insights into Pathophysiological Conditions

Ana Isabel Padrão[1], Rui Vitorino[1,], Rita Nogueira-Ferreira[1,2], Vanessa Almeida[1], Tiago Henriques-Coelho[2], José A. Duarte[3], Francisco Amado[1,4] and Rita Ferreira[1]*

[1]QOPNA, Mass Spectrometry Group,
University of Aveiro, Aveiro, Portugal
[2]Department of Physiology and Cardiothoracic Surgery,
Faculty of Medicine, University of Porto, Porto, Portugal
[3]CIAFEL, Faculty of Sports, University of Porto, Porto, Portugal
4School of Health Sciences, University of Aveiro, Portugal

Abstract

Striated muscle is highly reliant on mitochondria where essential mechanisms of energy production, signaling, biosynthesis and apoptosis are contained and so, the orchestration of these pathways plays a critical role in cell physiology. Mitochondrial proteome are adjusted to the specific needs of the muscle and to the environmental conditions that it faces. Cumulative evidences have linked mitochondrial dysfunction to the pathogenesis of cardiovascular diseases, diabetes mellitus, cancer and aging. Progress on mass spectrometry-based proteomics and bioinformatics made possibly the large-scale characterization of mitochondrial proteome and its dynamics under pathophysiological conditions. This chapter focuses on principles and technical considerations of state-of-the-art mitochondrial proteomics, and on its applications to the proteome characterization of mitochondria isolated from skeletal and cardiac muscles. The comparative analysis of mitochondrial proteome with mass spectrometry-based proteomics towards the

* Corresponding author: Rui Vitorino, Address: Department of Chemistry, University of Aveiro, 3810-193, Aveiro, Portugal. E-mail: rvitorino@ua.pt; Fax: +351234370084.

understanding of the relationship between mitochondrial dysfunction and diseases is explored as well as the challenges and future perspectives in this field.

Keywords: Mitochondrial plasticity; Proteomics; Pathophysiological conditions, Skeletal and cardiac muscles

1. Introduction

Mitochondria are the powerhouses of the cell, being the major site of ATP production via oxidative phosphorylation (OXPHOS) for tissue survival and functionality. Besides their key role in energy producing, mitochondria are critical organelles involved in multiple cell signalling cascades and in the regulation of cellular metabolism, development, calcium homeostasis and apoptosis (McBride et al. 2006; Benard et al. 2006; Benard and Rossignol 2008).

Some tissues like the striated muscles are highly reliant on mitochondria. Indeed, the mammalian heart is a hotspot of metabolic activity, consuming each day 100 times its own weight in ATP (Dorn 2013). Skeletal muscle accounts for approximately 40% of total body mass, being a major site of metabolic activity (Romanello and Sandri 2013).

In striated muscles there are two distinct populations of mitochondria that differ in their subcellular localization, morphology and biochemical properties. Subsarcolemmal (SS) mitochondria are located just beneath the sarcolemma whereas the smaller intermyofibrillar (IMF) mitochondria, which account for 80% of mitochondria, are interspersed in the contractile filaments (Riva et al. 2005; Hoppel et al. 2009; Hoppeler and Fluck 2003; Palmer et al. 1977; Hoppeler 1986). The plasticity of these mitochondrial populations was studied in response to distinct stimuli using proteomic approaches relied on mass spectrometry and expression differences were related to the location of mitochondria in the cell (Dabkowski et al. 2009; Baseler et al. 2011; Ferreira et al. 2010).

Proteomic cataloguing studies show that mitochondria contain about 1000 (yeast) to 1500 (human) different proteins (Pagliarini et al. 2008; Perocchi et al. 2006; Sickmann et al. 2003; Reinders et al. 2006). The vast majority (~98%) of the total protein composition of the organelle is encoded within the nucleus, synthesized upon cytosolic ribosomes and imported to submitochondrial destinations by an elaborate network of translocases and sorting machineries (Koehler 2000; Chacinska et al. 2009; Ryan and Hoogenraad 2007; Neupert and Herrmann 2007).

A remaining minority of the proteome is encoded by mitochondrion's own genome, which encodes only a few components of the respiratory chain complexes (13 in humans) (Duchen 2004). Therefore, the mitochondrial proteome should be viewed as a dynamic program generated by the cross-talk between the two genomes and able to adapt to tissue's needs or disease states (Ryan and Hoogenraad 2007; Dimmer and Rapaport 2008). Changes in the mitochondrial proteome exert influences over mitochondrial homeostasis, leading to disease in addition to disturb natural processes as development and ageing (Duchen 2004; Lowell and Shulman 2005; Wallace 1999).

2. Mitochondrial Proteome Characterization in Striated Muscle

The mitochondrial proteome is unique, complex and dynamically regulated as it adapts to the needs of tissues, like the striated muscles, or disease states (Johnson et al. 2007; Balaban 2010; Zhang et al. 2008). Besides alterations in protein levels, these conditions might regulate posttranslational modifications (PTMs) such as glycosylation, acetylation, phosphorylation and even ubiquitination (Witze et al. 2007; Kerner et al. 2011; Koc and Koc 2012). These modifications might affect protein folding and, consequently, organelle function. Such high complexity might explain the wide heterogeneity of mitochondrial proteome reported in different studies and highlights the technical challenges in proteome characterization (Warda et al. 2013). Over the last decade, great progresses were achieved in mass spectrometry-based approaches for the study of whole mitochondrial proteome, sub-proteome, mitochondrial protein complexes and their molecular organization, and for the characterization of mitochondrial protein PTMs in both physiological and pathological conditions.

The workflow for mitochondrial proteomics includes the isolation of mitochondria from tissues or cultured cells, large-scale separation of mitochondrial proteins, protein identification and quantification by MS, and bioinformatic analysis (figure 1).

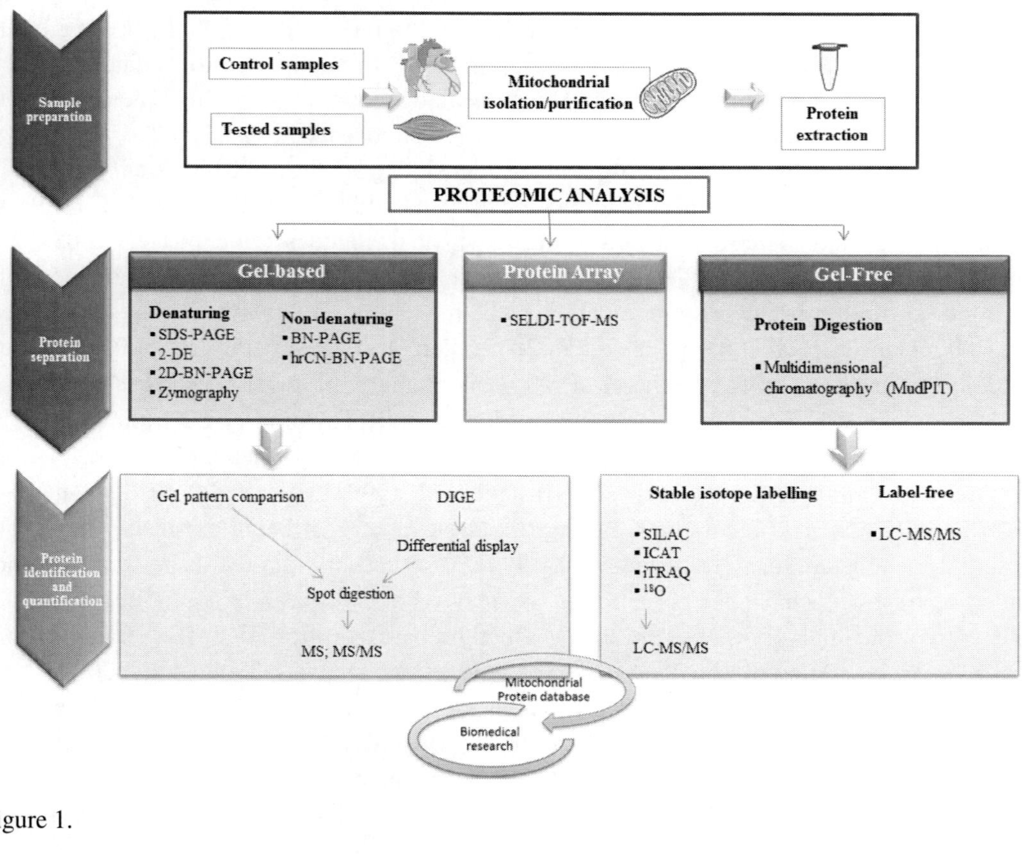

Figure 1.

Large-scale proteome profiling started with the classic two-dimensional electrophoresis (2-DE), which combines isoelectric focusing (IEF) and polyacrylamide gel electrophoresis in denaturating conditions (SDS-PAGE). This approach was optimized with the use of fluorescent dyes, Differential Gel Electrophoresis (DIGE), for comparative quantitative analysis. New approaches have been proposed and include gel-free analysis of mitochondrial proteins (MudPIT) with novel peptide labeling techniques for quantitation, and complementary functional analysis (e.g. Blue Native PAGE).

Finally, data integration with bioinformatic tools gives a broad range perspective of the mitochondrial proteome plasticity in response to pathophysiological conditions.

2.1. Mitochondrial Isolation and Purification Methods for Proteomics

Preparation of pure and functionally viable mitochondria is an important step to achieve reliable and reproducible proteomic outputs (Zhang et al. 2012). Mitochondrial fractionation consists of two major steps: disruption of the cellular organization (homogenization of tissue) and differential centrifugation of the homogenate (Huber et al. 2003). Homogenization buffers usually contain ethylene glycol tetraacetic acid (EGTA) and delipidated BSA to avoid mitochondrial swelling and the uncoupling of mitochondria respiration by free fatty acids, respectively (Graham 1993).

The most effective method for mitochondrial isolation is differential centrifugation, although several other techniques such as free-flow electrophoresis (Zischka et al. 2003), kit-based methods (Hartwig et al. 2009) have been applied to these types of studies (Kang et al. 2008; Jung et al. 2000). In 1948 it was proposed the first protocol to isolate rat liver mitochondria based on a differential centrifugation (Hogeboom et al. 1948). Then, isolation methods have been modified and tailored to tissues and animal species analyzed.

Compared with free-flow electrophoresis, density gradient centrifugation has great advantages both in obtaining functional mitochondria and good yields (Hartwig et al. 2009), it is easily set up, and ideally combined with analytical techniques (Huber et al. 2003). To reduce "contaminating" organelles like endoplasmic reticulum or lysosomes, mitochondria are further purified using density gradients, which are usually set up with sucrose (Meisinger et al. 2000; Hanson et al. 2001; Rezaul et al. 2005), Percoll (Ferreira et al. 2010; Pagliarini et al. 2008; Millar et al. 2001; Mootha et al. 2003; Werhahn et al. 2001; Fukada et al. 2004), Nycodenz (Jiang et al. 2004; Li et al. 2009) or Metrizamide (Taylor et al. 2003; Fountoulakis and Schlaeger 2003). Purification of mitochondria can also be performed by immunoisolation using mitochondria-specific antibodies (Herrnstadt et al. 1999; Lee et al. 2010). This method leads to highly purified preparations, but has the disadvantage of being expensive and large amounts of sample are lost (Herrnstadt et al. 1999). More recently, protocols performing mild proteolytic treatments have been reported to the analysis of cytoplasm-exposed mitochondrial proteins by proteomics (Forner et al. 2006), a method based on flow field-flow fractionation for size-based separation of mitochondria prior to proteomic analysis (Kang et al. 2008).

After purification, mitochondrial purity and intactness must be validated by immunoblotting using specific markers of mitochondria and of other common contaminating organelles (Zhang et al. 2008; Chen et al. 2010), by membrane potential measurements, respiratory control index assessments, and/or morphology evaluation by electron microscopy (Zhang et al. 2008).

Sample subfractionation might also be performed to reduce sample complexity before MS analyses and so increase the number of identified proteins; however, high amounts of sample are required and usually involves ultracentrifugation combined with density gradients (Ferreira et al. 2010; Forner et al. 2006).

2.2. Mitochondrial Protein Separation Approaches

The most widespread protein separation approaches used in mitochondrial proteomics can be divided in two main categories: gel-based and gel-free strategies (figure 1), which application to mitochondria proteome is reviewed in this section.

2.2.1. Gel-Based Approaches

Since it was first introduced in 1975 (O'Farrell 1975), 2-DE has evolved at different levels and became the workhorse of protein separation and the method of choice for differential protein expression analysis. It separates a complex protein mixture based upon two intrisic protein characteristics, net charge (isoelectric point, p*I*) and molecular weight (MW). This approach provides the highest resolution in soluble protein separation, and resolves hundreds or thousands of proteins at once (Mathy and Sluse 2008; Schulze and Usadel 2010; Lilley et al. 2002). In many cases it is possible to separate different isoforms of a given protein, which can be due to sequence variations or to PTMs (Roepstorff 2012). Nevertheless, 2-DE has lately come under assault due to its known limitations and in part to the development of alternative MS-based approaches. Any of the reasons behind this trend include issues related to reproducibility (Lilley et al. 2002), poor representation of low abundant proteins (Gygi et al. 2000), highly acidic/basic proteins, or proteins with extreme size or hydrophobicity (Ong and Pandey 2001; Gorg et al. 1997; Wilkins et al. 1998), and difficulties in automation of the gel-based techniques (Tonge et al. 2001). Moreover, the comigration of multiple proteins in a single spot renders comparative quantification rather inaccurate (Abdallah et al. 2012). Mitochondrial proteins appear mostly within the pH range of 3-11 and within the 15 to 100 kDa molecular mass range (Mathy and Sluse 2008). In fact, some proteins have a p*I* too alkaline to be well resolved on a typical wide range IPG strips (Kane et al. 2006), as for example, cytochrome c (p*I* = 10.3) (Flatmark 1966). A large percentage of mitochondrial proteins, such as the ones from electron transporter chain (ETC), which represent about 40% of inner membrane proteins (Schwerzmann et al. 1986), have high hydrophobic properties, making isoelectric focusing difficult (Hanson et al. 2001; Santoni et al. 2000). To avoid the limitations of 2-DE in the analysis of specific classes of mitochondrial proteins, SDS-PAGE coupled to liquid chromatography and to tandem MS (LC-MS/MS) has been applied in large-scale mitochondrial proteome research (Taylor et al. 2002; Padrao et al. 2012) since it takes advantage of the resolving power of electrophoresis in a single dimension.

Development of several fluorescent staining methods and protein labeling prior to 2-DE separation as the Difference Gel Electrophoresis (DIGE) method (Unlu et al. 1997), responded to specific concerns about the qualitative and quantitative limitations of protein spot detection (Van den Bergh and Arckens 2004). This technique is based on the incorporation of different fluorescent cyanide dyes (Cydye) in protein's lysine residues (Minden 2007; Marouga et al. 2005) and is therefore compatible with in-gel digestion and MS analysis (Lilley et al. 2002). After electrophoresis, the gel is scanned, fluorescent excitation from each distinct Cydye is captured and overlapped for the comparison of spot density between the other dyes and the standard signal, avoiding possible intra-gel variability (Mathy and Sluse 2008; Lilley and Friedman 2004). Comparing with coomassie- or silver-staining methods the sensitivity of the 2-DE-DIGE method is very high, with a detection limit of ~1 ng and a dynamic range of ~10^5 (Lilley et al. 2002; Tonge et al. 2001; D'Hertog et al. 2006).

DIGE offered the advantage of multiplexing, supressing inter-gel variability and reducing the number of gels required (Shaw et al. 2003), being strongly recommended for protein quantification and comparison (Mathy and Sluse 2008). So, 2-DE-DIGE has captured a lot of attention in the field of mitochondrial proteome research (Baseler et al. 2011; Meng et al. 2009; Egan et al. 2011).

Membrane-bound mitochondrial protein complexes may be analyzed from a functional perspective with special in-gel approaches. Schagger and von Jagow (1991) developed a method named Blue-Native Electrophoresis (BN-PAGE) that allow the separation of intact membrane protein complexes (e.g. mitochondrial respiratory chain complexes). A charge shifting agent, the dye Coomassie Brilliant Blue G-250, is added to the cathode buffer in order to stick to proteins conferring a uniform electric charge without unfold the protein. Thus, intact protein complexes can be separated on a nondenaturing gradient gel according to their MW, though the size and shape of each complex also influence how far that complex migrates into the gel. Moreover, 2D blue-native polyacrylamide gel electrophoresis (2D-BN-PAGE) has been successfully used to resolve mitochondrial membrane proteins (Schagger and von Jagow 1991; Devreese et al. 2002). Denaturing conditions are employed to separate complexes' individual subunits in the second dimension, based on their MW (Nijtmans et al. 2002). Several authors (Padrao et al. 2011; Brookes et al. 2002) have shown the great potential of this technique in functional proteomics for the analysis of mitochondrial molecular signaling, protein-to-protein interaction and posttranslational modifications, especially in respiratory chain proteins from the mitochondrial membrane (Abdallah et al. 2012). Among the limitations associated to BN-PAGE is its low resolving power for some proteins that cannot bind coomassie blue and so present a neutral or basic p*I*, and the interference of coomassie blue with some fluorimetric and catalytic assays (Wittig and Schagger 2008). In an attempt to overcome these disadvantages, a new methodology named high resolution clear-native electrophoresis (hrCN-PAGE) was developed (Wittig and Schagger 2008; Wittig et al. 2007; Wittig et al. 2007). This approach has a separation resolution as high as BN-PAGE, uses non-colored mixed micelles to induce a net negative charge on the proteins, and is compatible with fluorimetric and catalytic assays. Nevertheless, hrCN-PAGE favors the dissociation of labile proteins from protein complexes (Braun et al. 2007). Another variation of BN-PAGE is the Native DIGE technique (BN-DIGE) (Dani and Dencher 2008), which couples the fluorescent dyes labeling technique to the previously described BN-PAGE. BN-DIGE facilitates the (i) systematic and quantitative comparison of protein complexes of related protein fractions, (ii) structural investigation of protein complexes, (iii) assignment of protein complexes to subcellular fractions like organelles and (iv) electrophoretic mapping of isoforms of subunits from protein complexes with respect to a larger proteome (Dani and Dencher 2008; Heinemeyer et al. 2009; Wittig and Schagger 2009).

Zymography is a simple, sensitive, quantifiable, and functional approach for the analysis of proteolytic activity in cell and tissue extracts (Heussen and Dowdle 1980). The standard method is based on the use of SDS-PAGE co-polymerized with a protein substrate (e.g. gelatin, casein, or fibrin) of a protease (Wilkesman and Kurz 2009). Using non-reducing conditions (i.e. in the absence of heating and reducing agent), proteases activity is preserved and might be detected as white bands on a dark blue background resultant of coomassie blue staining (Wilkesman and Kurz 2009).

Zymography was already used to study mitochondrial proteolytic activity (Padrao et al. 2012), contributing to a better understanding of mitochondrial proteolysis in various pathophysiological conditions.

2.2.2. Gel-Free Approaches

Nowadays, gel-free proteomic techniques (Abdallah et al. 2012; Lefort et al. 2009; Roe and Griffin 2006) are considered an attractive alternative to 2-DE since they allow to overcome the disadvantages of gel-based approaches such as the laborious procedure involved, the large amount of samples required, the limited dynamic range, the difficulty in resolving low abundance proteins and the ones with extreme p*I*, MW and hydrophobicity (Tan et al. 2008). In fact, in the last years, a series of gel-free methods have provided complementary information to 2-DE, driving mitochondrial molecular investigation to a higher level of data production and accuracy.

Gel-free methods are based on the separation of complex protein samples by liquid chromatography (e.g. reversed-phase, strong cationic exchange) followed by direct MS analysis (Nagele et al. 2004; Gilmore and Washburn 2010). This separation strategy might involve more than one LC column, being referred as multidimensional protein identification technology (MudPIT) (Link et al. 1999; Washburn et al. 2001; Wolters et al. 2001; Yates et al. 1999; Yates et al. 2000). MudPIT has been recognized as a high throughput method with enhanced ability to identify thousands of proteins within a single experiment (Motoyama and Yates 2008). For some time, gel-free approaches have suffered from the disadvantage of being only qualitative, hence not useful for differential display analyses. In addition, most of these approaches fall short of the resolution achieved with 2-DE gels and are still biased with respect to p*I* and molecular weight. This is because very acidic proteins will be underrepresented in the mixture due to fewer tryptic cleavage sites, and low molecular weight proteins will be underrepresented due to the production of fewer peptides (Washburn et al. 2001). A clear advantage is that most of these methods are very easily automated (Lopez and Melov 2002).

In recent years, approaches in shotgun proteomics have been developed, and are being increasingly used (Gilmore and Washburn 2010). Gel-free strategies have also being enriched by new, parallel developments in mass spectrometry, HPLC systems [i.e., Ultra Performance Liquid Chromatography (UPLC) technologies] as well as by peptide fragmentation techniques (i.e., ETD), which increased the number of proteins identified by MS (Silvestri et al. 2010). Moreover, the combination of SDS-PAGE with HPLC-ESI-MS/ MS allowed the identification of 823 unique proteins in mitochondria isolated from human *vastus lateralis* biopsies, 487 of which were assigned to mitochondrion (Lefort et al. 2009).

2.2.3. Mitochondrial Protein Microarrays

A promising technique for proteome analysis is the surface-enhanced laser desorption (SELDI) (Kiehntopf et al. 2007; Tang et al. 2004), which is available for the high throughput analysis of complex protein samples. SELDI interfaces coupled to time-of-flight mass spectrometry (SELDI-TOF-MS) consists of binding proteins to a chip with selective surface chemistries followed by TOF analysis of the native proteins in order to obtain a protein profile of a proteome subset (Cadieux et al. 2004). This strategy significantly reduces the complexity of the analyzed samples because it selects specific proteins according to their biochemical properties (Seibert et al. 2004).

SELDI is widely used in biomedical studies to perform, above all, biomarker analysis, and it has also been applied to mitochondrial proteomics research (Nyblom et al. 2006).

2.3. Mitochondrial Protein Identification and Quantification

Most of the studies on mitochondrial proteomics relies on the gel-based separation of proteins, mainly by 2-DE, followed by the excision of a selected protein spot, in-gel digestion with trypsin and identification by MS (Aebersold and Mann 2003). In gel-free approaches, identification readily follows LC separation of digested peptides, being the separation step directly coupled to the mass spectrometer (D'Hertog et al. 2006). Regardless of the strategy used, one of the major problems in the analysis of mitochondrial plasticity is the proteome comparison of different samples towards the detection of expression differences related with the pathophysiological condition in study.

In gel-based assays, quantitative analysis relies on band/spot staining intensity and so, imaging software is required for (i) image alignment and spot matching across the gels, (ii) normalization, background adjustment and noise removal, (iii) spot detection, (iv) calculation of the spot volumes and (v) statistical analysis to highlight differentially present proteins. The software uses multivariate statistical packages, such as ANOVA (analysis of variance), apply FDR (false discovery rates) or *q*-values to avoid the wrongful assignment of significant changes and PCA (Principle Component Analysis) is also often carried out. However, some challenges are associated with computational 2-DE analysis such as experimental variation between gels and the high probability of piling several proteins under one spot (Abdallah et al. 2012).

To overcome these limitations, gel-free based quantitative strategies have been favored to compare differentially expressed mitochondrial proteins. The two main approaches include stable isotope-labeling and label-free for the comparison of LC-MS/MS data.

In stable isotope-labeling quantification two or more distinct protein samples are first differentially labeled, one with an isotopically “light” tag and the other with an isotopically “heavy” tag. After labeling, the samples are combined and the resulting pooled peptides are separated by LC, either in a single dimension or using multidimensional strategies, depending on the complexity of the mixture (Roe and Griffin 2006). Peptides common to both samples, although differentially isotopically labeled, retain the same chemical properties and behave similar during the purification and separation steps. As result, such peptides are detected as peak pairs, differing in mass determined by the isotope label used, and the mass spectral peak heights, or the peak areas, are compared to calculate relative abundance levels of peptides between samples. Proteins or peptides in different samples can be metabolically labeled with stable isotope labeling amino acids in cell culture (SILAC) (Ong et al. 2002), chemically with isotope-coded affinity tag (ICAT) (Gygi et al. 1999), labeled with isobaric tag for relative and absolute quantification (iTRAQ) (Ross et al. 2004) or enzymatically with ^{18}O during proteolytic digestion (Miyagi and Rao 2007). Each of these labeling techniques has its own peculiarities.

SILAC is based on the incorporation of "light" or "heavy" isotope agents (^{2}H, ^{13}C, ^{15}N) in arginine and lysine within distinct cell cultures, which will further synthetize labeled proteins for proteomic comparisons. It is also known as metabolic stable isotope labeling and has been used to investigate the mitochondrial proteome in distinct pathophysiological conditions, such as Parkinson's treatment (Jin et al. 2007), human cytomegalovirus infection (Zhang et al. 2011) and diabetic sensory neuropathy (Akude et al. 2011). The principal advantage of SILAC is its simplicity, because no *a posteriori* labeling is necessary to carry out the analysis; however, its application is limited to cell cultures (Ong et al. 2003). An alternative strategy assigned as SILAM (Stable Isotope Labeling of Mammals) was developed to metabolically label an entire mammal for quantitative MS analysis, being appropriate for animal models of disease (Geiger et al. 2013). However, no studies are known of its application to mitochondrial proteome analysis.

ICAT technique entails three functional elements: a specific chemical reactive group that binds to sulfhydryl groups of cysteinyl residues, an isotopicaly coded linker with light or heavy isotopes, and a biotin tag for affinity purification. The proteins containing cysteine residues are labeled either with light or heavy isotopes, where the latter form has eight ^{13}C atoms. Afterwards, light- and heavy-labeled samples are pooled and proteolytically cleaved. The complexity of the sample is reduced prior to MS analysis through the purification of tagged cysteine-containing peptides by affinity chromatography using biotin-avidin affinity columns. This technique was already applied to the study of mitochondrial proteome from primary neuron cultures (Lovell et al. 2005) but ICAT fails to identify non-cysteine containing proteins, being a major technical limitation. Moreover, ICAT has obviously limited use for the analysis of PTMs and splice isoforms (Bantscheff et al. 2007).

Another isotopic labeling approach involves the use of a protease (e.g. trypsin) and ^{18}O-enriched water (Yao et al. 2001), being the oxygen enzymatically incorporated in the C-terminal end of the peptide during the digestion procedure. Chromatographic and MS analyses are then performed to identify and quantify (relatively) the proteins from which the peptides were originated. This technique determines the ratio of individual protein expression levels between two samples, and can be used to quantitatively examine protein expression and PTMs, and to study protein-protein interactions (Miyagi and Rao 2007; Stewart et al. 2001; Mirgorodskaya et al. 2000). This quantitation method was used to evaluate the differential expression of cardiac mitochondrial proteins related with heart failure (Smith et al. 2008). Stable isotope-labeling experiments have great advantages in quantitative proteomics considering its accuracy and reproducibility in the study of complex samples. However, most of these stable isotope-labeling methods suffer potential limitations, including limited linear dynamic range, increased time and complexity of sample preparation and high cost of the labeling reagents (Chen et al. 2010). To overcome these limitations, a new amine reactive, isobaric tag named iTRAQ was developed (Ross et al. 2004). Within this method, proteins from different samples are digested and the resultant peptides are separately labeled with different isotopic variants of iTRAQ reagents, and then combined for LC-MS/MS analysis. A major improvement brought by this method is the ability to analyze up to eight different samples in the same experiment (Ross et al. 2004). iTRAQ is one of the most popular quantitation tools in mitochondrial proteome research (Baseler et al. 2011; Jullig et al. 2007; Kavazis et al. 2009; Glancy and Balaban 2011; Meany et al. 2007; Jullig et al. 2008).

Label-free LC-MS represents an attractive approach for quantitative mitochondrial proteomics (Lundgren et al. 2010; Lai et al. 2013; Zhu et al. 2010; Old et al. 2005) since it is simple, reproducible, cost effective, and less prone to errors and side reactions related to the labeling process. Protein quantification is generally based on two categories of measurements: i) ion intensity such as peptide peak areas or peak heights in chromatography (Old et al. 2005; Ono et al. 2006); ii) spectral counting of identified proteins after MS/MS analysis (Lundgren et al. 2010; Old et al. 2005; Gilchrist et al. 2006). Peptide peak intensity or spectral count is measured in individual LC-MS/MS or LC/LC-MS/MS runs and changes in protein abundance are calculated via a direct comparison between different analyses (Zhu et al. 2010). Though straightforward, this method has the disadvantage of include the variation related with all preparation steps as well as those due to variations in the performance of LC-MS/MS unit. Nevertheless, label-free quantitative proteomics was already successfully applied to the systematic analysis of tissue-specific mitochondrial proteome changes related with diabetes mellitus (Bugger et al. 2009) and cardiovascular diseases (Smith et al. 2008; Bugger et al. 2010).

2.3.1. Mitochondrial Protein Posttranslational Modifications

PTMs of mitochondrial proteins have been reported to regulate mitochondrial functions and so, numerous studies have been conducted using MS-based approaches. Phosphorylation is an important PTM that was reported to regulate mitochondrial dynamics (Koc and Koc 2012; Zhao et al. 2011). The identification of phosphorylated proteins and its specific phosphosites has been performed by MS-based strategies in combination with phospho-specific enrichment methods, such as immobilized metal ion affinity chromatography (IMAC) and titanium dioxide (TiO_2) chromatography (Eyrich et al. 2011). Besides phosphorylation, other PTMs in mitochondrial proteins, such as acetylation and oxidative modifications have been studied. Protein tyrosine nitration is a common modification that occurs in mitochondria under conditions of oxidative stress (Greenacre and Ischiropoulos 2001), being classically analyzed by 2-DE in combination with western blotting and MS (Turko et al. 2003). Protein oxidation has been detected by western blotting analysis of carbonylated proteins with an anti-DNP antibody followed by MS identification of the oxidized proteins (Padrao et al. 2012). Acetylation is another modification found in mitochondrial proteins, mainly in the ones involved in metabolic pathways (Kim et al. 2006). Its analysis usually involves immunoenrichment procedures with an acetyl lysine antibody. Despite all the achievements observed in the last years, many challenges remain in the PTMs characterization of mitochondrial proteins, but it is clear that MS-based approaches can make a unique contribution on how these modifications are modulated by distinct stimuli and what are their functional repercussions.

2.4. Mitochondrial Proteins Database

The need of cataloguing and archiving mitochondrial proteomic data already resulted in the development of several mitochondrial proteome databases that combine information on genetic, functional and pathogenic aspects of nuclear-encoded mitochondrial proteins and include MitoP2 (Elstner et al. 2009; Elstner et al. 2008), Mitoproteome (Cotter et al. 2004), Mito-Carta (Pagliarini et al. 2008), Human mitochondrial Protein database and MitoMiner (Smith and Robinson 2009), as well as the CardiacOrganellar Protein Atlas Knowledgebase (COPaKB) (table 1).

Table 1.

Database	Web link	Species	Information given
MitoP2	http://www.mitop.de:8080/mitop2/	yeast, mouse, human, Arabidopsis thaliana, and Neurospora crassa	Computational predictions of signalling sequences and summarize results from proteome mapping, mutant screening, expression profiling, protein-protein interaction and cellular sublocalization studies
Mitoproteome	http://www.mitoproteome.org/	human	Comprehensive curation of public databases and direct experimental evidence; contains both mitochondrial- and nuclear-encoded protein sequences.
MitoCarta	http://www.broadinstitute.org/pubs/MitoCarta/	mouse, human	Inventory of 1098 mouse genes encoding proteins with MS analysis of mitochondria isolated from fourteen tissues, protein localization through large-scale GFP tagging/microscopy, and integration with six other genome-scale datasets of mitochondrial localization, using a Bayesian approach. Inventory of 1013 human HomoloGene homologs to the mouse MitoCarta collection.
Human mitochondrial Protein database	http://bioinfo.nist.gov/	human	Inventory of proteins involved in mitochondrial biogenesis and function; consolidates information from SwissProt, LocusLink, Protein Data Bank, GenBank, Genome Database, Online Mendelian Inheritance in Man, Human Mitochondrial Genome Database, MITOMAP, Neuromuscular Disease Center and Human 2-D PAGE Databases. Tool to aid in the study of mitochondria-associated diseases.
MitoMiner	http://mitominer.mrc-mbu.cam.ac.uk/release-2.1/begin.do	not defined dataset but allows the analysis of rat, mouse, human, A. thaliana, B. taurus, D. melanogaster, G. lamblia ATCC50803, N. crassa OR74A, P. falciparum 3D7, S. cerevisiae S288c, S. pombae 972h-, T. thermophila SB210.	Integrates mitochondrial proteomics data for a range of organisms.
CardiacOrganellar Protein Atlas Knowledgebase	http://www.heartproteome.org/copa/default.aspx	human, mouse	Provides annotated peptide spectra, their association knowledge on proteome properties, cardiovascular biology, as well as tools to access this resource.

Table 2.

Proteomic approaches						Proteins differentially expressed		
Mitochondrial purification methods	Analytic method	Condition	Specie	Sample	Alterations of mitochondrial physiology	Up-regulated	Down-regulated	References
Type 1 Diabetes								
Differential centrifugation Percoll-density centrifugation	Isotopic labeling, 2-DE, MALDI-TOF	STZ-induced diabetes rats	Rat	Heart mitochondria	-	FAO	ETC; CK; VDAC1, Hsp60, Grp75	(Turko and Murad 2003)
Differential centrifugation	iTRAQ, MudPIT MS/MS	STZ-induced diabetes rats	Rat	Heart mitochondria	-	FAO (particularly long-chain fatty acids);	ETC; amino acid catabolism; biosynthesis and HSPs/chaperones	(Jullig, Hickey et al. 2007)
Differential centrifugation, density centrifugation	iTRAQ, DIGE; LC-MALDI-TOF/TOF	STZ-induced diabetic mice	Mice	Heart SS and IMF	-	FAO in IMF;	ETC in IMF; Protein import and substrate transport (mtHsp70, ANT1, mitochondrial phosphate carrier)	(Baseler, Dabkowski et al. 2011)
Differential centrifugation Percoll-density centrifugation	Label free	Akita diabetic mice	Mice	Heart mitochondria	↓State 3 respiration, ATP synthesis, mitochondrial cristae density in cardiac mitochondria	FAO	TCA	(Bugger, Chen et al. 2009)
Differential centrifugation Percoll-density centrifugation	BN-PAGE, zymography/LC-MALDI-TOF/TOF, Western blotting	STZ-induced diabetic rats	Rat	IMF gastrocnemius mitochondria	↓ complex V respiratory complex activity ↑ oxidative damage	-	Lon, paraplegin, MMP-9	(Padrao, Carvalho et al. 2012)

Proteomic approaches						Proteins differentially expressed		
Mitochondrial purification methods	Analytic method	Condition	Specie	Sample	Alterations of mitochondrial physiology	Up-regulated	Down-regulated	References
Type 1 Diabetes (Continued)								
Differential centrifugation	2DE, Western blotting; MALDI-TOF	ALS and ALR mice	mice	Heart mitochondria	# protein nitration levels	Nitrated SCOT, CK, peroxiredoxin 3, VDAC1	-	(Turko, Li et al. 2003)
Type 2 diabetes								
Differential centrifugation	2DE, LC-ESI-MS/MS; western-blotting	db/db mice	Mice	Heart mitochondria	# protein nitration levels	Nitrated SCOT, NADH dehydrogenase (ubiquinone) Fe-S protein 1, Trifunctional enzyme subunit alpha, dihydrolipoamide dehydrogenase, ATP synthase subunit alpha isoform 1, Aldehyde dehydrogenase family 6, Glutamate dehydrogenase 1, fumarate hydratase 1, Dihydrolipoamide succinyltransferase component of 2-oxoglutarate dehydrogenase complex, short-chain acyl-CoA dehydrogenase, Pyruvate hydrogenase E1 alpha, Isovaleryl-CoA dehydrogenase, mitochondrial, acetyl-Coenzyme A dehydrogenase medium chain, CK, glutamate oxaloacetate transaminase 2, Trifunctional enzyme subunit alpha (HADHB), 3-ketoacyl-CoA thiolase, isocitrate dehydrogenase 3 beta subunit	-	(Wang, Peng et al. 2010)

Table 2. (Continued)

Proteomic approaches						Proteins differentially expressed		
Mitochondrial purification methods	Analytic method	Condition	Specie	Sample	Alterations of mitochondrial physiology	Up-regulated	Down-regulated	References
Type 2 Diabetes (Continued)								
Differential centrifugation density centrifugation	iTRAQ; MudPIT	db/db mice	Mice	Heart SSM and IFM	↓Size and internal complexity of SS ↓State 3 respiration rates, ETC activities, and membrane potential of SS ↑oxidative damage in SS	FAO in both SSM and IMF of db/db hearts	ETC, ATP synthesis, and protein import machinery in SS	(Dabkowski, Baseler et al. 2010)
Differential centrifugation	2-DE; ESI-Q-TOF	Obese db/db mouse	Mice	Heart mitochondria	-	ATP synthase D chain; electron transfer flavoprotein subunit alpha	ubiquinol cytochrome-C reductase core protein 1	(Essop, Chan et al. 2011)
Cardiovascular diseases								
Gradient centrifugation	2-DE, MALDI-TOF	Chronic stress	Rat	Heart mitochondria	Mitochondrial swelling with vacuoles, ↓ ATP synthase activity, ↓ ATP content, ↓ respiratory control rate, ↓ oxidative phosphorylation	CK, prohibitin	TCA, FAO	(Liu, Qian et al. 2004)
Differential centrifugation	2-DE, MALDI-TOF	Desmin null	Mouse	Heart mitochondria	-	Ketone and acetate catabolism, ETC (complex I), VDCA2, mtHsp70, ubiquitin-like 1	Malate/aspartate shuttle, ketone utilization, PDH	(Fountoulakis, Soumaka et al. 2005)

Proteomic approaches						Proteins differentially expressed		
Mitochondrial purification methods	Analytic method	Condition	Specie	Sample	Alterations of mitochondrial physiology	Up-regulated	Down-regulated	References
Cardiovascular diseases (continued)								
Differential centrifugation and Percoll-density centrifugation	2-DE, MALDI-TOF	Ischemia-reperfusion	Rabbit	Heart mitochondria	-	PDH, ETC (complex I), FAO long-chain acyl-CoA dehydrogenase and 3-hydroxybutyrate dehydrogenase, mitofillin and translation elongation factor	TCA, ETC (complex III), VDCA2, enoyl-CoA hydratase, prohibitin	(Kim, Lee et al. 2006)
Differential centrifugation	2-DE, MALDI-TOF	Left anterior descending coronary artery ligation	Rat	Heart mitochondria	Small mitochondria with scattered abnormal vacuoles in failing hearts	TCA, PDH, glycolysis	FAO, ETC (complex I, III, IV, V)	(Wang, Bai et al. 2009)
Differential centrifugation	2-DE, ESI-MS/MS, BN-PAGE	Dyssynchronous or resynchronized heart failure	Dog	Heart mitochondria	↑ complex V activity, ↑ efficiency of oxidative phosphorylation in cardiac resynchronization	ETC (complexes I, II, III, V), PDH, VDAC3, protein synthesis and import, ROS scavenging enzymes	cytochrome c, VDAC2	(Agnetti, Kaludercic et al. 2010)
Differential centrifugation	iTRAQ, MudPIT MS/MS	Spontaneously hypertensive	Rat	Heart mitochondria	Enlarged mitochondria with irregularities of the cristae, ↓ respiratory capacity on succinate	Hsp60, Hsp10, mtHsp70, PDH, proapoptotic factors, cytochrome c, cyclophilin D, GPx1, MAO A, ETC (complexes I, III, IV)	ETC (complexes II, V), CK, FAO	(Jüllig, Hickey et al. 2008)

Table 2. (Continued)

Proteomic approaches						Proteins differentially expressed		
Mitochondrial purification methods	Analytic method	Condition	Specie	Sample	Alterations of mitochondrial physiology	Up-regulated	Down-regulated	References
Cardiovascular diseases (continued)								
Differential centrifugation and density gradient centrifugation	2D-DIGE, MALDI-TOF/TOF	Spontaneously hypertensive	Rat	Heart mitochondria	-	TCA, ETC (complexes I, II)	FAO, PDH, ETC (complex V), VDAC1	(Meng, Jin et al. 2009)
Differential centrifugation	LC-MS/MS, ^{18}O labeling	Dahl salt sensitive	Rat	Heart mitochondria	-	3-hydroxyacyl-CoA dehydrogenase type II, short-chain 3-hydroxyacyl-CoA dehydrogenase	Succinate dehydrogenase, fumarate hydratase, electron transfer flavoprotein alpha	(Smith, Matus et al. 2008)
Differential centrifugation	LC-MS/MS, Label-free	Transverse aortic constriction	Rat	Heart mitochondria	Mitochondria with disorganized cristae and/or reduced cristae density, i mitochondrial volume density, ↓ State 3 respiration	TCA, PDH, ETC (complex I, II, V)	FAO, ECT (complex I, III, IV, V)	(Bugger, Schwarzer et al. 2010)
Differential centrifugation	2-DE, liquid chromatography, ESI-MS/MS	Intermittent hypoxia	Rat	Heart mitochondria	↓ ATP content in intermittent hypoxia at ischemia compared with pro-ischemia and reperfusion	Aldehyde dehydrogenase, aconitase, malate dehydrogenase, ATP synthase β chain, electron transfer flavoprotein α subunit, sirtuin 5, methylmalonate-semialdehyde dehydrogenase	Ubiquinol-cytochrome c reductase iron-sulfur subunit, aspartate aminotransferase	(Zhu, Wu et al. 2012)

Proteomic approaches						Proteins differentially expressed		
Mitochondrial purification methods	Analytic method	Condition	Specie	Sample	Alterations of mitochondrial physiology	Up-regulated	Down-regulated	References
Aging								
Differential centrifugation	DIGE, LC-MS, Imunoblotting	Sarcopenia model	Rat	Skeletal muscle mitochondria	# activity of respiratory chain complexes	NADH dehydrogenase, prohibitin, VDAC2, Succinate dehydrogenase, Fis1, acyl-coA dehydrogenase, PRX-III, ubiquinol-cytochrome c reductase core I protein, Succinate CoA ligase, F1-ATPase	-	(O'Connell and Ohlendieck 2009)
Differential centrifugation	2-DE, MALDI-TOF, MALDI-TOF/TOF,HPLC-ESI/MS	24-25 months rats	Rat	Heart Mitochondria, Skeletal muscle mitochondria	-	Isocitrate dehydrogenase, malate dehydrogenase	-	(Chang et al. 2007)
Differential centrifugation	2-DE, Western blotting, MALDI-TOF, ESI-MS/MS, NSI-MS/MS	24-34 months rats	Rat	Skeletal muscle	# protein nitration levels	nitrated succinate dehydrogenase	-	(Kanski et al. 2003)
Differential centrifugation	2-DE, MALDI-TOF	30 months rats	Rat	Skeletal muscle	i protein phosphorylation levels	-	phosphorylated cytochrome c oxidase and aconitase	(Gannon et al. 2008)
Differential centrifugation	SDS-PAGE, LC-MALDI-TOF/TOF, Western blotting	25 months mice	Mice	Heart mitochondria	# protein carbonylation and nitration	carbonylated and nitrated ATP synthase and cytochrome *b-c1* complexes subunits; carbonylated NADH dehydrogenase [ubiquinone] iron-sulfur protein 2,	-	(Padrao et al. 2012)

Table 2. (Continued)

Proteomic approaches						Proteins differentially expressed		
Mitochondrial purification methods	Analytic method	Condition	Specie	Sample	Alterations of mitochondrial physiology	Up-regulated	Down-regulated	References
Aging (Continued)								
Differential centrifugation	SDS-PAGE, LC-MALDI-TOF/TOF, Western blotting	25 months mice	Mice	Heart mitochondria	# protein carbonylation and nitration	NADH dehydrogenase [ubiquinone] 1 alpha subcomplex subunit 10, ATP synthase subunits d and O,malate dehydrogenase, VDAC1, MnSOD	-	(Padrao et al. 2012)

Legend: FAO: Fatty acid oxidation, SCOT: succinyl-CoA:3-oxoacid CoA transferase, CK: creatine kinase, TCA: tricarboxylic acid cycle, ETC: electron transport chain, OXPHOS: oxidative phosphorylation, PDH: pyruvate dehydrogenase, VDAC: voltage dependent anion channel, ALS: alloxan-susceptible, ALR: alloxanresistant.

These databases list mitochondrial proteins identified via multiple approaches, including MS/MS analyses, literature references and bioinformatic evaluations. Among these, MitoP2 is one of the most comprehensive database, providing a list of mitochondrial proteins from yeast, mouse, human, *Arabidopsis thaliana*, and *Neurospora crassa*. It gives information about the functional annotation of proteins, as well as their subcellular location, their homologs, along with literature references.

MitoMiner is another mitochondrial protein database that integrates sets of proteomic data from several species (e.g. human, rat, mouse, *Drosophila melanogaster*, *Caenorhabditis elegans,* and *Saccharomyces cerevisiae*), thereby representing the most complete mitochondrial database known to date. It is based on MS experiments together with protein annotation data from Uniprot, metabolic pathway data from Kyoto Encyclopedia of Genes and Genomes (KEGG), protein homology from HomoloGene and disease information from Online Mendelian Inheritance in Man (OMIM).

Importantly, these data-sets have revealed that the number of mammalian mitochondrial proteins is much greater (> 3000) than the early studies have estimated (Meisinger et al. 2008; McDonald and Van Eyk 2003). COPaKB is an integrated resource of proteome biology configured to specifically focus on cardiovascular biology and medicine. It includes proteomic data from large-scale proteomic surveys of cardiac mitochondria (Wilkins 2009). Despite the advances achieved in mitochondrial proteomics research, data quality is still a major challenge in proteomics.

3. Mitochondrial Proteome Plasticity in Pathophysiological Conditions

Several inter-connected metabolic pathways are harbored in mitochondria, which themselves are connected to other important cellular functions. So, any local mitochondrial dysfunction can potentially lead to very complex cellular adaptations, which might be studied using comparative proteomic techniques.

Mitochondrial proteomics of tissues highly reliant on oxidative metabolism as striated muscles have been considered appropriate to increase the understanding of the molecular mechanisms underlying pathophysiological conditions related to heart or skeletal muscle dysfunction, and to find targets for novel therapies. So, in this section, the main findings on mitochondrial proteome alterations in cardiac and skeletal muscles are explored in cardiovascular diseases, diabetes mellitus, cancer and aging (summarized in table 2).

3.1. Cardiovascular Diseases

Mitochondrial proteomics have been applied to the study of cardiovascular diseases, the main "killer" in human beings (Jiang and Wang 2012). Most of the studies in this field aim to characterize heart mitochondrial dysfunction in order to give new insights into an early diagnosis and in the discovery of novel therapeutic targets (Hollander et al. 2011; McGregor and Dunn 2003; McGregor and Dunn 2006; Pieczenik and Neustadt 2007). For that, mitochondria isolated from heart or ventricles from different animal models of heart failure have been studied either with gel-based or gel-free approaches for protein separation, and mass spectrometry for protein identification (table 2).

For instance, using gel-based approaches (2-DE-MALDI-TOF), five proteins involved in tricarboxylic acid (TCA) cycle and lipid metabolism were found decreased and creatine kinase (CK) and prohibitin increased in the ventricles of stressed rats (Liu et al. 2004). More differentially expressed proteins were recognized in the analysis of heart from desmin null mice with proteins involved in the malate/aspartate shuttle, ketone utilization, and pyruvate dehydrogenase (PDH) subunits found down-regulated, and the electron transport chain (ETC) subunits, voltage dependent anion channel 2 (VDCA2), mtHsp70 and ubiquitin-like 1 were reported in significant higher levels (Fountoulakis et al. 2005). Changes in the abundance of metabolic proteins were also observed in ischemia-reperfusion (IR) and ischemia preconditioned (IPC) rabbit hearts (Kim et al. 2006). Whereas the expression of TCA cycle proteins decreased in IR hearts compared with control and IPC hearts, PDH increased. Respiratory chain proteins showed a variable pattern. The fatty acid oxidation (FAO) enzymes long-chain acyl-CoA dehydrogenase and 3-hydroxybutyrate dehydrogenase expression increased with ischemia, but enoyl-CoA hydratase decreased. Besides metabolism, alterations in the levels of proteins involved in transport and translation were also related with cardiac dysfunction. The mitochondrial proteome profiling of the left ventricle in a canine model of dyssynchronous or resynchronized heart failure performed with 2-DE-MS/MS, and BN-PAGE retrieved an overexpression of OXPHOS subunits (from complexes I, II, III and V), PDH complex proteins, VDAC3, proteins involved in protein synthesis and import and reactive oxygen species (ROS) scavenging enzymes. In contrast, cytochrome c and VDAC2 were down-regulated in cardiac resynchronization therapy, which was related with improved cardiac chamber mechanoenergetics, and decreased morbidity and mortality in patients with ventricular dyssynchrony (Agnetti et al. 2010).

MudPIT MS/MS analysis, combined with iTRAQ labeling, of mitochondria isolated from 20 month old spontaneously hypertensive rats evidenced the down-expression of several proteins from OXPHOS complexes II and V, CK, and proteins involved in FAO, and the overexpression of mitochondrial HSPs, PDH complex subunits, proapoptotic factors, proteins involved in permeability transition pore, ROS production and scavenging (Jüllig et al. 2008). Using LC-MS/MS with ^{18}O labeling for the analysis of heart mitochondrial proteome in a rat model of hypertension, cardiac hypertrophy and heart failure, the Dahl salt sensitive rat model, FAO dehydrogenases were found over-expressed whereas succinate dehydrogenase, fumarate hydratase and electron transfer flavoprotein alpha were down-regulated (Smith et al. 2008). LC-MS/MS with label-free quantitation also allowed the association of down-regulated FAO proteins, and over-expressed proteins from the PDH complex and TCA cycle with heart failure induced by transverse aortic constriction (Bugger et al. 2010).

There are no known studies that characterized the heart mitochondrial proteome in humans with cardiovascular diseases. However, to gain insight into mitochondrial proteome changes associated with these diseases, the analysis of the total heart tissue instead of isolated mitochondria was performed for the study of dilated cardiomyopathy (Knecht et al. 1994; Knecht et al. 1994) and of chronic reversibly and irreversibly dysfunctional myocardium, as well as in end-stage failing myocardium (Urbonavicius et al. 2009). The down-regulation of PDH, VDAC1, MnSOD, TCA cycle and ETC complex I subunits was verified by 2-DE and LC-MS/MS in irreversibly dysfunction and end-stage failing heart (Urbonavicius et al. 2009).

3.2. Diabetes Mellitus

Given their essential role in fatty acid and glucose metabolism (Baker and Haynes 2011; Duncan 2011), mitochondria from cardiac and skeletal muscles are likely to be impacted by diabetes mellitus (DM) and related complications (Baseler et al. 2011; Padrao et al. 2012; Jullig et al. 2007; Bugger et al. 2009; Duncan 2011; Romanello and Sandri 2010; Sivitz and Yorek 2010; Turko and Murad 2003; Dabkowski et al. 2010; Dabkowski et al. 2009; Essop et al. 2011). There are more studies in cardiac then in skeletal muscle mitochondria, focused on the understanding of the early events in DM-related cardiomyopathy. The majority of the mitochondrial proteome research has been conducted in animal models exhibiting T2DM (db/db mouse model) and T1DM, including streptozotocin (STZ)-induced T1DM mice/rats and diabetic Akita mouse.

Among the mitochondrial proteins modulated by STZ-induced T1DM, it was found by 2DE-MALDI-TOF/TOF an overexpression of FAO proteins and a selective down-regulation of OXPHOS and other mitochondrial proteins, such as CK, VDAC1, Hsp60, and Grp75 (Turko and Murad 2003). Employing iTRAQ labeling coupled with MudPIT, 65 proteins were identified in significantly different levels in cardiac mitochondria from STZ rats compared with control ones. Enzymes involved in long-chain fatty acids oxidation were found in increased amounts, further confirmed by enzyme assays, whereas multiple proteins involved in OXPHOS and catabolism of short-chain fatty acids and branched-chain amino acids were detected in decreased levels. Altered levels of several enzymes linked to oxidative stress were also observed (Jullig et al. 2007). Label-free proteome expression analyses of several tissues from wild-type and T1DM Akita mice (kidney, liver, brain, and heart) evidenced tissue-specific remodeling of the mitochondrial proteome, which in the case of the dysfunctional heart was manifested by an increase of FAO enzymes (Bugger et al. 2009).

Focusing on the role of mitochondrial protein quality control systems, zymography-LC-MALDI-MS/MS analysis was performed in order to identify the effect of T1DM on the expression and activity of mitochondrial proteases in *gastrocnemius* muscle. The AAA Lon protease, the metalloproteases PreP, LAP-3 and MIP, and cathepsin D were identified in muscle mitochondria and the content and activity of Lon protease was lower in the STZ animals and related with the accumulation of oxidized proteins (Padrao et al. 2012).

Besides changes in mitochondrial protein expression, PTMs seem to play an important role in DM-related mitochondria functionality. Protein tyrosine nitration was reported in the mitochondrial proteome from diabetic mouse heart (Turko and Murad 2003; Wang et al. 2010) and seems to affect energy production, antioxidant defense and apoptosis (Turko et al. 2003). Hollander's research group investigated thoroughly the differential response of the two populations of heart mitochondria in both T1DM and T2DM, using proteomic tools (Baseler et al. 2011; Dabkowski et al. 2010; Dabkowski et al. 2009). Based on iTRAQ and DIGE data, SS and IMF were analyzed to determine whether T1DM influenced their proteomic makeup in the heart. The decreased abundance of FAO and ETC proteins, of mitochondrial protein import and substrate transport (Hsp70, ANT1, and mitochondrial phosphate carrier) was more notorious in IMF mitochondria from diabetic subjects. Besides alterations in protein levels, PTMs including oxidations and deamidations were most prevalent in diabetic IMF mitochondria (Baseler et al. 2011). Using iTRAQ with MudPIT-based mass spectral analyses, proteins from FAO were found increased in both SS and IMF mitochondria of db/db hearts, though more pronounced in SS. Inner mitochondrial membrane proteins, including ETC, ATP synthase, and mitochondrial protein import machinery, were predominantly decreased in SS and related with impaired activity of this organelle (Dabkowski et al. 2010).

3.3. Cancer

Proteomic studies focused on mitochondria from striated muscle of individuals with cancer are scarce and mainly target cancer cachexia. Though recently emphasis has been given to mitochondrial dysfunction in cancer cachexia (Julienne et al. 2012; Lokireddy et al. 2012; White et al. 2012), no studies are known that performed protein profiling of mitochondria isolated from skeletal or cardiac muscles. Nevertheless, cancer-related muscle wasting was associated with altered expression of mitochondrial proteins measured by western blotting. For instance, cancer-related activation of TNF-alpha –induced NFκB was shown to decrease the expression of regulatory factors involved in mitochondrial biogenesis and consequently to disturb muscle oxidative capacity (Julienne et al. 2012; Wang et al. 2012). Moreover, the proteome analysis of whole *gastrocnemius*, *tibialis anterior*, *extensor digitorum longus* and *soleus,* and heart from rats inoculated with Yoshida AH-130 ascites hepatoma cells using 2-DE coupled to MALDI-TOF evidenced a higher susceptibility of proteins involved in ATP production and mitochondrial metabolism to oxidative damage (Marin-Corral et al. 2010).

3.4. Aging

The knowledge of mitochondrial proteome plasticity in skeletal and cardiac muscles of aged individuals have given new insights on the molecular mechanisms underlying this pathophysiological condition, which is characterized by a progressive, irreparable and time dependent deterioration of overall body functions (Figueiredo et al. 2008; Lenaz et al. 2000; Ventura et al. 2002). Proteomic profiling of mitochondria-enriched fractions from different animal's skeletal muscles has revealed a shift to a more aerobic oxidative metabolism in slower-twitching fibre populations during age-related muscle degeneration (O'Connell and Ohlendieck 2009; Ohlendieck 2011). Using DIGE with MS for protein identification, 11 mitochondrial proteins were found differentially expressed, with an aged-dependent elevation of ETC complexes I and III, mitofilin, peroxiredoxin isoform PRX-III, F1-ATPase, succinate dehydrogenase, fission protein Fis 1, succinate-coenzyme A ligase, acyl-coenzyme A dehydrogenase, VDAC2 and prohibitin (O'Connell and Ohlendieck 2009). The effect of aging was studied in mitochondria from 25-month-old rats and compared with 6-month-old ones using 2-DE coupled to MALDI-TOF-MS, MALDI-TOF/TOF-MS/MS and HPLC-ESI-MS and 3 heart proteins and 5 skeletal muscle proteins were found altered in an age dependent way, namely the metabolic enzymes isocitrate dehydrogenase and malate dehydrogenase (Chang et al. 2007).

Besides protein profiling, abnormal PTMs have been studied in aged skeletal and cardiac muscles' mitochondria (Schoneich 2006). For instance, higher levels of carbonylated ATP synthase, NADH dehydrogenase, pyruvate dehydrogenase and isocitrate dehydrogenase were reported in mitochondria from aged skeletal muscle based on data obtained from LC-ESI-MS/MS analysis with iTRAQ (Feng et al. 2008). Moreover, fast-twitch muscles seem to contain twice as many carbonylated mitochondrial proteins than slow-twitch muscles and 22 proteins showed significant changes in the carbonylation state with age, as for example trifunctional enzyme subunits α and β, acetyl-CoA acetyltransferase, NADH dehydrogenase (ubiquinone) Fe-S protein 1, Stress-70 protein, Hsp60, among others (Feng et al. 2008). The decreased capacity of aged mitochondria for ATP production was related with the increased carbonylation of specific proteins, mostly from TCA and OXPHOS (Alves et al. 2010). Using SDS-PAGE coupled to LC-MALDI-TOF/TOF and western blotting to identify the mitochondrial proteins more susceptibility to oxidation in mice cardiac muscle, it was verified an aged-related accumulation of oxidized OXPHOS subunits, some metabolic enzymes, transport proteins and antioxidant enzymes as MnSOD (Padrao et al. 2012).

3.5. Integrated Analysis of Mitochondrial Proteome Dynamics in Striated Muscle

Regardless the insult that triggers mitochondrial dysfunction, there are some similarities in the pattern of disease-related protein expression. Metabolic proteins, OXPHOS subunits, chaperones like HSPs and transport proteins as VDACs were found altered in all the pathophysiological conditions analyzed. Concerning metabolism, in diabetes mellitus, either type 1 or type 2, proteins from FAO were found up-regulated in cardiac mitochondria while ETC subunits were identified in significantly lower levels. Indeed, the metabolic phenotype of the diabetic heart is characterized by raises in fatty acid uptake and oxidation, a pattern largely mediated by competition between glucose and fatty acid metabolism (reviewed by Isfort et al. 2013). An opposite tendency was verified in distinct models of cardiovascular diseases, with FAO enzymes in lower levels in heart mitochondria. Though alterations in the expression of TCA and ETC proteins were also observed, distinct metabolic profiles were reported depending on the animal model studied. In contrast to the normal adult heart, the non-ischemic failing heart generally uses relatively more glucose at the expense of FAO. However, not all heart failure is metabolically the same. Comorbid conditions, degree of ischemia, degree of FAO metabolism machinery dysfunction, and plasma substrate levels also appear to modulate myocardial metabolism (reviewed by Kadkhodayan et al. 2012).

Besides metabolic proteins, the expression of heat shock proteins and VDACs was found to be modulated by these pathologies. The up-regulation of VDAC1 in T1DM and the altered expression of VDAC1-3 in cardiovascular diseases may result in disturbed energy production, metabolite cross-talk between cytosol and mitochondria, or apoptosis regulation (Shoshan-Barmatz and Ben-Hail 2012). The enhanced expression of HSPs might protect mitochondrial energetics, contributing to the recovery of myocardial function in ischaemia-reperfusion (Sammut and Harrison 2003).

More than protein profiling, several studies targeted the proteins more susceptible to PTMs in pathophysiological conditions, particularly in aging. Metabolic proteins are particularly prone to oxidative modifications in aged heart and skeletal muscle mitochondria. Indeed, increased carbonylated and nitrated TCA enzymes as aconitate hydratase and OXPHOS proteins like ATP synthase subunits α and β were previously reported (Turko et al. 2003; Alves et al. 2010). Because proteins that undergo oxidation are involved in major mitochondrial functions, such as energy production and antioxidant defense, increased levels of oxidized proteins have been associated with disease-related dysfunctional mitochondria.

4. Future Perspectives

The mitochondrion, as the biochemical nexus of the cell, is a critical consideration in the proteomics era. Great progress has already been made in mitochondrial proteomics to elucidate the role of mitochondria from striated muscles in the pathogenesis of diseases like diabetes mellitus or heart failure. However, mitochondria harbor a multitude of proteins and despite technical and methodological advances in mitochondrial protein profiling, there are many proteins yet to be characterized and their biological functions better understood. Standardizing experiments in terms of sample preparation, protein separation and identification as well as data analysis with bioinformatic tools is necessary for unbiased proteomic studies. Considering the increasingly recognized role of PTMs in the regulation of protein's function, further studies focused on the functional consequences of mitochondrial protein specific posttranslational modifications would provide a better understanding of the biological significance of signaling pathways centered on mitochondria.

Acknowledgments

This work was supported by Fundação para a Ciência e a Tecnologia (FCT, Portugal), European Union, QREN, FEDER and COMPETE for funding the QOPNA research unit (project PEst-C/QUI/UI0062/2011), research projects (PTDC/DES/114122/2009 and PTDC/DES/104567/2008) and post-graduation students (grant numbers SFRH/BD/66642/2009 to A.I.P, SFRH/BD/91067/2012 to R.N.F).

References

Abdallah, C., Dumas-Gaudot, E., Renaut, J., Sergeant, K. (2012) Gel-based and gel-free quantitative proteomics approaches at a glance. *International journal of plant genomics* 2012494572.

Aebersold, R., Mann, M. (2003) Mass spectrometry-based proteomics. *Nature* 422(6928): 198-207.

Agnetti, G., Kaludercic, N., Kane, L. A., Elliott, S. T., Guo, Y., Chakir, K., Samantapudi, D., Paolocci, N., Tomaselli, G. F., Kass, D. A., Van Eyk, J. E. (2010) Modulation of Mitochondrial Proteome and Improved Mitochondrial Function by Biventricular Pacing of Dyssynchronous Failing Hearts. *Circulation: Cardiovascular Genetics* 3(1):78-87.

Akude, E., Zherebitskaya, E., Chowdhury, S. K., Smith, D. R., Dobrowsky, R. T., Fernyhough, P. (2011) Diminished superoxide generation is associated with respiratory chain dysfunction and changes in the mitochondrial proteome of sensory neurons from diabetic rats. *Diabetes* 60(1):288-297.

Alves, R. M., Vitorino, R., Figueiredo, P., Duarte, J. A., Ferreira, R., Amado, F. (2010) Lifelong physical activity modulation of the skeletal muscle mitochondrial proteome in mice. *J. Gerontol. A Biol. Sci. Med. Sci.* 65(8):832-842.

Baker, B. M., Haynes, C. M. (2011) Mitochondrial protein quality control during biogenesis and aging. *Trends in biochemical sciences* 36(5):254-261.

Balaban, R. S. (2010) The mitochondrial proteome: a dynamic functional program in tissues and disease states. *Environmental and molecular mutagenesis* 51(5):352-359.

Bantscheff, M., Schirle, M., Sweetman, G., Rick, J., Kuster, B. (2007) Quantitative mass spectrometry in proteomics: a critical review. *Anal. Bioanal. Chem.* 389(4):1017-1031.

Baseler, W. A., Dabkowski, E. R., Williamson, C. L., Croston, T. L., Thapa, D., Powell, M. J., Razunguzwa, T. T., Hollander, J. M. (2011) Proteomic alterations of distinct mitochondrial subpopulations in the type 1 diabetic heart: contribution of protein import dysfunction. *Am. J. Physiol. Regul. Integr. Comp. Physiol.* 300(2):R186-200.

Benard, G., Faustin, B., Passerieux, E., Galinier, A., Rocher, C., Bellance, N., Delage, J. P., Casteilla, L., Letellier, T., Rossignol, R. (2006) Physiological diversity of mitochondrial oxidative phosphorylation. *American journal of physiology Cell physiology* 291(6):C11 72-1182.

Benard, G., Rossignol, R. (2008) Ultrastructure of the mitochondrion and its bearing on function and bioenergetics. *Antioxid. Redox. Signal* 10(8):1313-1342.

Braun, R. J., Kinkl, N., Beer, M., Ueffing, M. (2007) Two-dimensional electrophoresis of membrane proteins. *Analytical and bioanalytical chemistry* 389(4):1033-1045

Brookes, P. S., Pinner, A., Ramachandran, A., Coward, L., Barnes, S., Kim, H., Darley-Usmar, V. M. (2002) High throughput two-dimensional blue-native electrophoresis: a tool for functional proteomics of mitochondria and signaling complexes. *Proteomics* 2 (8):969-977.

Bugger, H., Chen, D., Riehle, C., Soto, J., Theobald, H. A., Hu, X. X., Ganesan, B., Weimer, B. C., Abel, E. D. (2009) Tissue-specific remodeling of the mitochondrial proteome in type 1 diabetic akita mice. *Diabetes* 58(9):1986-1997.

Bugger, H., Schwarzer, M., Chen, D., Schrepper, A., Amorim, P. A., Schoepe, M., Nguyen, T. D., Mohr, F. W., Khalimonchuk, O., Weimer, B. C., Doenst, T. (2010) Proteomic remodelling of mitochondrial oxidative pathways in pressure overload-induced heart failure. *Cardiovasc. Res.* 85(2):376-384.

Cadieux, P. A., Beiko, D. T., Watterson, J. D., Burton, J. P., Howard, J. C., Knudsen, B. E., Gan, B. S., McCormick, J. K., Chambers, A. F., Denstedt, J. D., Reid, G. (2004) Surface-enhanced laser desorption/ionization-time of flight-mass spectrometry (SELDI-TOF-MS): a new proteomic urinary test for patients with urolithiasis. *Journal of clinical laboratory analysis* 18(3):170-175.

Chacinska, A., Koehler, C. M., Milenkovic, D., Lithgow, T., Pfanner, N. (2009) Importing mitochondrial proteins: machineries and mechanisms. *Cell* 138(4):628-644.

Chang, J., Cornell, J. E., Van Remmen, H., Hakala, K., Ward, W. F., Richardson, A. (2007) Effect of aging and caloric restriction on the mitochondrial proteome. The journals of gerontology Series A, *Biological sciences and medical sciences* 62(3):223-234.

Chen, X., Li, J., Hou, J., Xie, Z., Yang, F. (2010) Mammalian mitochondrial proteomics: insights into mitochondrial functions and mitochondria-related diseases. *Expert review of proteomics* 7(3):333-345.

Cotter, D., Guda, P., Fahy, E., Subramaniam, S. (2004) MitoProteome: mitochondrial protein sequence database and annotation system. *Nucleic acids research* 32(Database issue): D4 63-467.

Dabkowski, E. R., Baseler, W. A., Williamson, C. L., Powell, M., Razunguzwa, T. T., Frisbee, J. C., Hollander, J. M. (2009) Mitochondrial dysfunction in the type 2 diabetic heart is associated with alterations in spatially distinct mitochondrial proteomes. *American journal of physiology Heart and circulatory physiology* 299(2):H529-540.

Dabkowski, E. R., Baseler, W. A., Williamson, C. L., Powell, M., Razunguzwa, T. T., Frisbee, J. C., Hollander, J. M. (2010) Mitochondrial dysfunction in the type 2 diabetic heart is associated with alterations in spatially distinct mitochondrial proteomes. *American journal of physiology Heart and circulatory physiology* 299(2):H529-540.

Dabkowski, E. R., Williamson, C. L., Bukowski, V. C., Chapman, R. S., Leonard, S. S., Peer, C. J., Callery, P. S., Hollander, J. M. (2009) Diabetic cardiomyopathy-associated dysfunction in spatially distinct mitochondrial subpopulations. *American journal of physiology Heart and circulatory physiology* 296(2):H359-369.

Dani, D., Dencher, N. A. (2008) Native-DIGE: a new look at the mitochondrial membrane proteome. *Biotechnology journal* 3(6):817-822.

Devreese, B., Vanrobaeys, F., Smet, J., Van Beeumen, J., Van Coster, R. (2002) Mass spectrometric identification of mitochondrial oxidative phosphorylation subunits separated by two-dimensional blue-native polyacrylamide gel electrophoresis. *Electrophoresis* 23(15):2525-2533.

D'Hertog, W., Mathieu, C., Overbergh, L. (2006) Type 1 diabetes: entering the proteomic era. *Expert review of proteomics* 3(2):223-236.

Dimmer, K. S., Rapaport, D. (2008) Proteomic view of mitochondrial function. *Genome biology* 9(2):209.

Dorn, G. W., 2nd (2013) Mitochondrial dynamics in heart disease. *Biochimica et biophysica acta* 1833(1):233-241.

Duchen, M. R. (2004) Mitochondria in health and disease: perspectives on a new mitochondrial biology. *Molecular aspects of medicine* 25(4):365-451.

Duncan, J. G. (2011) Mitochondrial dysfunction in diabetic cardiomyopathy. *Biochim. Biophys. Acta* 1813(7):1351-1359.

Egan, B., Dowling, P., O'Connor, P. L., Henry, M., Meleady, P., Zierath, J. R., O'Gorman, D. J. (2011) 2-D DIGE analysis of the mitochondrial proteome from human skeletal muscle reveals time course-dependent remodelling in response to 14 consecutive days of endurance exercise training. *Proteomics* 11(8):1413-1428.

Elstner, M., Andreoli, C., Ahting, U., Tetko, I., Klopstock, T., Meitinger, T., Prokisch, H. (2008) MitoP2: an integrative tool for the analysis of the mitochondrial proteome. *Molecular biotechnology* 40(3):306-315.

Elstner, M., Andreoli, C., Klopstock, T., Meitinger, T., Prokisch, H. (2009) The mitochondrial proteome database: MitoP2. *Methods in enzymology* 4573-20.

Essop, M. F., Chan, W. A., Hattingh, S. (2011) Proteomic analysis of mitochondrial proteins in a mouse model of type 2 diabetes. *Cardiovascular journal of Africa* 22(4):175-178.

Eyrich, B., Sickmann, A., Zahedi, R. P. (2011) Catch me if you can: mass spectrometry-based phosphoproteomics and quantification strategies. *Proteomics* 11(4):554-570.

Feng, J., Xie, H. W., Meany, D. L., Thompson, L. V., Arriaga, E. A., Griffin, T. J. (2008) Quantitative Proteomic Profiling of Muscle Type-Dependent and Age-Dependent Protein

Carbonylation in Rat Skeletal Muscle Mitochondria. *J. Gerontol. a-Biol.* 63(11):1137-1152.

Ferreira, R., Vitorino, R., Alves, R. M., Appell, H. J., Powers, S. K., Duarte, J. A., Amado, F. (2010) Subsarcolemmal and intermyofibrillar mitochondria proteome differences disclose functional specializations in skeletal muscle. *Proteomics* 10(17):3142-3154.

Figueiredo, P. A., Mota, M. P., Appell, H. J., Duarte, J. A. (2008) The role of mitochondria in aging of skeletal muscle. *Biogerontology* 9(2):67-84.

Flatmark, T. (1966) On the heterogeneity of beef heart cytochrome c. II. Some physico-chemical properties of the main subfractions (Cy I-Cy 3). *Acta chemica Scandinavica* 20 (6):1476-1486.

Forner, F., Arriaga, E. A., Mann, M. (2006) Mild protease treatment as a small-scale biochemical method for mitochondria purification and proteomic mapping of cytoplasm-exposed mitochondrial proteins. *Journal of proteome research* 5(12):3277-3287.

Fountoulakis, M., Schlaeger, E. J. (2003) The mitochondrial proteins of the neuroblastoma cell line IMR-32. *Electrophoresis* 24(1-2):260-275.

Fountoulakis, M., Soumaka, E., Rapti, K., Mavroidis, M., Tsangaris, G., Maris, A., Weisleder, N., Capetanaki, Y. (2005) Alterations in the heart mitochondrial proteome in a desmin null heart failure model. *J. Mol. Cell. Cardiol.* 38(3):461-474.

Fukada, K., Zhang, F., Vien, A., Cashman, N. R., Zhu, H. (2004) Mitochondrial proteomic analysis of a cell line model of familial amyotrophic lateral sclerosis. *Molecular and cellular proteomics : MCP* 3(12):1211-1223.

Geiger, T., Velic, A., Macek, B., Lundberg, E., Kampf, C., Nagaraj, N., Uhlen, M., Cox, J., Mann, M. (2013) Initial quantitative proteomic map of twenty-eight mouse tissues using the SILAC mouse. *Molecular and cellular proteomics : MCP.*

Gilchrist, A., Au, C. E., Hiding, J., Bell, A. W., Fernandez-Rodriguez, J., Lesimple, S., Nagaya, H., Roy, L., Gosline, S. J., Hallett, M., Paiement, J., Kearney, R. E., Nilsson, T., Bergeron, J. J. (2006) Quantitative proteomics analysis of the secretory pathway. *Cell* 127(6):1265-1281.

Gilmore, J. M., Washburn, M. P. (2010) Advances in shotgun proteomics and the analysis of membrane proteomes. *Journal of proteomics* 73(11):2078-2091.

Glancy, B., Balaban, R. S. (2011) Protein composition and function of red and white skeletal muscle mitochondria. *American journal of physiology Cell physiology* 300(6):C1280-1290.

Graham, J. M. (1993) Isolation of mitochondria, mitochondrial membranes, lysosomes, peroxisomes, and Golgi membranes from rat liver. *Methods Mol. Biol.* 1929-40

Greenacre, S. A., Ischiropoulos, H. (2001) Tyrosine nitration: localisation, quantification, consequences for protein function and signal transduction. *Free radical research* 34(6): 541-581

Gorg, A., Obermaier, C., Boguth, G., Csordas, A., Diaz, J. J., Madjar, J. J. (1997) Very alkaline immobilized pH gradients for two-dimensional electrophoresis of ribosomal and nuclear proteins. *Electrophoresis* 18(3-4):328-337.

Gygi, S. P., Corthals, G. L., Zhang, Y., Rochon, Y., Aebersold, R. (2000) Evaluation of two-dimensional gel electrophoresis-based proteome analysis technology. *Proceedings of the National Academy of Sciences of the US* 97(17):9390-9395.

Gygi, S. P., Rist, B., Gerber, S. A., Turecek, F., Gelb, M. H., Aebersold, R. (1999) Quantitative analysis of complex protein mixtures using isotope-coded affinity tags. *Nature biotechnology* 17(10):994-999.

Hanson, B. J., Schulenberg, B., Patton, W. F., Capaldi, R. A. (2001) A novel subfractionation approach for mitochondrial proteins: a three-dimensional mitochondrial proteome map. *Electrophoresis* 22(5):950-959.

Hartwig, S., Feckler, C., Lehr, S., Wallbrecht, K., Wolgast, H., Muller-Wieland, D., Kotzka, J. (2009) A critical comparison between two classical and a kit-based method for mitochondria isolation. *Proteomics* 9(11):3209-3214.

Heinemeyer, J., Scheibe, B., Schmitz, U. K., Braun, H. P. (2009) Blue native DIGE as a tool for comparative analyses of protein complexes. *Journal of proteomics* 72(3):539-544.

Herrnstadt, C., Clevenger, W., Ghosh, S. S., Anderson, C., Fahy, E., Miller, S., Howell, N., Davis, R. E. (1999) A novel mitochondrial DNA-like sequence in the human nuclear genome. *Genomics* 60(1):67-77.

Heussen, C., Dowdle, E. B. (1980) Electrophoretic analysis of plasminogen activators in polyacrylamide gels containing sodium dodecyl sulfate and copolymerized substrates. *Analytical biochemistry* 102(1):196-202.

Hogeboom, G. H., Schneider, W. C., Pallade, G. E. (1948) Cytochemical studies of mammalian tissues; isolation of intact mitochondria from rat liver; some biochemical properties of mitochondria and submicroscopic particulate material. *The Journal of biological chemistry* 172(2):619-635.

Hollander, J. M., Baseler, W. A., Dabkowski, E. R. (2011) Proteomic remodeling of mitochondria in heart failure. *Congest. Heart Fail.* 17(6):262-268.

Hoppel, C. L., Tandler, B., Fujioka, H., Riva, A. (2009) Dynamic organization of mitochondria in human heart and in myocardial disease. *Int. J. Biochem. Cell. Biol.* 41 (10):1949-1956.

Hoppeler, H. (1986) Exercise-induced ultrastructural changes in skeletal muscle. *International journal of sports medicine* 7(4):187-204.

Hoppeler, H., Fluck, M. (2003) Plasticity of skeletal muscle mitochondria: structure and function. *Medicine and science in sports and exercise* 35(1):95-104.

Huber, L. A., Pfaller, K., Vietor, I. (2003) Organelle proteomics: implications for subcellular fractionation in proteomics. *Circulation research* 92(9):962-968.

Isfort, M., Stevens, S. C., Schaffer, S., Jong, C. J., Wold, L. E. (2013) Metabolic dysfunction in diabetic cardiomyopathy. *Heart failure reviews.*

Jiang, X. S., Zhou, H., Zhang, L., Sheng, Q. H., Li, S. J., Li, L., Hao, P., Li, Y. X., Xia, Q. C., Wu, J. R., Zeng, R. (2004) A high-throughput approach for subcellular proteome: identification of rat liver proteins using subcellular fractionation coupled with two-dimensional liquid chromatography tandem mass spectrometry and bioinformatic analysis. *Molecular and cellular proteomics : MCP* 3(5):441-455.

Jiang, Y., Wang, X. (2012) Comparative mitochondrial proteomics: perspective in human diseases. *Journal of hematology and oncology* 511.

Jin, J., Davis, J., Zhu, D., Kashima, D. T., Leroueil, M., Pan, C., Montine, K. S., Zhang, J. (2007) Identification of novel proteins affected by rotenone in mitochondria of dopaminergic cells. *BMC neuroscience* 867.

Julienne, C. M., Dumas, J. F., Goupille, C., Pinault, M., Berri, C., Collin, A., Tesseraud, S., Couet, C., Servais, S. (2012) Cancer cachexia is associated with a decrease in skeletal muscle mitochondrial oxidative capacities without alteration of ATP production efficiency. *Journal of cachexia, sarcopenia and muscle.*

Jüllig, M., Hickey, A. J. R., Chai, C. C., Skea, G. L., Middleditch, M. J., Costa, S., Choong, S. Y., Philips, A. R. J., Cooper, G. J. S. (2008) Is the failing heart out of fuel or a worn engine running rich? A study of mitochondria in old spontaneously hypertensive rats. *Proteomics* 8(12):2556-2572.

Jullig, M., Hickey, A. J., Chai, C. C., Skea, G. L., Middleditch, M. J., Costa, S., Choong, S. Y., Philips, A. R., Cooper, G. J. (2008) Is the failing heart out of fuel or a worn engine running rich? A study of mitochondria in old spontaneously hypertensive rats. *Proteomics* 8(12):2556-2572.

Jullig, M., Hickey, A. J., Middleditch, M. J., Crossman, D. J., Lee, S. C., Cooper, G. J. (2007) Characterization of proteomic changes in cardiac mitochondria in streptozotocin-diabetic rats using iTRAQ isobaric tags. *Proteomics Clinical applications* 1(6):565-576.

Johnson, D. T., Harris, R. A., Blair, P. V., Balaban, R. S. (2007) Functional consequences of mitochondrial proteome heterogeneity. *American journal of physiology Cell physiology* 292 (2):C698-707.

Jung, E., Heller, M., Sanchez, J. C., Hochstrasser, D. F. (2000) Proteomics meets cell biology: the establishment of subcellular proteomes. *Electrophoresis* 21(16):3369-3377.

Kadkhodayan, A., Coggan, A. R., Peterson, L. R. (2012) A "PET" area of interest: myocardial metabolism in human systolic heart failure. *Heart failure reviews.*

Kane, L. A., Yung, C. K., Agnetti, G., Neverova, I., Van Eyk, J. E. (2006) Optimization of paper bridge loading for 2-DE analysis in the basic pH region: application to the mitochondrial subproteome. *Proteomics* 6(21):5683-5687.

Kang, D., Oh, S., Reschiglian, P., Moon, M. H. (2008) Separation of mitochondria by flow field-flow fractionation for proteomic analysis. *The Analyst* 133(4):505-515.

Kavazis, A. N., Alvarez, S., Talbert, E., Lee, Y., Powers, S. K. (2009) Exercise training induces a cardioprotective phenotype and alterations in cardiac subsarcolemmal and intermyofibrillar mitochondrial proteins. *American journal of physiology Heart and circulatory physiology* 297(1):H144-152.

Kerner, J., Lee, K., Hoppel, C. L. (2011) Post-translational modifications of mitochondrial outer membrane proteins. *Free radical research* 45(1):16-28.

Kiehntopf, M., Siegmund, R., Deufel, T. (2007) Use of SELDI-TOF mass spectrometry for identification of new biomarkers: potential and limitations. *Clinical chemistry and laboratory medicine : CCLM / FESCC* 45(11):1435-1449.

Kim, N., Lee, Y., Kim, H., Joo, H., Youm, J. B., Park, W. S., Warda, M., Cuong, D. V., Han, J. (2006) Potential biomarkers for ischemic heart damage identified in mitochondrial proteins by comparative proteomics. *Proteomics* 6(4):1237-1249.

Kim, S. C., Sprung, R., Chen, Y., Xu, Y., Ball, H., Pei, J., Cheng, T., Kho, Y., Xiao, H., Xiao, L., Grishin, N. V., White, M., Yang, X. J., Zhao, Y. (2006) Substrate and functional diversity of lysine acetylation revealed by a proteomics survey. *Molecular cell* 23(4):607-618.

Knecht, M., Regitz-Zagrosek, V., Pleissner, K. P., Emig, S., Jungblut, P., Hildebrandt, A., Fleck, E. (1994) Dilated cardiomyopathy: computer-assisted analysis of endomyocardial biopsy protein patterns by two-dimensional gel electrophoresis. *Eur. J. Clin. Chem. Clin. Biochem.* 32(8):615-624.

Knecht, M., Regitz-Zagrosek, V., Pleissner, K. P., Jungblut, P., Steffen, C., Hildebrandt, A., Fleck, E. (1994) Characterization of myocardial protein composition in dilated cardiomyopathy by two-dimensional gel electrophoresis. *Eur. Heart J.* 15 Suppl. D37-44.

Koc, E. C., Koc, H. (2012) Regulation of mammalian mitochondrial translation by post-translational modifications. *Biochim. Biophys. Acta* 1819(9-10):1055-1066.

Koehler, C. M. (2000) Protein translocation pathways of the mitochondrion. *FEBS letters* 476 (1-2):27-31.

Lai, X., Wang, L., Witzmann, F. A. (2013) Issues and applications in label-free quantitative mass spectrometry. *International journal of proteomics* 2013756039.

Lee, Y. H., Tan, H. T., Chung, M. C. (2010) Subcellular fractionation methods and strategies for proteomics. *Proteomics* 10(22):3935-3956.

Lefort, N., Yi, Z., Bowen, B., Glancy, B., De Filippis, E. A., Mapes, R., Hwang, H., Flynn, C. R., Willis, W. T., Civitarese, A., Hojlund, K., Mandarino, L. J. (2009) Proteome profile of functional mitochondria from human skeletal muscle using one-dimensional gel electrophoresis and HPLC-ESI-MS/MS. *J. Proteomics* 72(6):1046-1060.

Lenaz, G., D'Aurelio, M., Merlo Pich, M., Genova, M. L., Ventura, B., Bovina, C., Formiggini, G., Parenti Castelli, G. (2000) Mitochondrial bioenergetics in aging. *Biochimica et biophysica acta* 1459(2-3):397-404.

Li, J., Cai, T., Wu, P., Cui, Z., Chen, X., Hou, J., Xie, Z., Xue, P., Shi, L., Liu, P., Yates, J. R., 3rd, Yang, F. (2009) *Proteomic analysis of mitochondria from Caenorhabditis elegans. Proteomics* 9(19):4539-4553.

Lilley, K. S., Friedman, D. B. (2004) All about DIGE: quantification technology for differential-display 2D-gel proteomics. *Expert review of proteomics* 1(4):401-409.

Lilley, K. S., Razzaq, A., Dupree, P. (2002) Two-dimensional gel electrophoresis: recent advances in sample preparation, detection and quantitation. *Current opinion in chemical biology* 6(1):46-50.

Link, A. J., Eng, J., Schieltz, D. M., Carmack, E., Mize, G. J., Morris, D. R., Garvik, B. M., Yates, J. R., 3rd (1999) Direct analysis of protein complexes using mass spectrometry. *Nature biotechnology* 17(7):676-682.

Liu, X. H., Qian, L. J., Gong, J. B., Shen, J., Zhang, X. M., Qian, X. H. (2004) Proteomic analysis of mitochondrial proteins in cardiomyocytes from chronic stressed rat. *Proteomics* 4(10):3167-3176.

Lokireddy, S., Wijesoma, I. W., Bonala, S., Wei, M., Sze, S. K., McFarlane, C., Kambadur, R., Sharma, M. (2012) Myostatin is a novel tumoral factor that induces cancer cachexia. *The Biochemical journal.*

Lopez, M. F., Melov, S. (2002) Applied proteomics: mitochondrial proteins and effect on function. *Circ. Res.* 90(4):380-389.

Lovell, M. A., Xiong, S., Markesbery, W. R., Lynn, B. C. (2005) Quantitative proteomic analysis of mitochondria from primary neuron cultures treated with amyloid beta peptide. *Neurochemical research* 30(1):113-122.

Lowell, B. B., Shulman, G. I. (2005) Mitochondrial dysfunction and type 2 diabetes. *Science* 307 (5708):384-387.

Lundgren, D. H., Hwang, S. I., Wu, L., Han, D. K. (2010) Role of spectral counting in quantitative proteomics. *Expert review of proteomics* 7(1):39-53.

Marin-Corral, J., Fontes, C. C., Pascual-Guardia, S., Sanchez, F., Olivan, M., Argiles, J. M., Busquets, S., Lopez-Soriano, F. J., Barreiro, E. (2010) Redox balance and carbonylated proteins in limb and heart muscles of cachectic rats. *Antioxid. Redox. Signal* 12(3):365-380.

Marouga, R., David, S., Hawkins, E. (2005) The development of the DIGE system: 2D fluorescence difference gel analysis technology. *Analytical and bioanalytical chemistry* 382(3):669-678.

Mathy, G., Sluse, F. E. (2008) Mitochondrial comparative proteomics: strengths and pitfalls. *Biochim. Biophys. Acta* 1777(7-8):1072-1077.

McBride, H. M., Neuspiel, M., Wasiak, S. (2006) Mitochondria: more than just a powerhouse. *Curr. Biol.* 16(14):R551-560.

McDonald, T. G., Van Eyk, J. E. (2003) Mitochondrial proteomics. Undercover in the lipid bilayer. *Basic research in cardiology* 98(4):219-227.

McGregor, E., Dunn, M. J. (2003) Proteomics of heart disease. *Human Molecular Genetics* 12(suppl. 2):R135-R144.

McGregor, E., Dunn, M. J. (2006) Proteomics of the Heart: Unraveling Disease. *Circulation Research* 98(3):309-321.

Meany, D. L., Xie, H., Thompson, L. V., Arriaga, E. A., Griffin, T. J. (2007) Identification of carbonylated proteins from enriched rat skeletal muscle mitochondria using affinity chromatography-stable isotope labeling and tandem mass spectrometry. *Proteomics* 7 (7): 1150-1163.

Meisinger, C., Sickmann, A., Pfanner, N. (2008) The mitochondrial proteome: from inventory to function. *Cell* 134(1):22-24.

Meisinger, C., Sommer, T., Pfanner, N. (2000) Purification of Saccharomcyes cerevisiae mitochondria devoid of microsomal and cytosolic contaminations. *Analytical biochemistry* 287(2):339-342.

Meng, C., Jin, X., Xia, L., Shen, S.-M., Wang, X.-L., Cai, J., Chen, G.-Q., Wang, L.-S., Fang, N.-Y. (2009) Alterations of Mitochondrial Enzymes Contribute to Cardiac Hypertrophy before Hypertension Development in Spontaneously Hypertensive Rats. *Journal of Proteome Research* 8(5):2463-2475

Minden, J. (2007) Comparative proteomics and difference gel electrophoresis. *BioTechniques* 43(6):739, 741, 743 passim.

Millar, A. H., Sweetlove, L. J., Giege, P., Leaver, C. J. (2001) Analysis of the Arabidopsis mitochondrial proteome. *Plant physiology* 127(4):1711-1727.

Mirgorodskaya, O. A., Kozmin, Y. P., Titov, M. I., Korner, R., Sonksen, C. P., Roepstorff, P. (2000) Quantitation of peptides and proteins by matrix-assisted laser desorption/ ionization mass spectrometry using (18)O-labeled internal standards. *Rapid communications in mass spectrometry : RCM* 14(14):1226-1232.

Miyagi, M., Rao, K. C. (2007) Proteolytic 18O-labeling strategies for quantitative proteomics. *Mass spectrometry reviews* 26(1):121-136.

Mootha, V. K., Bunkenborg, J., Olsen, J. V., Hjerrild, M., Wisniewski, J. R., Stahl, E., Bolouri, M. S., Ray, H. N., Sihag, S., Kamal, M., Patterson, N., Lander, E. S., Mann, M. (2003) Integrated analysis of protein composition, tissue diversity, and gene regulation in mouse mitochondria. *Cell* 115(5):629-640.

Motoyama, A., Yates, J. R., 3rd (2008) Multidimensional LC separations in shotgun proteomics. *Analytical chemistry* 80(19):7187-7193.

Nagele, E., Vollmer, M., Horth, P., Vad, C. (2004) 2D-LC/MS techniques for the identification of proteins in highly complex mixtures. *Expert Rev. Proteomics* 1(1):37-46.

Neupert, W., Herrmann, J. M. (2007) Translocation of proteins into mitochondria. *Annual review of biochemistry* 76723-749.

Nijtmans, L. G., Henderson, N. S., Holt, I. J. (2002) Blue Native electrophoresis to study mitochondrial and other protein complexes. *Methods* 26(4):327-334.

Nyblom, H. K., Thorn, K., Ahmed, M., Bergsten, P. (2006) Mitochondrial protein patterns correlating with impaired insulin secretion from INS-1E cells exposed to elevated glucose concentrations. *Proteomics* 6(19):5193-5198.

O'Connell, K., Ohlendieck, K. (2009) Proteomic DIGE analysis of the mitochondria-enriched fraction from aged rat skeletal muscle. *Proteomics* 9(24):5509-5524.

O'Farrell, P. H. (1975) High resolution two-dimensional electrophoresis of proteins. *The Journal of biological chemistry* 250(10):4007-4021.

Ohlendieck, K. (2011) Skeletal muscle proteomics: current approaches, technical challenges and emerging techniques. *Skeletal muscle* 1(1):6.

Old, W. M., Meyer-Arendt, K., Aveline-Wolf, L., Pierce, K. G., Mendoza, A., Sevinsky, J. R., Resing, K. A., Ahn, N. G. (2005) Comparison of label-free methods for quantifying human proteins by shotgun proteomics. *Molecular and cellular proteomics : MCP* 4(10): 1487-1502.

Ong, S. E., Blagoev, B., Kratchmarova, I., Kristensen, D. B., Steen, H., Pandey, A., Mann, M. (2002) Stable isotope labeling by amino acids in cell culture, SILAC, as a simple and accurate approach to expression proteomics. *Molecular and cellular proteomics: MCP* 1 (5): 376-386.

Ong, S. E., Foster, L. J., Mann, M. (2003) Mass spectrometric-based approaches in quantitative proteomics. *Methods* 29(2):124-130.

Ong, S. E., Pandey, A. (2001) An evaluation of the use of two-dimensional gel electrophoresis in proteomics. *Biomolecular engineering* 18(5):195-205.

Ono, M., Shitashige, M., Honda, K., Isobe, T., Kuwabara, H., Matsuzuki, H., Hirohashi, S., Yamada, T. (2006) Label-free quantitative proteomics using large peptide data sets generated by nanoflow liquid chromatography and mass spectrometry. *Molecular and cellular proteomics : MCP* 5(7):1338-1347.

Padrao, A. I., Carvalho, T., Vitorino, R., Alves, R. M., Caseiro, A., Duarte, J. A., Ferreira, R., Amado, F. (2012) Impaired protein quality control system underlies mitochondrial dysfunction in skeletal muscle of streptozotocin-induced diabetic rats. *Biochim. Biophys. Acta* 1822(8):1189-1197.

Padrao, A. I., Ferreira, R., Vitorino, R., Alves, R. M., Figueiredo, P., Duarte, J. A., Amado, F. (2012) Effect of lifestyle on age-related mitochondrial protein oxidation in mice cardiac muscle. *Eur. J. Appl. Physiol.* 112(4):1467-1474.

Padrao, A. I., Ferreira, R. M., Vitorino, R., Alves, R. M., Neuparth, M. J., Duarte, J. A., Amado, F. (2011) OXPHOS susceptibility to oxidative modifications: the role of heart mitochondrial subcellular location. *Biochimica et biophysica acta* 1807(9):1106-1113.

Pagliarini, D. J., Calvo, S. E., Chang, B., Sheth, S. A., Vafai, S. B., Ong, S. E., Walford, G. A., Sugiana, C., Boneh, A., Chen, W. K., Hill, D. E., Vidal, M., Evans, J. G., Thorburn, D. R., Carr, S. A., Mootha, V. K. (2008) A mitochondrial protein compendium elucidates complex I disease biology. *Cell* 134(1):112-123.

Palmer, J. W., Tandler, B., Hoppel, C. L. (1977) Biochemical properties of subsarcolemmal and interfibrillar mitochondria isolated from rat cardiac muscle. *The Journal of biological chemistry* 252(23):8731-8739.

Perocchi, F., Jensen, L. J., Gagneur, J., Ahting, U., von Mering, C., Bork, P., Prokisch, H., Steinmetz, L. M. (2006) Assessing systems properties of yeast mitochondria through an interaction map of the organelle. *PLoS genetics* 2(10):e170.

Pieczenik, S. R., Neustadt, J. (2007) Mitochondrial dysfunction and molecular pathways of disease. *Experimental and Molecular Pathology* 83(1):84-92.

Reinders, J., Zahedi, R. P., Pfanner, N., Meisinger, C., Sickmann, A. (2006) Toward the complete yeast mitochondrial proteome: multidimensional separation techniques for mitochondrial proteomics. *J. Proteome. Res.* 5(7):1543-1554.

Rezaul, K., Wu, L., Mayya, V., Hwang, S. I., Han, D. (2005) A systematic characterization of mitochondrial proteome from human T leukemia cells. *Molecular and cellular proteomics: MCP* 4(2):169-181.

Romanello, V., Sandri, M. (2013) Mitochondrial biogenesis and fragmentation as regulators of protein degradation in striated muscles. *Journal of molecular and cellular cardiology* 5564-72.

Riva, A., Tandler, B., Loffredo, F., Vazquez, E., Hoppel, C. (2005) Structural differences in two biochemically defined populations of cardiac mitochondria. *American journal of physiology Heart and circulatory physiology* 289(2):H868-872.

Roe, M. R., Griffin, T. J. (2006) Gel-free mass spectrometry-based high throughput proteomics: tools for studying biological response of proteins and proteomes. *Proteomics* 6(17):4678-4687.

Roepstorff, P. (2012) Mass spectrometry based proteomics, background, status and future needs. *Protein and cell* 3(9):641-647.

Romanello, V., Sandri, M. (2010) Mitochondrial biogenesis and fragmentation as regulators of muscle protein degradation. *Curr. Hypertens. Rep.* 12(6):433-439.

Ross, P. L., Huang, Y. N., Marchese, J. N., Williamson, B., Parker, K., Hattan, S., Khainovski, N., Pillai, S., Dey, S., Daniels, S., Purkayastha, S., Juhasz, P., Martin, S., Bartlet-Jones, M., He, F., Jacobson, A., Pappin, D. J. (2004) Multiplexed protein quantitation in Saccharomyces cerevisiae using amine-reactive isobaric tagging reagents. *Molecular and cellular proteomics : MCP* 3(12):1154-1169.

Ryan, M. T., Hoogenraad, N. J. (2007) Mitochondrial-nuclear communications. *Annual review of biochemistry* 76701-722.

Santoni, V., Molloy, M., Rabilloud, T. (2000) Membrane proteins and proteomics: un amour impossible? *Electrophoresis* 21(6):1054-1070.

Schagger, H., von Jagow, G. (1991) Blue native electrophoresis for isolation of membrane protein complexes in enzymatically active form. *Anal. Biochem.* 199(2):223-231.

Schoneich, C. (2006) Protein modification in aging: An update. *Exp. Gerontol.* 41(9):807-812.

Schulze, W. X., Usadel, B. (2010) Quantitation in mass-spectrometry-based proteomics. *Annual review of plant biology* 61491-516

Schwerzmann, K., Cruz-Orive, L. M., Eggman, R., Sanger, A., Weibel, E. R. (1986) Molecular architecture of the inner membrane of mitochondria from rat liver: a combined biochemical and stereological study. *The Journal of cell biology* 102(1):97-103

Seibert, V., Wiesner, A., Buschmann, T., Meuer, J. (2004) Surface-enhanced laser desorption ionization time-of-flight mass spectrometry (SELDI TOF-MS) and ProteinChip technology in proteomics research. *Pathology, research and practice* 200(2):83-94.

Shaw, J., Rowlinson, R., Nickson, J., Stone, T., Sweet, A., Williams, K., Tonge, R. (2003) Evaluation of saturation labelling two-dimensional difference gel electrophoresis fluorescent dyes. *Proteomics* 3(7):1181-1195.

Shoshan-Barmatz, V., Ben-Hail, D. (2012) VDAC, a multi-functional mitochondrial protein as a pharmacological target. *Mitochondrion* 12(1):24-34.

Sickmann, A., Reinders, J., Wagner, Y., Joppich, C., Zahedi, R., Meyer, H. E., Schonfisch, B., Perschil, I., Chacinska, A., Guiard, B., Rehling, P., Pfanner, N., Meisinger, C. (2003) The proteome of Saccharomyces cerevisiae mitochondria. *Proceedings of the National Academy of Sciences of the US* 100(23):13207-13212.

Silvestri, E., Lombardi, A., Glinni, D., Senese, R., Cioffi, F., Lanni, A., Goglia, F., Moreno, M., Lange, P. D. (2010) *Mammalian Mitochondrial Proteome And Its Functions: Current Investigative Techniques And Future Perspectives On Ageing And Diabetes.*

Sivitz, W. I., Yorek, M. A. (2010) Mitochondrial dysfunction in diabetes: from molecular mechanisms to functional significance and therapeutic opportunities. *Antioxidants and redox signaling* 12(4):537-577.

Smith, A. C., Robinson, A. J. (2009) MitoMiner, an integrated database for the storage and analysis of mitochondrial proteomics data. *Molecular and cellular proteomics : MCP* 8 (6):1324-1337.

Smith, J. R., Matus, I. R., Beard, D. A., Greene, A. S. (2008) Differential expression of cardiac mitochondrial proteins. *Proteomics* 8(3):446-462.

Stewart, I. I., Thomson, T., Figeys, D. (2001) 18O labeling: a tool for proteomics. *Rapid communications in mass spectrometry : RCM* 15(24):2456-2465.

Sammut, I. A., Harrison, J. C. (2003) Cardiac mitochondrial complex activity is enhanced by heat shock proteins. *Clin. Exp. Pharmacol.* P 30(1-2):110-115

Tan, S., Tan, H. T., Chung, M. C. (2008) Membrane proteins and membrane proteomics. *Proteomics* 8(19):3924-3932.

Tang, N., Tornatore, P., Weinberger, S. R. (2004) Current developments in SELDI affinity technology. *Mass spectrometry reviews* 23(1):34-44.

Tonge, R., Shaw, J., Middleton, B., Rowlinson, R., Rayner, S., Young, J., Pognan, F., Hawkins, E., Currie, I., Davison, M. (2001) Validation and development of fluorescence two-dimensional differential gel electrophoresis proteomics technology. *Proteomics* 1(3): 377-396.

Taylor, S. W., Fahy, E., Zhang, B., Glenn, G. M., Warnock, D. E., Wiley, S., Murphy, A. N., Gaucher, S. P., Capaldi, R. A., Gibson, B. W., Ghosh, S. S. (2003) Characterization of the human heart mitochondrial proteome. *Nature biotechnology* 21(3):281-286.

Taylor, S. W., Warnock, D. E., Glenn, G. M., Zhang, B., Fahy, E., Gaucher, S. P., Capaldi, R. A., Gibson, B. W., Ghosh, S. S. (2002) An alternative strategy to determine the mitochondrial proteome using sucrose gradient fractionation and 1D PAGE on highly purified human heart mitochondria. *Journal of proteome research* 1(5):451-458.

Turko, I. V., Li, L., Aulak, K. S., Stuehr, D. J., Chang, J. Y., Murad, F. (2003) Protein tyrosine nitration in the mitochondria from diabetic mouse heart. Implications to dysfunctional mitochondria in diabetes. *The Journal of biological chemistry* 278(36): 33972-33977.

Turko, I. V., Murad, F. (2003) Quantitative protein profiling in heart mitochondria from diabetic rats. *The Journal of biological chemistry* 278(37):35844-35849.

Unlu, M., Morgan, M. E., Minden, J. S. (1997) Difference gel electrophoresis: a single gel method for detecting changes in protein extracts. *Electrophoresis* 18(11):2071-2077.

Urbonavicius, S., Wiggers, H., Botker, H. E., Nielsen, T. T., Kimose, H. H., Ostergaard, M., Lindholt, J. S., Vorum, H., Honore, B. (2009) Proteomic analysis identifies mitochondrial metabolic enzymes as major discriminators between different stages of the failing human myocardium. *Acta Cardiol.* 64(4):511-522.

Van den Bergh, G., Arckens, L. (2004) Fluorescent two-dimensional difference gel electrophoresis unveils the potential of gel-based proteomics. *Current opinion in biotechnology* 15(1):38-43.

Ventura, B., Genova, M. L., Bovina, C., Formiggini, G., Lenaz, G. (2002) Control of oxidative phosphorylation by Complex I in rat liver mitochondria: implications for aging. *Biochimica et biophysica acta* 1553(3):249-260.

Wallace, D. C. (1999) Mitochondrial diseases in man and mouse. *Science* 283(5407):1482-1488.

Wang, X., Pickrell, A. M., Zimmers, T. A., Moraes, C. T. (2012) Increase in muscle mitochondrial biogenesis does not prevent muscle loss but increased tumor size in a mouse model of acute cancer-induced cachexia. *PloS one* 7(3):e33426.

Wang, Y., Peng, F., Tong, W., Sun, H., Xu, N., Liu, S. (2010) The nitrated proteome in heart mitochondria of the db/db mouse model: characterization of nitrated tyrosine residues in SCOT. *Journal of proteome research* 9(8):4254-4263.

Warda, M., Kim, H. K., Kim, N., Ko, K. S., Rhee, B. D., Han, J. (2013) A matter of life, death and diseases: mitochondria from a proteomic perspective. *Expert Rev. Proteomics* 10(1):97-111.

Washburn, M. P., Wolters, D., Yates, J. R., 3rd (2001) Large-scale analysis of the yeast proteome by multidimensional protein identification technology. *Nature biotechnology* 19(3):242-247.

Werhahn, W., Niemeyer, A., Jansch, L., Kruft, V., Schmitz, U. K., Braun, H. (2001) Purification and characterization of the preprotein translocase of the outer mitochondrial membrane from Arabidopsis. Identification of multiple forms of TOM20. *Plant physiology* 125(2):943-954.

White, J. P., Puppa, M. J., Sato, S., Gao, S., Price, R. L., Baynes, J. W., Kostek, M. C., Matesic, L. E., Carson, J. A. (2012) IL-6 regulation on skeletal muscle mitochondrial remodeling during cancer cachexia in the ApcMin/+ mouse. *Skeletal muscle* 2(1):14.

Wilkesman, J., Kurz, L. (2009) Protease analysis by zymography: a review on techniques and patents. *Recent patents on biotechnology* 3(3):175-184.

Wilkins, M. (2009) Proteomics data mining. *Expert review of proteomics* 6(6):599-603.

Wilkins, M. R., Gasteiger, E., Sanchez, J. C., Bairoch, A., Hochstrasser, D. F. (1998) Two-dimensional gel electrophoresis for proteome projects: the effects of protein hydrophobicity and copy number. *Electrophoresis* 19(8-9):1501-1505.

Wittig, I., Carrozzo, R., Santorelli, F. M., Schagger, H. (2007) Functional assays in high-resolution clear native gels to quantify mitochondrial complexes in human biopsies and cell lines. *Electrophoresis* 28(21):3811-3820.

Wittig, I., Karas, M., Schagger, H. (2007) High resolution clear native electrophoresis for in-gel functional assays and fluorescence studies of membrane protein complexes. *Molecular and cellular proteomics: MCP* 6(7):1215-1225.

Wittig, I., Schagger, H. (2008) Features and applications of blue-native and clear-native electrophoresis. *Proteomics* 8(19):3974-3990.

Wittig, I., Schagger, H. (2009) Native electrophoretic techniques to identify protein-protein interactions. *Proteomics* 9(23):5214-5223.

Witze, E. S., Old, W. M., Resing, K. A., Ahn, N. G. (2007) Mapping protein post-translational modifications with mass spectrometry. *Nature methods* 4(10):798-806.

Wolters, D. A., Washburn, M. P., Yates, J. R., 3rd (2001) An automated multidimensional protein identification technology for shotgun proteomics. *Analytical chemistry* 73(23): 5683-5690.

Yao, X., Freas, A., Ramirez, J., Demirev, P. A., Fenselau, C. (2001) Proteolytic 18O labeling for comparative proteomics: model studies with two serotypes of adenovirus. *Analytical chemistry* 73(13):2836-2842.

Yates, J. R., 3rd, Carmack, E., Hays, L., Link, A. J., Eng, J. K. (1999) Automated protein identification using microcolumn liquid chromatography-tandem mass spectrometry. *Methods Mol. Biol.* 112553-569.

Yates, J. R., 3rd, Link, A. J., Schieltz, D. (2000) Direct analysis of proteins in mixtures. Application to protein complexes. *Methods Mol. Biol.* 14617-26.

Zhang, A., Williamson, C. D., Wong, D. S., Bullough, M. D., Brown, K. J., Hathout, Y., Colberg-Poley, A. M. (2011) Quantitative proteomic analyses of human cytomegalovirus-induced restructuring of endoplasmic reticulum-mitochondrial contacts at late times of infection. *Molecular and cellular proteomics : MCP* 10(10):M111 009936.

Zhang, J., Lin, A., Powers, J., Lam, M. P., Lotz, C., Liem, D., Lau, E., Wang, D., Deng, N., Korge, P., Zong, N. C., Cai, H., Weiss, J., Ping, P. (2012) Perspectives on: SGP symposium on mitochondrial physiology and medicine: mitochondrial proteome design: from molecular identity to pathophysiological regulation. *The Journal of general physiology* 139(6):395-406.

Zhang, J., Li, X., Mueller, M., Wang, Y., Zong, C., Deng, N., Vondriska, T. M., Liem, D. A., Yang, J. I., Korge, P., Honda, H., Weiss, J. N., Apweiler, R., Ping, P. (2008) Systematic characterization of the murine mitochondrial proteome using functionally validated cardiac mitochondria. *Proteomics* 8(8):1564-1575.

Zhao, X., Leon, I. R., Bak, S., Mogensen, M., Wrzesinski, K., Hojlund, K., Jensen, O. N. (2011) Phosphoproteome analysis of functional mitochondria isolated from resting human muscle reveals extensive phosphorylation of inner membrane protein complexes and enzymes. *Molecular and cellular proteomics : MCP* 10(1):M110 000299.

Zhu, W., Smith, J. W., Huang, C. M. (2010) Mass spectrometry-based label-free quantitative proteomics. *Journal of biomedicine and biotechnology* 2010840518.

Zischka, H., Weber, G., Weber, P. J., Posch, A., Braun, R. J., Buhringer, D., Schneider, U., Nissum, M., Meitinger, T., Ueffing, M., Eckerskorn, C. (2003) Improved proteome analysis of Saccharomyces cerevisiae mitochondria by free-flow electrophoresis. *Proteomics* 3(6):906-916.

In: Recent Advances in Proteomics Research
Editor: Ming D. Li
ISBN: 978-1-62948-218-7

Chapter 7

Structural Mass Spectrometry of Halophilic Proteasomal Nanomachines Using Chemical Cross-Linking

***Julie Maupin-Furlow*[1,*], *Ivanka Karadžić*[2] *and David Goodlett*[3]**
[1]Department of Microbiology and Cell Science, University of Florida, Gainesville, Florida, US
[2]Department of Chemistry, School of Medicine, University of Belgrade, Belgrade, Serbia
[3]School of Pharmacy, University of Maryland, Baltimore, US

Abstract

Structural mass spectrometry is a powerful technology that can provide direct determination of protein structure through three overall approaches: covalent labeling based on footprinting technology, chemical cross-linking, and hydrogen-deuterium exchange. Of these three, chemical cross-linking coupled with mass spectrometry (CXMS) can provide intra- and inter-molecular distance constraints that can be used to resolve protein folding in monomeric protein subunits and interactions at molecular interfaces in protein complexes. Challenges associated with CXMS include detection of low abundant cross-linked peptides, analysis of complex spectra, and computational tools to facilitate the identification of cross-linked peptides. A simple strategy based on data from a single LC-MS/MS analysis and an open modification search strategy to interpret the data has been previously reported using cytochrome c and validated using a much more complex protein machine, the bacterial T3SS needle complex. We recently applied the same CXMS pipeline process to examine the structure of unusual, acidic proteins of proteasomal 20S core particles (CPs) from *Haloferax volcanii* and demonstrated that

* Corresponding author: Julie Maupin-Furlow, PhD. Professor, University of Florida, Department of Microbiology and Cell Science, PO Box 110700, Bldg 981, Museum Rd., Gainesville, FL 32611-0700. Tel: (352) 392-4095; Fax: (352) 392-5922; E-mail: jmaupin@ufl.edu.

even a small number of intermolecular cross-link constraints can be sufficient for the topology prediction of CPs protein complex. Successful application of CXMS data in validation of complex protein structures generated by computational methods implies that routine analysis of protein structure based on acquisition of cross-linking distance constraints is a realistic goal.

Several topics related to structural mass spectrometry of halophilic proteasomal nanomachines using chemical cross-linking will be discussed including the general principle of CXMS, characteristics of cross-linkers, properties of CPs, known strategies of CXMS (particularly those applied to microbial proteins), bioinformatics in cross-link identification, and specificity of halophilic proteins of haloarchaea. Finally, the relationship of CXMS and comparative modeling, use of distance constraints in validation of inter- and intramolecular distances, and molecular modeling software, especially those that allow efficient, but simple use of cross-linking distance constraints to create and analyze protein assemblies on a routine basis will be described.

Keywords: Foot printing technology (FP), chemical cross-linking (CX), hydrogen-deuterium exchange (HDX), bottom-up proteomics, top-down proteomics, halophilic proteins, Archaea, proteasomes, N-hydroxysuccinimide (NHS) esters, ethyl-3-(3-dimethylamino-propyl)carbodiimide (EDC), intrapeptide cross-link, interpeptide cross-link, xQuest, Popitam, xComb, ModBase, MSX-3D, DaliLite, VEGA ZZ

Introduction

Functionally relevant biological nanomachines of protein structure are often macromolecular complexes composed of protein monomers. However, of the total number of PDB deposited structures of proteins, which is almost 90000, only a small number represent protein complexes. Obviously, in spite of great progress in technologies to determine protein structures, analysis of multiprotein complexes is still challenging. Mapping structural details of functional protein complexes in solution by chemical cross-linking mass spectrometry (CXMS) has been used for different individual applications, including homology modeling. Over the last ten years, several protein complexes have been successfully studied by chemical cross-linking and mass spectrometry, indicating the increased potential for use of CXMS methods in the structural determination of protein complexes (Rappsilber 2011; Rappsilber 2012). Here we review our simple, fast and accurate method to generate distance constraints between juxtaposed amino acids and discuss methods used to validate molecular models of multisubunit complexes that are halophilic and of low lysine with no cysteine residues. Proteasomal 20S core particles (CPs) of the haloarchaeon *Haloferax volcanii* were used to investigate the quaternary structure of halophilic proteins based on their symmetrical, yet distinct subunit composition. Proteasomal CPs are cylindrial barrel-like structures of four-stacked homoheptameric rings of α- and β-type subunits organized in $\alpha_7\beta_7\beta_7\alpha_7$ stoichiometry. The CPs of *Hfx. volcanii* are formed from a single type of β subunit associated with α1 and/or α2 subunits (Wilson et al. 1999). Distance constraints obtained by CXMS, together with the available structural data of non-halophilic CPs, facilitated the selection of accurate models of *Hfx. volcanii* proteasomal CPs composed of α1-, α2-, and β-homoheptameric rings from among several different possible PDB structures (Karadzic et al. 2012). Leitner et al. were one the first groups to suggest that identification of even a low number of cross-linking distance constraints can facilitate probing complex protein structures (Leitner et al... 2010).

Moreover, chemical cross-linking together with molecular modeling is recently reported to be a strategy that could be useful to rapidly differentiate solution structures from the crystal structures of proteins (Rozbesky et al. 2013). Thus, acquisition of cross-linking distance constraints of protein assemblies is now a realistic method to be used on a routine basis to improve modeling accuracy.

CXMS technology provides improved opportunities for modeling by mapping structural details of protein complexes in solution, but some efficient platforms to link CXMS and modeling are still lacking. Once the technological limits of CXMS have been addressed, CXMS data is likely to be incorporated into modeling algorithms that will allow for faster development of integrated structural biology (Rappsilber 2011).

Structural Mass Spectrometry

Proteins are the most diverse macromolecules in living systems and serve crucial functions in essentially all biological processes. Proteins spontaneously fold into three-dimensional or tertiary structures, with many still organized into larger assemblies composed of multiple protein molecules or polypeptide chains (subunits) giving complex quaternary structures. The determination of protein structure is critical to understand protein function at a molecular level.

Significant progress has been made toward determination of the spatial and topological organization of protein and protein complexes at atomic level resolution, by traditional methods, such as nuclear magnetic resonance (NMR) spectroscopy and X-ray crystallography, which provide detailed information on the structure of highly purified proteins, but have limitations when analyzing protein complexes (Singh et al. 2010; Lee 2008; Sinz 2006). In this post-genomic era, high-throughput techniques that are sensitive, such as mass spectrometry (MS), can play a significant role in the large-scale analysis of protein structure. More generally applicable, albeit less direct, MS techniques typically rely on some sort of chemical labelling reaction prior to MS analysis to reveal available functional groups that are chemically modified.

Recent advances in MS have facilitated the development of structural MS techniques as powerful technologies that can provide direct insight into protein structure. Three overall approaches are used in structural MS including covalent labeling based on chemical foot printing technologies (FP), chemical cross-linking (CX), and hydrogen-deuterium exchange (HDX). Before reviewing our CXMS work we will briefly review the other two methods as they often are used in conjunction with CXMS and modeling to solve structures.

Chemical FP has been used to map the interfaces between macromolecules as well to map conformational changes caused by ligand binding (Wang and Chance 2011). In FP experiments, the solvent accessible surface of the macromolecule is examined after modification or cleavage of macromolecules (Takamoto and Chance 2006).

Hydroxyl radicals are the most popular reagents used in FP and can react with side-chain sites in proteins that are solvent accessible.Several methods of generating hydroxyl radicals for FP experiments have been developed, such as chemical methods (Tullius and Dombroski

1986), radiolysis of water (Sclavi et al. 1998) or photolysis of hydrogen peroxide (Sharp et al.. 2004). Hydroxyl radicals easily oxidize side chains of surface reactive amino acid residues in aqueous solution and react with protein functional groups depending on their specific chemistry and solvent accessibility. The relative reactivity of amino acids to hydroxyl radicals follows the order: Cys > Met > Trp > Tyr > Phe > Cystine > His > Leu ~ Ile > Arg ~ Lys~ Val > Ser ~ Thr ~ Pro > Gln ~ Glu > Asp ~ Asn > Ala > Gly (Wang and Chance 2011). Proteins irreversibly modified by FP are exposed to proteolysis and can be further analyzed by LC/MS that provides details of surface accessibility for specific side chains in the protein structure (Wang and Chance 2011).

Hydrogen-deuterium exchange (HDX) is an isotope exchange reaction in which hydrogen atoms on a protein molecule exchange with deuterium from the solvent. The HDX method relies on exchange of the amide hydrogen atoms attached to the backbone forming amides with deuterium (Brock 2012).

After the solution conformational information has been captured, the protein is either directly analyzed with liquid chromatography–mass spectrometry (LC-MS), or digested with acid proteases prior to LC/MS analysis to allow for the localization of deuterium incorporation along the primary structure of the studied protein (Brock 2012). H/D exchange rates have been used to explore mechanisms of protein folding (Konermann and Simmons 2003), determine the allosteric impact of post-translational modifications and ligand binding (Betts et al. 2008), and define truncation points for enhancing crystallization success (Pantazatos et al. 2004). HDX has also found a role in mapping interactions between proteins (Ling et al. 2010). FP and HDX are methods that can be considered as complementary. While HDX has the ability to probe backbone secondary structure conformation in solution, hydroxyl radical-mediated FP primarily labels side chains of proteins in solution, providing no information on the backbone structure. Thus, HDX, as a method that probes protein backbone, is a complementary approach to CX in mapping the interfaces between macromolecules.

General Principle and Strategies of CXMS

Amongst the already mentioned MS techniques that are used for structural analysis of proteins, chemical cross-linking coupled with mass spectrometry (CXMS) has emerged as a powerful method, which can yield site-specific low resolution structure information with sample quantities of 2-3 order of magnitude less than and faster experimental turn-around times than required for X-ray and NMR experiments (Singh et al. 2010). The general principle of CXMS is the covalent capture of juxtaposed amino acids using a variety of cross-linking reagents and subsequent analysis of cross-linked proteins/peptides by MS. The sequence and structure of cross-linked peptides are assigned from tandem mass spectra using various bioinformatics tools some of which we will describe further below. Regardless of the tools used, it is the intent that detailed analysis of these data will provide intra- and inter-molecular distance constraints that can in turn be used to resolve protein folding in monomeric protein subunits and interactions at molecular interfaces in protein complexes (Petrotchenko and Borchers 2010).

Advantages associated with CXMS analysis are summarized as follows (Sinz 2006):

- Fast, sensitive and requires only femtomole amounts of protein sample,
- No limitation in size or mass of proteins or protein complexes,
- Structure of proteins in solution and flexible region can be studied and identified,
- Transient interactions can be investigated,
- Can be used for *in vivo* studies,
- Provides information on distances between particular residues in proteins or complexes, and
- Provides experimental flexibility due to availability of a wide variety of chemical cross-linkers.

Two principal strategies have been developed to identify cross-link modifications by MS including bottom-up (peptide level) and top-down (intact protein level) approaches (Sinz 2006; Renzone et al. 2007). High resolution and high accuracy MS methods are necessary prerequisites for confident analysis of cross-linked proteins in both cases. In the bottom-up approach, cross-linked proteins are digested prior to MS analysis. Electrospray ionization tandem mass spectrometry (ESI-MS/MS) is the method of choice to analyze cross-linked peptides generated by this bottom-up approach.

In a top-down analytical approach, cross-linked proteins without digestion are subjected to MS, thus, requiring a MS method of ultra-high resolution and mass accuracy, as well as robust instrumentation. In spite of these stringent requirements, top-down analysis of proteins has developed rapidly in recent years (Li et al. 2011; Xie et al. 2006; Breuker et al. 2008; Novak et al. 2003). Furthermore, a combined top-down bottom-up approach has been recently used to elucidate the structure of regulatory β subunits in the (αβγδ)4 phosphorylase kinase complex by CXMS (Nadeau et al. 2012).

Despite success, bottom-up CXMS approaches remain challenging, because the mixtures of cross-linked peptides generated by protease digestion that are analyzed by LC-MS/MS can be complex. Namely, the detection of low abundant cross-linked peptides in complex spectra and a need to use robust computer algorithms hampers routine application of this method. The problem of finding the proverbial "needle in a haystack" that is the typical low stoichiometry cross-linked peptide product has been creatively solved by several sophisticated detection strategies [reviewed by (Singh et al. 2010; Sinz 2006)]. Cross-linkers with stable isotopes, radioactive isotopes, fluorophores, and affinity tags that can be cleaved from the peptide have been developed to facilitate the purification and identification of cross-linked products (Leitner et al. 2010).

The most straightforward bottom-up protocol typically involves three steps

I protein reaction with a certain cross-linker,
II enzymatic digestion of the cross-linked proteins into peptides, and
III separation of peptides by reverse phase liquid chromatography (LC) coupled with MS/MS analysis.

In complex CXMS analysis, the first step is proper selection of a chemical cross-linker that is adequate for the given protein complex. A large number of cross-linking reagents have been used to generate chemical bridges between juxtaposed amino acid side chain groups. Only the most common cross-linkers that target amino groups are reviewed.

Characteristics of Cross-Linkers

A large and increasing set of chemical cross-linking reagents has been investigated for use as a molecular ruler to provide information on distances between cross-linked amino acid residues that are relevant to both the tertiary and quaternary structure of proteins. One of the most important steps in CXMS is selection of an appropriate cross-linker. Cross-linker characteristics and principles of selection have been extensively studied and recently reviewed (Leitner et al. 2010; Lee 2008; Sinz 2006; Back et al. 2002). Important aspects to consider when choosing a cross-linker are its specificity (amino acid specific, photoreactive, nonspecific), arm type (homo/hetero bifunctional, trifunctional), arm length (zero length, carbon chain), water solubility (soluble, membrane permeable), presence of label (unlabeled, isotope labeled, affinity labeled) and cleavability (non-cleavable, chemically cleavable, CID cleavable). Aside from the large number of described and commercially available cross-linkers that are utilized, only a few organic reactions based on nucleophilicity of the functional groups of proteins are used to attach the cross-linkers. While amino and sulfhydryl are two the most often targeting groups in side chains of amino acids, cross-linking reagents with functionalities targeting amino groups are by far the most common reagents (Leitner et al. 2010; Sinz 2006). Based on the reaction of acylation of ε-amino group of lysine, a list of homobifunctional and heterobifunctional cross-linkers containing N-hydroxysuccinimide esters has been synthesized and used in CXMS protocols (Sinz 2006). Homobifunctional N-hydroxysuccinimide (NHS) esters primarily target the ε-amino group of lysine residues of proteins and can differ in the length of the spacer arm from 6.4 to 35.8 Å. Broadly used commercially available NHS esters are listed in Table 1. Listed homobifunctional N-hydroxysuccinimide esters differ not only in spacer arm length, but also in solubility and cleavability, giving a broad frame for proper selection of a cross-linker for a specific problem.

N-hydroxysuccinimide (NHS) esters react with nucleophiles to release the NHS or sulfo-NHS group and to create stable amide and imide bonds with primary or secondary amines, such as free N-termini and ε-amino groups of lysine side chains of proteins (Figure 1). Although the most commonly used NHS esters are often described as reactive exclusively towards primary amines, side reactions with other amino acid residues have also been reported. In particular, NHS reactivity with serine, threonine and tyrosine residues is significant but highly dependent on the pH and on adjacent amino acids within the protein (Mädler et al. 2009). In contrast to most bifunctional reagents (such as the NHS esters), which introduce a bridge between cross-linked residues, the zero length carbodiimide cross-linkers [such as 1-ethyl-3-(3-dimethylaminopropyl)carbodiimide or EDC] create a covalent bond between carboxylate and amine groups of peptides or proteins without an intervening linker, allowing a direct evaluation of contact interactions between protein surfaces (Figure 2) (Sinz 2006).

As presented in Figure 3, a diversity of products can be created by the cross-linking reagent, which further complicates the analysis. To describe the distinct cross-linking products in complex mixtures, a uniform nomenclature has been created. According to the classification of cross-linking products, three types of single modifications are considered. Type 0 is defined as dead-end, type 1 is defined as intrapeptide, and type 2 is defined as interpeptide with the two peptides denoted as α and β chains, where α is the longer of the two peptides (Schilling et al. 2003).

Table 1. Commercial homobifunctional N-hydroxysuccinimide (NHS) esters

Abbreviation	Chemical name	Spacer arm length, Å	Reference
TSAT	Tris-succinimidyl aminotriacetate	4.2	(Nowak et al.. 2005)
DST	Disuccinimidyl tartarate	6.4	(Green et al.. 2001; Sinz 2003)
DSG BS2G-d_0	Disuccinimidyl glutarate Bis[sulfosuccinimidyl] glutarate-d_0/ d_4	7.7	(Lee 2008; Green et al.. 2001)
BS3-d_0 DSS	Bis[sulfosuccinimidyl] suberate Disuccinimidyl suberate	11.4	(Green et al.. 2001; Sinz 2003)
DTSSP DSP	3,3´-Dithiobis[sulfosuccinimidylpropionate] Dithiobis[succinimidyl propionate]	12.0	(Lee 2008; Green et al.. 2001)
BSCOES	Bis[2-(succinimidooxycarbonyloxy)-ethyl]sulfone	13.0	(Green et al.. 2001)
EGS sulfo EGS	Ethylene glycol bis(succinimidylsuccinate) Ethylene-glycolbis(sulfosuccinimidylsuccinate)	16.1	(Green et al.. 2001; Sinz 2003)
BS(PEG)5	Bis(succinimidyl)penta(ethylene glycol)	21.7	(Hermanson 2008)
BS(PEG)9	Bis(succinimidyl)nona(ethylene glycol)	35.8	(Hermanson 2008)

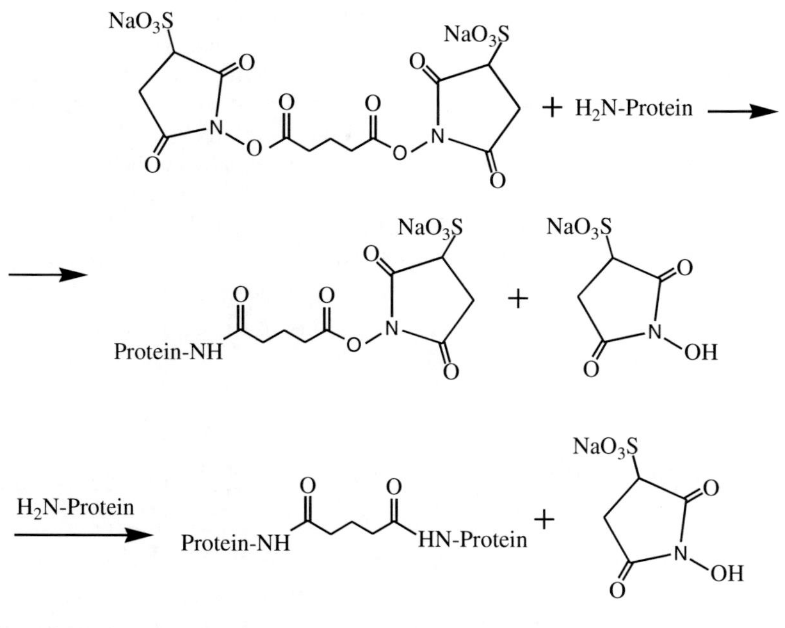

Figure 1. Reaction of N-hydroxysuccinimide (NHS) ester with amines. The NHS ester reacts with an amine group of a lysine residue from a protein creating an intrapeptide or an interpeptide cross-link.

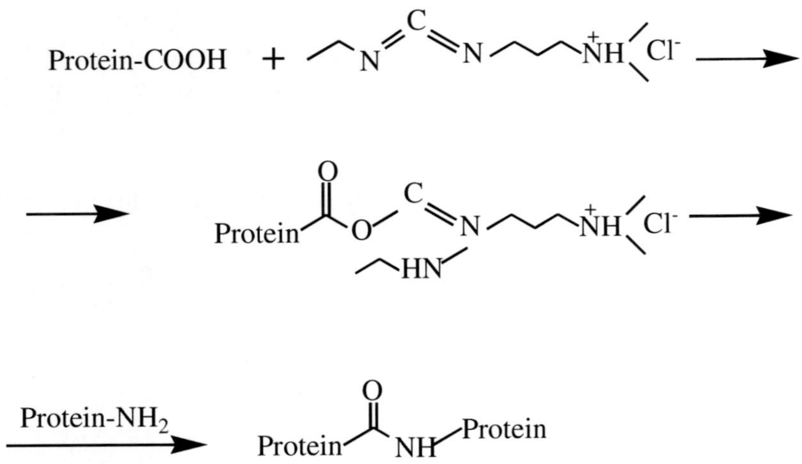

Figure 2. The zero length 1-Ethyl-3-(3-dimethylaminopropyl)carbodiimide (EDC) reaction. The common EDC reagent reacts with a carboxyl group on protein molecule 1, forming an amine-reactive O-acylisourea intermediate which in turn reacts with an amine on protein molecule 2, yielding a conjugate of the two molecules joined by a stable amide bond and unlike most other reagents leaves no chemical trace of the EDC reagent in the cross-linked product.

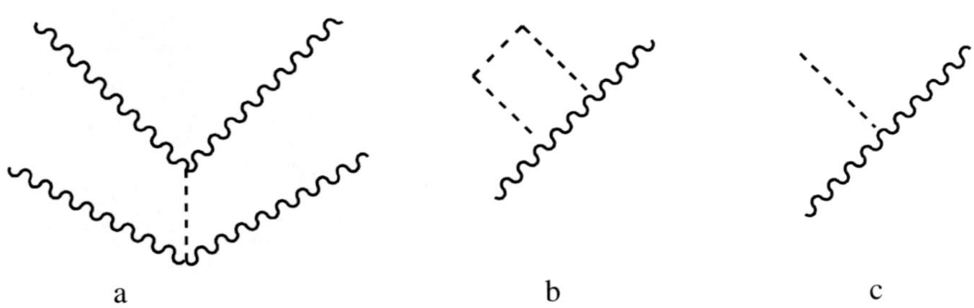

Figure 3. Nomenclature of common products of chemical cross-linking reactions. Typical products include: a) Type 2 or interpeptide crosslink connects two peptides, b) Type 1 or intrapeptide crosslink is within a single peptide, and c) Type 0 or dead end is a covalent bond that joins the crosslinking agent to a single peptide; (--) indicates a chemical crosslink and (—) indicates peptide chain.

The enormous complexity of the cross-linking reaction mixture makes identification of the cross-linked peptides difficult and requires support of robust computer programs to facilitate downstream analysis of the datasets. Although a large number of algorithms have been described to date, only those convenient for routine use in any biological laboratory are reviewed in the next section.

Bioinformatics in Cross-Link Identification

Broad and routine use of chemical cross-linking combined with mass spectrometry has been limited largely by the difficulty of identifying the sequences of cross-linked peptide(s) from their fragmentation spectra with high confidence. Extracting structural information from highly complex MS data obtained from cross-linked reaction mixtures is beyond the capacity of general, MS-based proteomic analysis software. Several computer programs have been developed for analysis of CXMS data. All major software tools with their applicability to different workflows and specific experimental requirements have recently been reviewed (Mayne and Patterton 2011).

The first generation of CXMS analytical software was based on precursor ion data, whereas the next generation had dual functionality based on use of precursor ion data to predict cross-linked peptide candidates and their validation by MS/MS (Gao et al. 2006; Seebacher et al. 2006). Finally, the latest generation of programs such as X!Link and xQuest, directly identify cross-linked peptides in tandem mass spectra (Rinner et al. 2008; Lee et al. 2007). Advantages and shortcomings of cross-linking software developed and used from 2000-2010 have been reviewed in detail (Singh et al. 2010; Lee 2008). In addition, new software pLink, for data analysis of cross-linked proteins compatible with multiple homo- or hetero-bifunctional cross-linkers, has recently been reported (Yang et al. 2012).

For samples of limited availability, a cross-linking strategy that utilizes data from a single LC-MS/MS run based on the open-modification search engine Popitam has been reported (Singh et al. 2008; Hernandez et al. 2003). The designed method can be used to map both chemically and naturally occurring cross-links in proteins. Popitam considers each peptide in the pair of cross-linked peptides to be post-translationally modified by the other peptide in the pair such that each has an unknown mass at an unknown amino acid. A peptide is identified as cross-linked if two sequence tags corresponding to the unique peptides are found in the same tandem mass spectrum matching the precursor mass. High mass accuracy of precursors and fragment ions, together with the specificity of derived sequence tag provide a small number of cross-linked candidates from a large data set, as was successfully demonstrated in analysis of complex CYP2E1-b5 (Singh, 2008) and a type III secretion system (T3SS) from *Samonella enterica* (Sanowar et al. 2010).

To facilitate further identification of cross-linked peptides and to make CXMS data analysis more generally applicable across laboratories, a database creation tool referred to as xComb has been developed (Panchaud et al. 2010). xComb strategy comprises four steps:

1 Creation of a database of theoretical cross-linked peptides created from a list of analyzed proteins,
2 Generation of a cross-linked peptides FASTA database with each entry corresponding to one cross-linking combination,
3 Construction of a peaklist file that considers greater than or equal to +4 precursors of only high mass accuracy, and
4 Search of tandem mass spectra against the cross-linked peptides database using any standard search engine (*e.g.,* Sequest, Mascot and Phenyx).

The FASTA database created by xComb may be used with any existing protein sequence database search software. This is convenient because the scoring schemas with which a user is already familiar enable immediate identification from tandem mass spectra of cross-linked peptide pairs using their preferred search engine. The search is highly specific, particularly with high mass accuracy data, and restricted to only cross-linked sequences. The fact that the database search takes place against the xComb generated FASTA files of peptides, not proteins, makes it exceptionally fast. In the end, publically available and easy to use, the xComb database processing tool is compatible with standard search engines and, thus, greatly facilitates CXMS by making data analysis rapid and routine.

CXMS and Comparative Modeling

As summarized by (Leitner et al. 2010), cross-linking data contain valuable information and have been used to:

1 Detect transient protein-protein interactions in a biological system,
2 Determine the sites of protein interaction in complex protein assemblies,
3 Construct protein-protein interaction networks,
4 Measure spatial distances between juxtaposed amino acids on the surface of folded proteins.

The first results, which showed that cross-linking experiments can provide low resolution interatomic distance information, were published almost forty years ago (Lutter and Kurland 1975; Clegg and Hayes 1974). Soon after that, a possibility to solve tertiary structures of proteins, based on enough distance information, was indicated (Cohen and Sternberg 1980). However, a rapid method for identification of cross-linked residues was lacking until recent advances in MS and an extensive development of proteomics technology.

As of January 2013, the number of deposited structures of proteins and protein complexes in the PDB Current Holdings Breakdown was reported to be 87,838, significantly lower than the number of known protein sequences (538,849 according to UniProtKB/Swiss-Prot protein knowledgebase release 2013_01 statistics). Thus, three-dimensional structures of most proteins are still unknown. However, only a few thousand overall protein structures or folds are estimated to exist, so predicting and refining structure based on solved ones, using molecular modeling, is considered a realistic goal (Back et al. 2003). Nevertheless, experimental validation of proposed models is necessary to improve modeling accuracy. In that sense, the use of experimentally generated distance constraints obtained by CXMS is recognized as an emerging approach to validate protein structures generated by computational methods and to address the gap in protein structure identification. Relatively recently, cross-linking of a native protein, bovine basic fibroblast growth factor (FGF)-2, was reported to provide in a single experiment, a significant number of intermolecular cross-links that were interpreted as distance constraints and used to reject poor and select a correct structural model from a set of homology based models (Young et al. 2000). Later on, by testing several cross-linkers with a range of arm length, chemical cross-linking was confirmed to indeed provide reliable three-dimensional structural information on dark-state bovine rhodopsin (Jacobsen et al. 2006). CXMS associated with molecular modeling was shown to be powerful, even when only partial structural information with a limited number of approximate distance constraints was available (Mouradov et al. 2008).

A set of rules to select between numerous three-dimensional protein models using CXMS is now well developed. From an early implementation of CXMS, (Havel et al.. 1979) proposed the following rules to accurately select among protein models: 1) numerous imprecise distance constraints are preferred over a few constraints with precise distance information, 2) distance constraints from remote residues in the primary sequence are preferred over sequentially close residues, and 3) distance constraints should involve as many different residues as possible. Soon after that study, several new observations related to accuracy of distance information, as well as quality of modeling software were reported by (Cohen and Sternberg 1980). Recently, Leitner et al. have further addressed the number of experimentally observed cross-links necessary for a reasonable prediction of protein complex topology (Leitner et al. 2010). Analysis of a large benchmark data set validated the conclusion that even for less accurate distance constraints three or four cross-links could give adequate information using appropriate modeling software (Leitner et al. 2010). In agreement with this conclusion, several models of protein complexes based on a low number of distance constraints have been successfully proposed to date. Included in these models is one developed by Tang and Clore, who showed that the prediction of the topology of two bacterial phosphotransferase complexes was possible with only three intermolecular cross-link constraints (Tang and Clore 2006). Likewise, Schulz et al. managed to predict a model of an annexin complex with only four constraints (Schulz et al. 2007). Furthermore, a complex of the mitogen-activated protein kinase pathway enzyme ERK2 and its regulator PTP-SL was modeled on three constraints (Balasu et al. 2009). All of these results indicate that identification of even a low number of cross-linking distance constraints can essentially facilitate probing the structures of multiprotein complexes. Thus, acquisition of cross-linking distance constraints of protein assemblies became a realistic method to be used on a routine basis to improve modeling accuracy (Leitner et al. 2010). A recently published interpretation of the in solution structure of an activating immune receptor, C-type lectin-like domain of NKR-P1C, using chemical cross-linking and molecular modeling indicates that this rapid strategy could be useful to differentiate solution structures from the crystal structures of proteins (Rozbesky et al. 2013).

Investigation of Multiprotein Complexes by Methods of Structural Proteomics

Many protein complexes have molecular masses in the mega-Dalton range and are not amenable to X-ray crystallographic or NMR analysis but, in some cases, can be analyzed by lower resolution methods such as electron microscopy. On the other hand, these low resolution methods are not applicable on smaller protein complexes. Thus, in spite of great progress of proteomics technology in protein structure determination, analyses of multiprotein complexes are still challenging. However, for the last ten years, several protein complexes have been successfully studied by chemical cross-linking and mass spectrometry, indicating a high potential for the CXMS toolbox in the structural determination of protein complexes (Rappsilber 2011; Rappsilber 2012). Within the last ten years, multiprotein complexes analyzed by CXMS include the Ffh.FtsY complex (Chu et al. 2004), Hsp90 chaperone GRP94 (Chu et al. 2006), a heterotetrameric Ndc80 complex (Maiolica et al. 2007), the annexin A2/p11 complex (Schulz et al. 2007), the calmodulin–Munc13 complex (Dimova et al. 2009), the phi29 connector/scaffolding complex (Fu et al. 2010), the GroEL–GroES chaperonin complex (Trnka and Burlingame 2010), RNA polymerase II in complex with transcription factor TF (Chen et al. 2010), a T3SS needle complex (Sanowar et al. 2010), chaperonin TriC/CCT (Leitner et al. 2012), and a 20S proteasomal core particle (Karadzic et al. 2012). Mapping structural details of the listed functional complexes in solution by CXMS have been used for different individual applications, but typically are used to support homology modeling. Furthermore, an intention and need to develop a fast and accurate CXMS method to generate distance constraints between juxtaposed amino acids and to validate molecular models has been documented (Karadzic et al. 2012; Kao et al. 2011).

Software to Incorporate CXMS Distance Constraints for Validation of Theoretical Models

As already stated earlier, distance determinations obtained by the CXMS technique can lead to important advances in mapping the protein topology in low resolution structures refined by computational methods. *Ab initio* and comparative modeling of protein complexes result in a number of possible structures. Although the number of actual proteins is vast, most proteins belong to a limited set of tertiary structural motifs (Back et al. 2003; Mouradov et al. 2008). Thus, comparative protein modeling, which uses previously solved structures/ templates as starting point, can be very effective in providing a preliminary estimate of protein structure. Distance constraints generated by CXMS can facilitate the evaluation and/or construction of the modeled protein structure. Alongside with growing experimental data related to validation of protein models, web-accessible resources that assist in analyzing protein structures and structural models and evaluating their reliability become particularly important. In that sense, a publically available ModBase database that allows users to request modeling of certain sequences, as well as to update comparative models on demand, meets the highest criterion and needs of the scientific community to easily access, download and further analyze protein models.

Thanks to the ModBase tool, any lab in the world now has a chance to obtain a structural model of a protein with known amino acid sequence in a simple and fast approach. Albeit complex, comparative modeling is comprised of four sequential steps: i) identifying known structures (templates) related to the sequence to be modeled (target), ii) aligning the target sequence with the templates, iii) building models and iv) assessing the models (Pieper et al. 2011). We next review four published software solutions to the problem of modeling using CXMS data.

MODBASE. ModBase (http://salilab.org/modbase) is database of annotated comparative protein structure models. The base currently contains over 10 million reliable models for domains in over 2 million proteins with unique sequences (Pieper et al. 2011). ModBase models are available through the ModBase interface as well as the Protein Model Portal (http://www.proteinmodelportal.org/). The models are calculated by ModPipe, an automated modeling pipeline that relies on Modeller (Eswar et al. 2006) for fold assignment, sequence–structure alignment, model building and model assessment. By default, for each target–template alignment, 10 models are calculated, and the model with the best value of the Discrete Optimized Protein Energy (DOPE) statistical potential (Shen and Sali 2006) is selected and evaluated by several additional quality criteria (Pieper et al. 2011).

MSX-3D. *In silico* generated models need to be efficiently validated and filtered by experimental data. Molecular modeling is complicated for numerous reasons. In particular, approximations needed to generate a theoretical model can be too numerous, protein structure for close homologs may be unknown, the target may be distantly related to known proteins, and the template proteins sharing the same fold may be lacking. For its analytical sensitivity, accuracy and speed, CXMS data appears to be the method of choice to validate structures obtained by modeling. The MSX-3D software platform is a tool specifically designed to confirm theoretical models using CXMS data (Heymann et al. 2008). The tool allows for classical peptide identification in addition to an interactive 3D visualization of the distance constraints obtained by cross-linking experiments. An important point for the scientific community is that the tool is freely available. By using MSX-3D, observed cross-links can be easily and rapidly compared with the linkages predicted from homology modeled structures and peptide candidates that fit the length of used cross-linker spacer arm can be identified.

Functionally relevant biological nanomachines of protein structure are often macromolecular complexes composed of protein monomers. By contrast, most structures deposited in PDB are single chain proteins. Multimeric protein complexes are usually difficult to isolate and crystalize, so very often structural information is obtained from subcomplexes with a limited number of monomers. Because of such difficulties and limits, computational methods for predicting structures of protein complexes are of a great importance. To predict the structure of protein complexes, protein-protein docking methods defined within a Rosetta protein modeling suite have become very popular (Kaufmann et al. 2010), and the efficiency of this type of application has been demonstrated (Sanowar et al. 2010). Furthermore, the positions, orientations, and conformations of the Rpn subunits and AAA-ATPase hexamer of Rpt subunits of 26S proteasomes, was determined using an open source Integrative Modeling Platform (IMP) package (http://salilab.org/imp/) (Lasker et al. 2010; Alber et al. 2007). The approach was also used to determine the positions of 456 subunits in the eightfold symmetric ring of the yeast nuclear pore complex (Alber et al. 2007; Alber et al. 2007).

As an alternative approach, prediction based on comparison with known strucures of protein complexes based on the Dali method (Holm and Park 2000) has recently been reported (Karadzic et al. 2012). In brief, DaliLite was designed as a standalone package of the Dali method to compare all chains of investigated structure against all chains in the known structure, generating superimposed coordinate files that can be downloaded or viewed interactively (Holm et al. 2008). As a result of comparison, DaliLite generates atomic coordinates of an entire complex, allowing smaller protein subunits to be properly assembled in the final complex structure.

DaliLite. DaliLite software is a freely available network service for comparing protein structures (http://www.ebi.ac.uk/Tools/structure/dalilite/). The Dali method was routinely used to scan and to compare newly solved structures against those in the PDB, or to compare predicted structures to the X-ray crystal structure based on sensitive measure of geometrical similarity between protein structures (Holm and Park 2000). Comparing 3D structures may sometimes reveal biologically interesting similarities that are not detectable by comparing primary and secondary amino acid sequences. Aside from the described usage, DaliLite can be used to construct protein assembles from subunits. The principle Holm and Park (2000) used to create a complex assembly from subunits is simple. Atomic coordinates of every single monomer from protein assembly are compared against coordinates of a chosen monomer from known complex protein structure in the Protein Data Bank (PDB), to generate coordinates of single monomer in the complex structures. A strong match between analyzed subunits and templates is considered to assemble protein subunits into a complex. The last step, thus, requires molecular modeling software to manage, display, and analyze the three dimensional structure of the complexes whose coordinates are generated by DaliLite.

Vega ZZ. While the computational chemistry and some important molecular modeling packages are still complicated, an easy-to-use Windows-based molecular modeling software with a complete 3D graphic interface has been developed to do the most common molecular modeling operations in a fast and simple way. VEGA ZZ software is free to download and easy to use on a common PC. In spite of its relative simplicity, VEGA ZZ is a complete molecular modeling software, which can be successfully used to solve many computational chemistry problems including homology modeling of proteins, as well as creation of protein complexes (Pedretti et al. 2002).

Main features of described bioinformatics tools for analysis of data from cross-linking experiments and for validation of theoretical models based on CXMS distance constraints are summarized in Table 2.

Table 2. Main features of bioinformatics tools for cross-linking data analysis and for validation of molecular models based on CXMS distance constraints

Bioinformatics tool	Main features	Reference
Cross-link identification		
xQuest	Software that directly identifies cross-linked peptides in MS[1]and tandem mass spectra; can search large sequence databases	(Rinner et al. 2008)
Popitam	Open mass modification search software that uses tandem mass spectra, can map both chemically and naturally occurring cross-links in proteins	(Hernandez et al. 2003)
xComb	Database creation tool compatible with standard tandem mass spectra search engines	(Panchaud et al. 2010)
CXMS and comparative modeling		
ModBase	Database that allows users to request modeling of certain protein sequences	(Pieper et al. 2011)
MSX-3D	Software platform specifically designed to confirm theoretical models using CXMS data	(Heymann et al. 2008)
DaliLite	Software for scanning and comparing newly solved protein structures against those in the PDB, can be used to construct protein assembles from subunits	(Holm and Park 2000; Holm et al. 2008)
VEGA ZZ	Complete Windows-based molecular modeling software	(Pedretti et al. 2002)

No doubt, CXMS technology provides improved opportunities for modeling by mapping structural details of protein complexes in solution, but some efficient platforms to link CXMS and modeling are still lacking. Once the technological limits of CXMS are addressed, CXMS data is likely to be incorporated into modeling algorithms that will allow for faster development of integrated structural biology (Rappsilber 2011).

Case Study: The Multiprotein 20S Proteasome Core Particles of Haloarchaea

Archaea are one of three major evolutionary lineages of life. Recent advances in the genome sequencing of halophilic archaea (or haloarchaea) have made this group of microbes ideal for providing insight into cell physiology (Hartman *et* al. 2010; Leigh et al. 2011). Haloarchaea have developed into model organisms that are used to study many biological processes, including the function of the ubiquitin-proteasome system (Humbard et al. 2010). Proteasomes from archaea are highly related to those of eukaryotes. Hfx. volcanii is the only archaeon that has been demonstrated to synthesize three different proteasomal subtypes composed of various combinations of α1, α2 and β subunits [encoded by psmA (HVO_1091), psmC (HVO_2923) and psmB (HVO_1562), respectively] classified in the 20S proteasome superfamily (Wilson et al. 1999; Karadzic et al. 2012; Kaczowka and Maupin-Furlow 2003).

Proteasomes are self-compartmentalized proteinaceous nanomachines, composed of triple A-ATPases (Rpts) and 20S core particles (CPs), that are important in proteolysis and often essential for growth (Zhou et al. 2008). The proteasomal CP is a cylindrical complex of four-stacked heptameric rings of α- and β-type subunits organized in $\alpha_7\beta_7\beta_7\alpha_7$ stoichiometry. The *Hfx. volcanii* CPs are of a 600 kDa molecular mass corresponding to a 28 subunit complex of 14 α-type and 14 β-type subunits. The α-type subunits form the outer rings and the β-type subunits form the two inner rings that harbor the proteolytic active sites (Wilson et al. 1999; Kaczowka and Maupin-Furlow 2003). As *Hfx. volcanii* synthesizes two different α subunits (α1 and α2), this organism has the potential to make three different CPs: two symmetric (α1ββα1, α2ββα2) and one asymmetric (α1ββα2). Since the α1 and α2 proteins share only 55.5% identity, significant structural differences in the homoheptameric rings formed by α1 and α2 are predicted (Kaczowka and Maupin-Furlow 2003).

Structural characterization of eukaryotic proteasomes by CXMS has been recently published (Kao et al. 2011; Lasker et al. 2012). Kao et al. (2011) developed a method based on a newly designed MS-cleavable cross-linker, disuccinimidyl sulfoxide (DSSO), and an integrated use of MS, MS/MS, and MS^3 to characterize the yeast 20S proteasome complex. A total of 13 non-redundant interlinked peptides of the 20S proteasome were identified, representing the first application of an MS-cleavable cross-linker for the characterization of a multisubunit protein complex.

Lasker et al. (2012) determined the structure of the complete 26S proteasome from *Schizosaccharomyces pombe* using an integrative approach based on residue-specific intersubunit cross-links, 8.4-Å cryo-EM map, crystallographic structures of RP subunit homologs, and known protein–protein interaction data from a variety of species.

Amongst prokaryotic proteasomes, only the topology of CPs from haloarchaeon *Hfx. volcanii* has been determined by CXMS (Karadzic et al. 2012). A fast and accurate method allowed the generation of distance constraints between juxtaposed amino acids and the validation of molecular models of the halophilic proteasomal 20S core particles (CPs) from *Hfx. volcanii.* Chemically cross-linked subunits of the *Hfx. volcanii* CPs were analyzed by high-performance MS. Distance constraints obtained by CXMS, together with available structural data of non-halophilic CPs deposited in the RCSB protein databank, facilitated the selection of accurate models of *Hfx. volcanii* proteasomal CPs.

Specificity of Halophilic Haloarchaeal CP Nanomachines and Strategy of Its Investigation

Halophilic proteins are quite specific with regard to amino acid composition often having an extremely high content of acidic amino acids and unusual abundance of surface exposed acidic residues. With regard to CXMS investigation of *Hfx. volcanii* CPs, several points must be emphasized. First, haloarchaeal proteins are typically halophilic (salt-loving) comprising a high content of acidic residues that are often surface exposed and a low number of basic amino acid residues (*e.g.*, lysine, a target of many cross-linkers). In addition, *Hfx. volcanii* CPs do not contain any cysteine residues. All of these factors limit and complicate the investigation of the halophilic CPs by CXMS, since lysine and cysteine residues are often targets of commercial cross-linkers. Furthermore, only a few crystal structures of haloarchaeal proteins have been resolved to date, while proteasomal CPs have been derived for a few non-halophilic archaea and eukarya. PDB entries of non-halophilic archaeal proteasomes include: 1j2q of *Archaeoglobus fulgidus* (Groll et al. 2003), 3h4p of *Methanocaldococcus jannaschii* (Zhang et al. 2009), and 1pma of *Thermoplasma acidophilum* (Löwe et al. 1995). In spite of these limitations and in light of the above review of the field, CXMS coupled with protein modeling, has recently emerged as a powerful approach to predict the structure of these unusual, acidic proteins found in halophilic proteasomes which we will next review further.

Taking into account advantages of CXMS to elucidate the tertiary and quaternary structure of proteins, a symmetric α1ββα1 and α2ββα2 core particle of *Hfx. volcanii* proteasomes was studied to determine the juxtaposed amino acids in the CP and to validate the theoretical 3D protein structures predicted by comparative modeling (Karadzic et al. 2012). To analyze the two symmetric CP subtypes of *Hfx. volcanii* by CXMS the following relatively simple strategy was used. First, the complexes (α1ββα1 and α2ββα2) were treated with the homobifunctional lysine reactive bis(sulfosuccinimidyl) glutarate or BS2G and zero length EDC cross-linkers.

Tandem mass spectra were generated from tryptic peptides of the cross-linked CPs and analyzed to confirm the identity of the peptides detected and to evaluate the type of cross-linkages using an approach similar to Singh et al. (2008).

Briefly, according to this method, all data were first generated at high mass accuracy on LTQ-Orbitrap mass spectrometer, xComb was used to generate a cross-linking database from the MS/MS data that were uploaded, data were searched with respect to the type of cross-linker with Phenyx (Panchaud et al. 2010), and, finally, the tandem mass spectra of cross-linked peptides were annotated using MS2Assign (Schilling et al. 2003).

According to the classification of cross-linking products interpeptide, intrapeptide and dead-end modifications were considered. The cross-linked peptides of the α1ββα1 and α2ββα2 CP subtype that were identified after treatment with BS2G and EDC are listed in Table 3 [for further details see (Karadzic et al. 2012)].

Table 3. Cross-linked peptides of α1ββα1 and α2ββα2 proteasomal CPs identified after treatment with homobifunctional lysine reactive BS2G and zero-length EDC

Measured precursor mass (Da)	Charge state	MMA (ppm)	Crosslinker	Subunit, aa residue no.	Sequence 1 — Sequence 2	Cross-linked sites	Distance intra-molecular (Å)[a]	Distance inter-molecular (Å)[b]
α1ββα1 CP:								
2945.56	5	7.6	BS2G	α1, 44-55 α1, 58-70	TPEGVVLAADKR — SPLMEPTSVEKIHK	K54-K68	9.4	37.1 (1j2p) 37.8 (1fnt) 36.3 (1ryp)
2453.28	4	5.7	EDC	α1,44-55 α1, 58-68	TPEGVVLAADKR — SPLMEPTSVEK	K54-E67	6.9	37.9 (1j2p) 36.9 (1fnt) 37.9 (1ryp)
3652.74	4	1.7	EDC	α1,150-171 α1, 58-68	LYETDPSGTPYEWKAVSIGADR—SPLMEPTSVEK	K163-E62	39.3	4.8 (1j2p)[b] 6.7 (1fnt) 9.7 (1ryp)
4418.12 (1 ox)	4	3.9	BS2G	β, 38-78, intrapeptide	ADELGDKETKTGTTTVGIKTEEGVVLATDMRASMGYMVSSK	K44-K47	6.1	/
2072.03	4	4.9	EDC	β, 69-82 β, 209-213	ASMGYMVSSKDVQK — SAVER	K78-E212	30.9	29.4/4.6[c] (1fnt) 16.4 (1j2q) 26.5/22.1[c] (1ryp)
α2ββα2 CP:								
4164.12	4	2.4	EDC	α2,116-148 α2, 88-94	TITDNIQESTQSGGTRPYGASLLIGGVENGSGR — KLVDFAR	K88- D119	18.5	3.1 (1j2p) 4.8 (1fnt) 5.5 (1ryp)
2115.06	4	3.5	EDC	α2, 68-88, intrapeptide	LHKLDDALGAATAGHVADARK	K70-D72	3.2	/
4418.12 (1 ox)	4	4.2	BS2G	β, 38-78, intrapeptide	ADELGDKETKTGTTTVGIKTEEGVVLATDMRASMGYMVSSK	K44-K47	6.1	/
2072.03	4	4.9	EDC	β, 69-82 β, 209-213	ASMGYMVSSKDVQK — SAVER	K78-E212	30.9	29.4/4.6[c] (1fnt) 16.4 (1j2q) 26.5/22.1[c] (1ryp)

[a]Distance between cross-linked amino acid residues within the subunit calculated by MSX-3D.

[b]Intermolecular distance of cross-linked amino acid residues between subunits calculated as average using modeled structures based on 1j2p, 1fnt, 1ryp, and 1j2q PDB templates.

[c]Intermolecular distance between cross-linked amino acid residues from subunits within one ring β7/distance between cross-linked amino acid residues from subunits within two rings β7β7, calculated as average using modeled structures based on 1j2p, 1fnt, 1ryp, and 1j2q PDB templates.

Use of Cross-Linking Distance Constraints to Optimize Models of Halophilic Multisubunit Complexes on a Routine Basis

Distance constraints obtained by CXMS were used for validation of theoretical models from ModBase of α1, α2 and β subunits of *Hfx. volcanii* 20S proteasomal CPs (Karadzic et al.. 2012). MSX-3D and the VEGA ZZ programs were used to display, analyze, and manage the three dimensional (3D) structure of the protein complexes. The same tools were used to validate 3D structures of complexes of α1, α2 and β subunits (α1ββα1 and α2ββα2 CP subtypes) obtained by homology modeling according to the PDB templates: 1j2p (α-ring of the proteasome from *A. fulgidus*), 1ryp (20S proteasome from the yeast *Saccharomyces cerevisiae*), and 1fnt (20S proteasome from the yeast *S. cerevisiae* in complex with the proteasomes activator PA26 from *Trypanosome brucei*), Figure 4 Based on biochemical suppositions a stoichiometry of seven α1 and β subunits assembled in homoheptameric rings was considered, and the coordinates of this α- and β-ring were obtained by DaliLite software. The heptameric α1-ring was best modeled using 1j2p as a template (Figure 4a and b), giving the DaliLite Z-score of 41 with a sequence identity of 51% and suggesting a high geometrical similarity between the *Hfx. volcanii* α1 homoheptameric ring and template 1j2p. The homoheptameric ring of α2 modeled on coordinates of the 1j2p, 1fnt, and 1ryp PDB structures (as above for α1) gave the best Z-score of 29.3 and the highest sequence identity of 42 %, for the 1j2p structure, too. Using both cross-linkers, the same interpeptide and intrapeptide cross-linkages of subunit β were identified in both the α1ββα1 and α2ββα2 CP subtypes (Table 3). Intermolecular distances in the homoheptameric β_7-ring and doublet of $\beta_7\beta_7$-rings, modeled according to the 1ryp, 1fnt, and 1j2q (20S proteasome from *A. fulgidus*) PDB templates, were calculated and found to differ significantly suggesting the model based on 1fnt was more accurate for the packing of the $\beta_7\beta_7$-rings than the models based on 1ryp and 1j2q (Figure 4c).

Conclusion

CXMS provides a powerful, and with recent advances in software and instrumentation, a relatively facile approach for analysis of protein structure. These methods offer the biochemist tremendous opportunities to hypothesize structures from relatively few constraints and when combined with modeling even to confirm or refute the proposed structures. Over the past ten years, the size and complexity of the proteins investigated with cross-linking and mass-spectrometry have increased. Based on CXMS the architecture of multiprotein complexes of 12 subunits (Pol II complex), 15 subunits (Pol II-TFIIF complex), 21 subunits (GroEL-GroES chaperonin complex) and even 28 subunits (CPs from *Hfx. volcanii*) were successfully resolved. However, until recently CXMS was performed only for non-halophilic (*i.e.*, non-acidic) proteins of relatively neutral pI.

Using the newly available tools described herein, CXMS analysis of halophilic proteins when coupled with atomic structural data (determined experimentally for non-halophilic protein homologs) provided useful distance constraints for modeling the 3D structures of extremely acidic and halophilic protein complexes of the proteasomal CPs of *Hfx. volcanii*, an archaeon isolated from the Dead Sea. As cross-linking reactions were carried out, under conditions that are close to physiological (neutral pH and a high ionic strength), the results show that the set of distance constraints generated by a CXMS approach can be a tool to fine-tune existing models or to create accurate structures of multiprotein complexes in solution, including specific, high saline solutions. Specifically, the α1ββα1 and α2ββα2 CPs of *Hfx. volcanii* are presumed to be associated as four-stacked heptameric rings in an $\alpha_7\beta_7\beta_7\alpha_7$ symmetry based on analogy to non-halophilic archaeal CPs for which detailed X-ray crystal structures are known. Several observed cross-links were used to validate the predicted 3D-structures of the *Hfx. volcanii* proteasomal α- and β-type subunits, and intermolecular distance constraints were used to assemble these proteins (α1, α2, and β) into the quaternary structure of the CP complexes. Distance constraints obtained by CXMS have facilitated selection of an accurate model from several possible PDB proteasomal models and assisted in the determination of the arrangement of the α1- and β- homoheptameric rings in the complex protein structures.

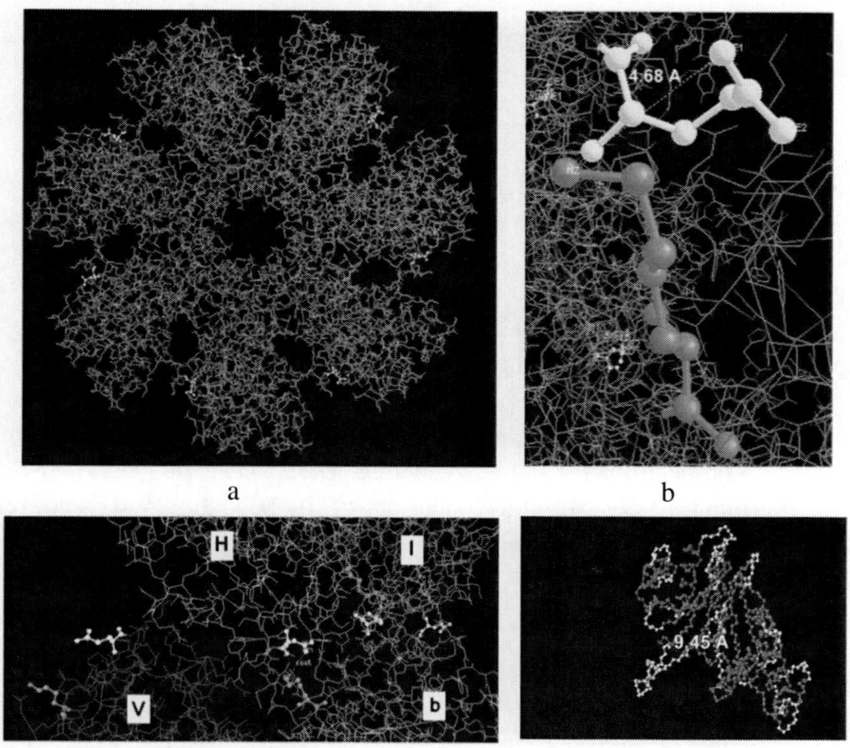

Figure 4. Structural models of the $\alpha1_7$-ring and $\beta_7\beta_7$-rings of *Hfx. volcanii* 20S core particles (CPs). Structural models shown are: a) heptameric $\alpha1_7$-ring modeled according 1j2p PDB template, b) distances between the NZ of K163 and OE1 of E-62 from the two α1 subunits estimated at 4.7 Å, c) $\beta_7\beta_7$-ring modeled according to the PDB template 1fnt, as H, I, V, and b. The measured distance between K-78 from the b chain and E-212 from the H chain was 4.64 Å, d) MSX-3D predicts that lysine K54 from TPEGVVLAADKRSR cross-linked with lysine K68 from SPLMEPTSVEKIHK is at a distance of 9.45 Å in the model structure of the α1 subunit (see Table 3).

Finally, given its efficiency, accuracy and simplicity, the CXMS strategy reported for the CPs of *Hfx. volcanii* can be applied by the broad scientific community for structural elucidation of even these less commonly studied protein complexes in a rapid and routine way available.

Acknowledgments

Thank you to the editors and reviewers for their critical evaluation of the manuscript. We apologize for not citing many important references that have contributed to this review owing to space limitations. Work is funded in part by grants awarded to JM-F from the National Institutes of Health (GM57498) and the Division of Chemical Sciences, Geosciences and Biosciences, Office of Basic Energy Sciences of the US Department of Energy (DE-FG02-05ER15650) and to IK from Ministry of Education and Science of the Republic of Serbia (Project No. III 43004).

References

Alber, F., Dokudovskaya, S., Veenhoff, L. M., Zhang, W., Kipper, J., Devos, D., Suprapto, A., Karni-Schmidt, O., Williams, R., Chait, B. T., Rout, M. P., Sali, A. (2007) Determining the architectures of macromolecular assemblies. *Nature* 450(7170):683-694.

Alber, F., Dokudovskaya, S., Veenhoff, L. M., Zhang, W., Kipper, J., Devos, D., Suprapto, A., Karni-Schmidt, O., Williams, R., Chait, B. T., Sali, A., Rout, M. P. (2007) The molecular architecture of the nuclear pore complex. *Nature* 450(7170):695-701.

Back, J. W., de Jong, L., Muijsers, A. O., de Koster, C. G. (2003) Chemical cross-linking and mass spectrometry for protein structural modeling. *J. Mol. Biol.* 331(2):303-313.

Back, J. W., Notenboom, V., de Koning, L. J., Muijsers, A. O., Sixma, T. K., de Koster, C. G., de Jong, L. (2002) Identification of cross-linked peptides for protein interaction studies using mass spectrometry and 18O labeling. *Anal. Chem.* 74(17):4417-4422.

Balasu, M. C., Spiridon, L. N., Miron, S., Craescu, C. T., Scheidig, A. J., Petrescu, A. J., Szedlacsek, S. E. (2009) Interface analysis of the complex between ERK2 and PTP-SL. *PLoS One* 4(5):e5432.

Betts, G. N., van der Geer, P., Komives, E. A. (2008) Structural and functional consequences of tyrosine phosphorylation in the LRP1 cytoplasmic domain. *J. Biol. Chem.* 283(23): 15656-15664.

Breuker, K., Jin, M., Han, X., Jiang, H., McLafferty, F. W. (2008) Top-down identification and characterization of biomolecules by mass spectrometry. *J. Am. Soc. Mass. Spectrom.* 19(8):1045-1053.

Brock, A. (2012) Fragmentation hydrogen exchange mass spectrometry: a review of methodology and applications. *Protein Expr. Purif.* 84(1):19-37.

Chen, Z. A., Jawhari, A., Fischer, L., Buchen, C., Tahir, S., Kamenski, T., Rasmussen, M., Lariviere, L., Bukowski-Wills, J. C., Nilges, M., Cramer, P., Rappsilber, J. (2010) Architecture of the RNA polymerase II-TFIIF complex revealed by cross-linking and mass spectrometry. *EMBO J.* 29(4):717-726.

Chu, F., Maynard, J. C., Chiosis, G., Nicchitta, C. V., Burlingame, A. L. (2006) Identification of novel quaternary domain interactions in the Hsp90 chaperone, GRP94. *Protein Sci.* 15 (6):1260-1269.

Chu, F., Shan, S. O., Moustakas, D. T., Alber, F., Egea, P. F., Stroud, R. M., Walter, P., Burlingame, A. L. (2004) Unraveling the interface of signal recognition particle and its receptor by using chemical cross-linking and tandem mass spectrometry. *Proc. Natl. Acad. Sci. US* 101(47):16454-16459.

Clegg, C., Hayes, D. (1974) Identification of neighbouring proteins in the ribosomes of *Escherichia coli.* A topographical study with the cross-linking reagent dimethyl suberimidate. *Eur. J. Biochem.* 42(1):21-28.

Cohen, F. E., Sternberg, M. J. (1980) On the use of chemically derived distance constraints in the prediction of protein structure with myoglobin as an example. *J. Mol. Biol.* 137(1):9-22.

Dimova, K., Kalkhof, S., Pottratz, I., Ihling, C., Rodriguez-Castaneda, F., Liepold, T., Griesinger, C., Brose, N., Sinz, A., Jahn, O. (2009) Structural insights into the calmodulin-Munc13 interaction obtained by cross-linking and mass spectrometry. *Biochemistry* 48(25):5908-5921.

Eswar, N., Webb, B., Marti-Renom, M. A., Madhusudhan, M. S., Eramian, D., Shen, M. Y., Pieper, U., Sali, A. (2006) Comparative protein structure modeling using Modeller. *Curr. Protoc. Bioinformatics* Chapter 5 Unit 5.6.

Fu, C. Y., Uetrecht, C., Kang, S., Morais, M. C., Heck, A. J., Walter, M. R., Prevelige, P. E. (2010) A docking model based on mass spectrometric and biochemical data describes phage packaging motor incorporation. *Mol. Cell. Proteomics* 9(8):1764-1773.

Gao, Q., Xue, S., Doneanu, C. E., Shaffer, S. A., Goodlett, D. R., Nelson, S. D. (2006) Pro-CrossLink. Software tool for protein cross-linking and mass spectrometry. *Anal. Chem.* 78(7):2145-2149.

Green, N. S., Reisler, E., Houk, K. N. (2001) Quantitative evaluation of the lengths of homobifunctional protein cross-linking reagents used as molecular rulers. *Protein Sci.* 10 (7):1293-1304.

Groll, M., Brandstetter, H., Bartunik, H., Bourenkow, G., Huber, R. (2003) Investigations on the maturation and regulation of archaebacterial proteasomes. *J. Mol. Biol.* 327(1):75-83.

Hartman, A., Norais, C., Badger, J., Delmas, S., Haldenby, S., Madupu, R., Robinson, J., Khouri, H., Ren, Q., Lowe, T., Maupin-Furlow, J., Pohlschroder, M., Daniels, C., Pfeiffer, F., Allers, T., Eisen, J. (2010) The complete genome sequence of *Haloferax volcanii* DS2, a model archaeon. *Plos One* e9605.

Havel, T., Crippen, G., Kuntz, I. (1979) Effects of distance constraints on macromolecular conformation. II. Simulation of experimental results and theoretical predictions. *Biopolymers* 18(1):73-81.

Hermanson, G. T. (2008) *Bioconjugate Techniques.* Academic Press (an imprint of Elsevier), US.

Hernandez, P., Gras, R., Frey, J., Appel, R. D. (2003) Popitam: towards new heuristic strategies to improve protein identification from tandem mass spectrometry data. *Proteomics* 3(6):870-878.

Heymann, M., Paramelle, D., Subra, G., Forest, E., Martinez, J., Geourjon, C., Deléage, G. (2008) MSX-3D: a tool to validate 3D protein models using mass spectrometry. *Bioinformatics* 24(23):2782-2783.

Holm, L., Kääriäinen, S., Rosenström, P., Schenkel, A. (2008) Searching protein structure databases with DaliLite v.3. *Bioinformatics* 24(23):2780-2781.

Holm, L., Park, J. (2000) DaliLite workbench for protein structure comparison. *Bioinformatics* 16(6):566-567.

Humbard, M., Miranda, H., Lim, J., Krause, D., Pritz, J., Zhou, G., Chen, S., Wells, L., Maupin-Furlow, J. (2010) Ubiquitin-like small archaeal modifier proteins (SAMPs) in *Haloferax volcanii*. *Nature* 54-60.

Jacobsen, R. B., Sale, K. L., Ayson, M. J., Novak, P., Hong, J., Lane, P., Wood, N. L., Kruppa, G. H., Young, M. M., Schoeniger, J. S. (2006) Structure and dynamics of dark-state bovine rhodopsin revealed by chemical cross-linking and high-resolution mass spectrometry. *Protein Sci.* 15(6):1303-1317.

Kaczowka, S., Maupin-Furlow, J. (2003) Subunit topology of two 20S proteasomes from *Haloferax volcanii*. *J. Bacteriol.* 185(1):165-174.

Karadzic, I., Maupin-Furlow, J., Humbard, M., Prunetti, L., Singh, P., Goodlett, D. R. (2012) Chemical cross-linking, mass spectrometry, and *in silico* modeling of proteasomal 20S core particles of the haloarchaeon *Haloferax volcanii*. *Proteomics* 12(11):1806-1814.

Kao, A., Chiu, C. L., Vellucci, D., Yang, Y., Patel, V. R., Guan, S., Randall, A., Baldi, P., Rychnovsky, S. D., Huang, L. (2011) Development of a novel cross-linking strategy for fast and accurate identification of cross-linked peptides of protein complexes. *Mol. Cell. Proteomics* 10(1):M110.002212.

Kaufmann, K. W., Lemmon, G. H., Deluca, S. L., Sheehan, J. H., Meiler, J. (2010) Practically useful: what the Rosetta protein modeling suite can do for you. *Biochemistry* 49(14): 2987-2998.

Konermann, L., Simmons, D. A. (2003) Protein-folding kinetics and mechanisms studied by pulse-labeling and mass spectrometry. *Mass. Spectrom. Rev.* 22(1):1-26.

Lasker, K., Förster, F., Bohn, S., Walzthoeni, T., Villa, E., Unverdorben, P., Beck, F., Aebersold, R., Sali, A., Baumeister, W. (2012) Molecular architecture of the 26S proteasome holocomplex determined by an integrative approach. *Proc. Natl. Acad. Sci. US* 109(5):1380-1387.

Lasker, K., Phillips, J. L., Russel, D., Velázquez-Muriel, J., Schneidman-Duhovny, D., Tjioe, E., Webb, B., Schlessinger, A., Sali, A. (2010) Integrative structure modeling of macromolecular assemblies from proteomics data. *Mol. Cell. Proteomics* 9(8):1689-1702.

Lee, Y. J. (2008) Mass spectrometric analysis of cross-linking sites for the structure of proteins and protein complexes. *Mol. Biosyst.* 4(8):816-823.

Lee, Y. J., Lackner, L. L., Nunnari, J. M., Phinney, B. S. (2007) Shotgun cross-linking analysis for studying quaternary and tertiary protein structures. *J. Proteome. Res.* 6(10): 3908-3917.

Leigh, J. A., Albers, S. V., Atomi, H., Allers, T. (2011) Model organisms for genetics in the domain Archaea: methanogens, halophiles, *Thermococcales* and *Sulfolobales*. *FEMS Microbiol. Rev.* 35(4):577-608.

Leitner, A., Joachimiak, L. A., Bracher, A., Mönkemeyer, L., Walzthoeni, T., Chen, B., Pechmann, S., Holmes, S., Cong, Y., Ma, B., Ludtke, S., Chiu, W., Hartl, F. U., Aebersold, R., Frydman, J. (2012) The molecular architecture of the eukaryotic chaperonin TRiC/CCT. *Structure* 20(5):814-825.

Leitner, A., Walzthoeni, T., Kahraman, A., Herzog, F., Rinner, O., Beck, M., Aebersold, R. (2010) Probing native protein structures by chemical cross-linking, mass spectrometry, and bioinformatics. *Mol. Cell. Proteomics* 9(8):1634-1649.

Li, H., Lin, T. Y., Van Orden, S. L., Zhao, Y., Barrow, M. P., Pizarro, A. M., Qi, Y., Sadler, P. J., O'Connor, P. B. (2011) Use of top-down and bottom-up Fourier transform ion cyclotron resonance mass spectrometry for mapping calmodulin sites modified by platinum anticancer drugs. *Anal. Chem.* 83(24):9507-9515.

Ling, J. M., Shima, C. H., Schriemer, D. C., Schryvers, A. B. (2010) Delineating the regions of human transferrin involved in interactions with transferrin binding protein B from Neisseria meningitidis. *Mol. Microbiol.* 77(5):1301-1314.

Löwe, J., Stock, D., Jap, B., Zwickl, P., Baumeister, W., Huber, R. (1995) Crystal structure of the 20S proteasome from the archaeon *T. acidophilum* at 3.4 Å resolution. *Science* 268 (5210):533-539.

Lutter, L. C., Kurland, C. G. (1975) Chemical determination of protein neighbourhoods in a cellular organelle. *Mol. Cell. Biochem.* 7(2):105-116.

Mädler, S., Bich, C., Touboul, D., Zenobi, R. (2009) Chemical cross-linking with NHS esters: a systematic study on amino acid reactivities. *J. Mass. Spectrom.* 44(5):694-706.

Maiolica, A., Cittaro, D., Borsotti, D., Sennels, L., Ciferri, C., Tarricone, C., Musacchio, A., Rappsilber, J. (2007) Structural analysis of multiprotein complexes by cross-linking, mass spectrometry, and database searching. *Mol. Cell. Proteomics* 6(12):2200-2211.

Mayne, S. L., Patterton, H. G. (2011) Bioinformatics tools for the structural elucidation of multi-subunit protein complexes by mass spectrometric analysis of protein-protein cross-links. *Brief Bioinform.* 12(6):660-671.

Mouradov, D., King, G., Ross, I. L., Forwood, J. K., Hume, D. A., Sinz, A., Martin, J. L., Kobe, B., Huber, T. (2008) Protein structure determination using a combination of cross-linking, mass spectrometry, and molecular modeling. *Methods Mol. Biol.* 426459-474.

Nadeau, O. W., Lane, L. A., Xu, D., Sage, J., Priddy, T. S., Artigues, A., Villar, M. T., Yang, Q., Robinson, C. V., Zhang, Y., Carlson, G. M. (2012) Structure and location of the regulatory β subunits in the (αβγδ)4 phosphorylase kinase complex. *J. Biol. Chem.* 287 (44):36651-36661.

Novak, P., Young, M. M., Schoeniger, J. S., Kruppa, G. H. (2003) A top-down approach to protein structure studies using chemical cross-linking and Fourier transform mass spectrometry. *Eur. J. Mass. Spectrom.* (Chichester, Eng.) 9(6):623-631.

Nowak, D. E., Tian, B., Brasier, A. R. (2005) Two-step cross-linking method for identification of NF-kappaB gene network by chromatin immunoprecipitation. *Biotechniques* 39(5):715-725.

Panchaud, A., Singh, P., Shaffer, S. A., Goodlett, D. R. (2010) xComb: a cross-linked peptide database approach to protein-protein interaction analysis. *J. Proteome. Res.* 9(5):2508-2515.

Pantazatos, D., Kim, J. S., Klock, H. E., Stevens, R. C., Wilson, I. A., Lesley, S. A., Woods, V. L. (2004) Rapid refinement of crystallographic protein construct definition employing enhanced hydrogen/deuterium exchange MS. *Proc. Natl. Acad. Sci. US* 101(3):751-756.

Pedretti, A., Villa, L., Vistoli, G. (2002) VEGA: a versatile program to convert, handle and visualize molecular structure on Windows-based PCs. *J. Mol. Graph. Model* 21(1):47-49.

Petrotchenko, E. V., Borchers, C. H. (2010) Crosslinking combined with mass spectrometry for structural proteomics. *Mass. Spectrom. Rev.* 29(6):862-876.

Pieper, U., Webb, B. M., Barkan, D. T., Schneidman-Duhovny, D., Schlessinger, A., Braberg, H., Yang, Z., Meng, E. C., Pettersen, E. F., Huang, C. C., Datta, R. S., Sampathkumar, P., Madhusudhan, M. S., Sjölander, K., Ferrin, T. E., Burley, S. K., Sali, A. (2011) ModBase, a database of annotated comparative protein structure models, and associated resources. *Nucleic Acids Res.* 39(Database issue):D465-474.

Rappsilber, J. (2011) The beginning of a beautiful friendship: cross-linking/mass spectrometry and modelling of proteins and multi-protein complexes. *J. Struct. Biol.* 173 (3):530-540.

Rappsilber, J. (2012) Cross-linking/mass spectrometry as a new field and the proteomics information mountain of tomorrow. *Expert Rev. Proteomics* 9(5):485-487.

Renzone, G., Salzano, A. M., Arena, S., D'Ambrosio, C., Scaloni, A. (2007) Mass spectrometry-based approaches for structural studies on protein complexes at low-resolution. *Current Proteomics* 4:1-16.

Rinner, O., Seebacher, J., Walzthoeni, T., Mueller, L. N., Beck, M., Schmidt, A., Mueller, M., Aebersold, R. (2008) Identification of cross-linked peptides from large sequence databases. *Nat. Methods* 5(4):315-318.

Rozbesky, D., Sovova, Z., Marcoux, J., Man, P., Ettrich, R., Robinson, C. V., Novak, P. (2013) Structural model of lymphocyte receptor NKR-P1C revealed by mass spectrometry and molecular modeling. *Anal. Chem.* 85(3):1597-1604.

Sanowar, S., Singh, P., Pfuetzner, R. A., André, I., Zheng, H., Spreter, T., Strynadka, N. C., Gonen, T., Baker, D., Goodlett, D. R., Miller, S. I. (2010) Interactions of the transmembrane polymeric rings of the *Salmonella enterica* serovar Typhimurium type III secretion system. *MBio* 1(3).

Schilling, B., Row, R. H., Gibson, B. W., Guo, X., Young, M. M. (2003) MS2Assign, automated assignment and nomenclature of tandem mass spectra of chemically crosslinked peptides. *J. Am. Soc. Mass. Spectrom.* 14(8):834-850.

Schulz, D. M., Kalkhof, S., Schmidt, A., Ihling, C., Stingl, C., Mechtler, K., Zschörnig, O., Sinz, A. (2007) Annexin A2/P11 interaction: new insights into annexin A2 tetramer structure by chemical crosslinking, high-resolution mass spectrometry, and computational modeling. *Proteins* 69(2):254-269.

Sclavi, B., Woodson, S., Sullivan, M., Chance, M., Brenowitz, M. (1998) Following the folding of RNA with time-resolved synchrotron X-ray footprinting. *Methods Enzymol.* 295:379-402.

Seebacher, J., Mallick, P., Zhang, N., Eddes, J. S., Aebersold, R., Gelb, M. H. (2006) Protein cross-linking analysis using mass spectrometry, isotope-coded cross-linkers, and integrated computational data processing. *J. Proteome. Res.* 5(9):2270-2282.

Shen, M. Y., Sali, A. (2006) Statistical potential for assessment and prediction of protein structures. *Protein Sci.* 15(11):2507-2524.

Singh, P., Panchaud, A., Goodlett, D. R. (2010) Chemical cross-linking and mass spectrometry as a low-resolution protein structure determination technique. *Anal. Chem.* 82(7):2636-2642.

Singh, P., Shaffer, S. A., Scherl, A., Holman, C., Pfuetzner, R. A., Larson Freeman, T. J., Miller, S. I., Hernandez, P., Appel, R. D., Goodlett, D. R. (2008) Characterization of protein cross-links via mass spectrometry and an open-modification search strategy. *Anal. Chem.* 80(22):8799-8806.

Sharp, J. S., Becker, J. M., Hettich, R. L. (2004) Analysis of protein solvent accessible surfaces by photochemical oxidation and mass spectrometry. *Anal. Chem.* 76(3):672-683.

Sinz, A. (2006) Chemical cross-linking and mass spectrometry to map three-dimensional protein structures and protein-protein interactions. *Mass. Spectrom. Rev.* 25(4):663-682.

Sinz, A. (2003) Chemical cross-linking and mass spectrometry for mapping three-dimensional structures of proteins and protein complexes. *J. Mass. Spectrom.* 38(12): 1225-1237.

Takamoto, K., Chance, M. R. (2006) Radiolytic protein footprinting with mass spectrometry to probe the structure of macromolecular complexes. *Annu. Rev. Biophys. Biomol. Struct.* 35:251-276.

Tang, C., Clore, G. M. (2006) A simple and reliable approach to docking protein-protein complexes from very sparse NOE-derived intermolecular distance restraints. *J. Biomol. NMR* 36(1):37-44.

Trnka, M. J., Burlingame, A. L. (2010) Topographic studies of the GroEL-GroES chaperonin complex by chemical cross-linking using diformyl ethynylbenzene: the power of high resolution electron transfer dissociation for determination of both peptide sequences and their attachment sites. *Mol. Cell. Proteomics* 9(10):2306-2317.

Tullius, T. D., Dombroski, B. A. (1986) Hydroxyl radical "footprinting": high-resolution information about DNA-protein contacts and application to lambda repressor and Cro protein. *Proc. Natl. Acad. Sci. US* 83(15):5469-5473.

Wang, L., Chance, M. R. (2011) Structural mass spectrometry of proteins using hydroxyl radical based protein footprinting. *Anal. Chem.* 83(19):7234-7241.

Wilson, H., Aldrich, H., Maupin-Furlow, J. (1999) Halophilic 20S proteasomes of the archaeon *Haloferax volcanii*: Purification, characterization, and gene sequence analysis. *J. Bacteriol.* 181(18):5814-5824.

Xie, Y., Zhang, J., Yin, S., Loo, J. A. (2006) Top-down ESI-ECD-FT-ICR mass spectrometry localizes noncovalent protein-ligand binding sites. *J. Am. Chem. Soc.* 128(45):14432-14433.

Yang, B., Wu, Y. J., Zhu, M., Fan, S. B., Lin, J., Zhang, K., Li, S., Chi, H., Li, Y. X., Chen, H. F., Luo, S. K., Ding, Y. H., Wang, L. H., Hao, Z., Xiu, L. Y., Chen, S., Ye, K., He, S. M., Dong, M. Q. (2012) Identification of cross-linked peptides from complex samples. *Nat. Methods* 9(9):904-906.

Young, M. M., Tang, N., Hempel, J. C., Oshiro, C. M., Taylor, E. W., Kuntz, I. D., Gibson, B. W., Dollinger, G. (2000) High throughput protein fold identification by using experimental constraints derived from intramolecular cross-links and mass spectrometry. *Proc. Natl. Acad. Sci. US* 97(11):5802-5806.

Zhang, F., Hu, M., Tian, G., Zhang, P., Finley, D., Jeffrey, P. D., Shi, Y. (2009) Structural insights into the regulatory particle of the proteasome from *Methanocaldococcus jannaschii*. *Mol. Cell.* 34(4):473-484.

Zhou, G., Kowalczyk, D., Humbard, M., Rohatgi, S., Maupin-Furlow, J. (2008) Proteasomal components required for cell growth and stress responses in the haloarchaeon *Haloferax volcanii*. *J. Bacteriol.* 190(24):8096-8105.

In: Recent Advances in Proteomics Research
Editor: Ming D. Li

ISBN: 978-1-62948-218-7

Chapter 8

Introduction to Protein Arrays and their Application to Nicotine Research in Macrophages

Wen-Yan Cui[1]* and Ming D. Li[2]

[1]State Key Laboratory for Diagnosis and Treatment of Infections Diseases, The First Affiliated Hospital, Zhejiang University School of Medicine, Hangzhou, Zhejiang, China
[2]Department of Psychiatry and Neurobehavioral Sciences, University of Virginia, Charlottesville, VA, US

Abstract

Protein array provides a powerful tool for high-throughput and precise measurement of proteins involved in various biological events. In contract to commonly used DNA microarray, a protein array can directly determine the quantities of hundreds to thousands of proteins, which are the functional effecters of cell activities, and of their post-translational modified forms, making it an effective and promising approach in the proteomics era. This chapter provides an overview of this relatively new functional genomics technology, which includes different designs of protein arrays, how to conduct an array experiment, what kind of data can be generated, and how to analyze data at both single protein and pathway levels properly. Finally, a set of real protein array data is provided as an example showing how to do research by using this technology. Specifically, for this example, we describe how we designed the experiment, how we analyzed the data, what hypothesis we drew, and how to verify the array result and draw scientific conclusions. By using this example, we hope to make this attractive approach more practical for researchers.

* Corresponding author: Wenyan Cui, PhD, State Key Laboratory for Diagnosis and Treatment of Infections Diseases, The First Affiliated Hospital, Zhejiang University School of Medicine, Hangzhou, Zhejiang, China. Tel: +86-0571-87232417; E-mail: cuiwy@zju.edu.cn.

Keywords: Protein array, pathway analysis, high-throughput, proteomics, genomics, nicotine, inflammation, Toll-like Receptors (TLRs), data analysis, phosphorylation, post-translational modification, gene expression

Introduction

Proteomics study enables detection of complex protein mixtures with relatively small sample quantities. Proteomics technology initially relied on two-dimensional electrophoresis and subsequent mass spectrometry (Tyers and Mann 2003). With the rapid development of mass spectrometry, direct mass spectrometric identification became capable of handling proteomics research needs directly. However, this approach is limited by its depth of coverage, which depends on the abundance of the most enriched protein in the sample mixture (Tyers and Mann 2003). Later, other approaches, such as two-hybrid screening, emerged to complement the mass spectrometric method (Tyers and Mann 2003; Fields and Song 1989). Miniaturized and multiplexed protein/antibody chips emerged in the 1980s (Ekins 1998). This technology permits simultaneous and sensitive measurement of multiple targets with low sample consumption. Since then, the application of planar microarrays using antibodies for detection of proteins and peptides has become widespread in life science research (Angenendt 2005; Joos et al. 2002; Kricka et al. 2003; Templin et al. 2003; MacBeath 2002; Min and Mrksich 2004; Stoll et al. 2005; Zhu and Snyder 2003).

DNA microarray technology has enabled high-throughput quantitation of the abundance of various mRNAs simultaneously in various systems (Schena et al. 1995). The data are processed using different strategies, such as single-gene identification and pathway or network analysis, with the goal of predicting the biological state of each subject of interest. However, the ultrafunctional unit in mediating cellular activities is protein rather than mRNA. Given the concern about potentially poor correlation between mRNA and protein expression levels, at least for some genes (Gygi et al. 1999), the development of a multiplexed protein measurement system has become a logical requirement for monitoring biological changes. Since early 20th Century, great attention has been focused on protein array development (Service 2000; Service 2001; Walter et al. 2000). Although there still exist some limitations, this strategy has been improved greatly during recent years by boosting of knowledge of conditional expression of proteins and pathways and their post-transcriptional modification.

Considering that protein represents the effecter influencing almost every aspect of cell life, the usage of protein array and other proteomics approaches such as 2D PAGE and mass spectrometry (see other chapters of this book for details) could spread into a vast area of both life science research and clinical application (Tyers and Mann 2003). Although the purpose of applications differ, they all can be categorized into the following two groups: 1) Conditional modeling: testing the changes of protein expression profile or pathways/networks under physiological conditions, in disease states, or after drug application. This method can be used, not only the mechanism determination for a disease of interest, but also for pharmacology study of drug toxicity and action. 2) Clinical practice: by monitoring different biomarkers in biological samples, protein arrays can provide a comprehensive view helping physicians access the "real-time" condition of the patient during disease development.

Although as a new high-throughput protein quantitating approach, a protein array sounds fancy and high-tech, it is based on the standard antigen-binding strategy, such as has been widely used in single-protein measurement for years. Through interaction with monoclonal or polyclonal antigen-specific antibodies printed on the array surface, different proteins are located in different physical positions and can be identified according to the "map" provided by the manufacturer of the array. By measuring the enhanced signal from each spot, the relative protein abundance is recorded. This signal can be either fluorescent or chemiluminescent, generated from different dyes. Given that all the reactions take place on a single slide with at most ten thousand antibodies (to date), the necessary sample volumes are reduced dramatically compared with traditional approaches such as ELISA or Western blotting. On the other hand, as all these reactions are using the same sample and taking place simultaneously, the likelihood of generating systemic error becomes lower, and thus, it would make more sense to integrate them during data analysis, such as marking them on the same pathway or predicting a disease state. In this chapter, we introduce several of the most commonly used protein arrays, which are commercially available, and then discuss approaches to data analysis. Finally, we present an example to illustrate how we determine genes and pathways modulated by nicotine with the protein array approach.

Designs of Protein Arrays

The most common protein arrays are antibodies printed on a planar surface. Theoretically, there is a wide choice of the supporting material, such as glass, silicon, plastic, membrane, etc.; but to date, nitrocellulose membrane and nitrocellulose-coated glass slides are the most popular (Table 1). Almost all the commercial products involve one of these two platforms, which have the advantages of low cost and high affinity for proteins. The nitrocellulose-coated glass also stands out for its high signal-to-noise ratio and flatness. A disadvantage is that such arrays need specialized equipment for reading, whereas the nitrocellulose membrane system requires only a Western blot-developing station, which is available in most molecular biology laboratories.

Although they use the same planar platform, the array design principle differs. The most widely used principles are "direct labeling" or "sandwich," which are different in the labeling and hybridization strategies. The direct labeling design uses antibodies printed on the platform for protein capture. The captured proteins have been tagged. In the example shown in Figure 1A, the sample proteins have been ligated with biotin as a tag. Then a fluorophor such as cyanine (Cy3 in the example) is labeled on the protein through the tag. The sandwich design also uses antibodies printed on the platform to capture the targeted proteins (Figure 1B). After that, the captured proteins are incubated with a cocktail of tagged antibodies. Then the tagged antibodies are labeled with fluorophor (Cy3 in the example). The tagging strategy (biotin-streptavidin in this example) is just for sensitivity enhancement.

In some products, the direct labeling design allows incubation of two samples with the array at the same time, which are ligated with different tags and then conjugated with different dyes. These two samples interact with the antibodies on the array in a competitive manner. Compared with the direct labeling design, the sandwich design has higher specificity and sensitivity.

Table 1. Different types of arrays used for protein quantification

Design Principle	Array Provider	Signal Output	Solid Support	Equipment Needed
Direct labeling	Sigma-Aldrich Co. LLC.	Fluorescent	Glass slide	Microarray scanner
	Full Moon Biosystems, Inc.	Fluorescent	Glass slide	Microarray scanner
Sandwich-based	Cell Signaling Technology, Inc.	Chemiluminescent or fluorescent	Membrane or glass slide	Western blot developing station or microarray scanner
	R and D Systems, Inc.	Chemiluminescent	Membrane	Western blot developing station
	Abcam plc.	Chemiluminescent	Membrane	Western blot developing station
Variable	RayBiotech, Inc.	Chemiluminescent or fluorescent	Membrane or glass slide	Western blot developing station or microarray scanner

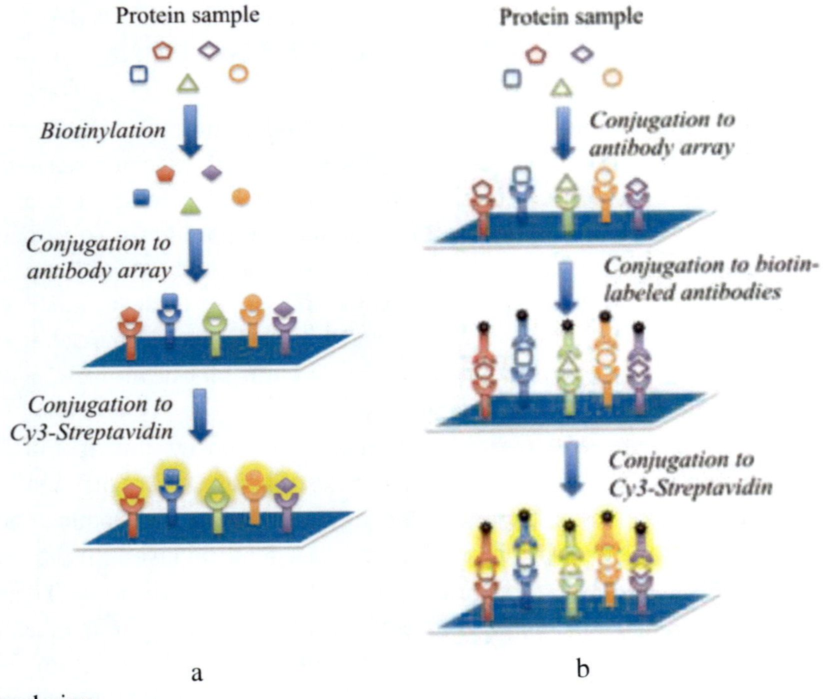

[a]Direct labeling design.
[b]Sandwich-based design.

Figure 1. Protein array designs. (a) The "direct labeling" design uses only one layer of antibodies printed on the array. The biotin-tagged samples bind specifically to these antibodies and then are labeled with a dye. (b) The "sandwich-based" design uses two layers of antibodies: one printed on the array, and the other bridging the sample and the dye. After incubating the sample with the array and thorough washing, cocktail antibody buffer was applied to the array. The antibodies in this cocktail are tagged with biotin in this example. Then the second layer of antibodies is labeled with a dye binding to the tag.

By introducing a standard curve with known concentrations of analytes, this approach can be quantitative. However, the need for paired antibodies limits its usage. And to avoid cross-reaction among cocktail antibodies, it is limited to only ~50 analytes per array to preserve specificity, whereas the direct labeling design can do tens of thousands of targets in one array, being limited only by its dimensions and the availability of antibodies.

Besides the planar platform arrays using the direct labeling or sandwich design, which are most common, there are some other forms of arrays, such as the bead-based, which is also a choice for protein detection (Elshal and McCoy 2006; Morgan et al. 2004; Nolan and Sklar 2002). These tools use antibody-coated beads for protein capture and magnifies the signal in a manner similar to that used with the planar array. Then the array signal is determined by a flow cytometer.

Array Data Processing and Analysis

Yes, protein array is a powerful tool to do parallel quantitation of multiple proteins; however, how to measure the intensity of each spot correctly, to compare the values of experimental groups appropriately, and use the data effectively to reveal underling mechanisms represent key challenges. This section provides some guidelines on how to do data analysis, as indicated by the array providers and the literature. Because the membrane arrays contain only limited numbers of spots, which can be quantified easily using the same strategy as Western blot, this part will focus primarily on the glass slide array to show how to do high-throughput quantitation and data analysis.

1. Spots on Array

Generally speaking, the glass slide array surface can be divided into two areas: the lower one is labeled with a barcode for its identification, and the body part contains one or more blocks holding antibody spots. There are usually duplicate spots for each antibody. For each slide, beside the antibodies for the targeting proteins, there are spots for controls, such as positive, negative, and internal controls and empty spots (Table 2).

These controls are essential for quality control and following statistical analysis, although one type of array product may not contain all of them. Positive controls for a fluorescent array are usually monoclonal anti-Cy3/Cy5 antibodies or Cy3/Cy5-labeled bovine serum albumin (BSA). They can be used for locating the blocks on the array.

Table 2. Controls commonly used on antibody array

Control type	Content	Usage
Positive control	Biotin-conjugated IgG protein, Cy3/Cy5- conjugated BSA, etc.	Array orientation and/or normalization
Negative control	BSA or sampling buffer	Represent the background
Empty	Blank	Represent the background
Internal control	House-keeping proteins	Normalization

For some chemilumiescent arrays, positive control spots can also be used for normalization. Negative control and empty spots are for background checking and unspecific hybridization surveillance. The negative control contains BSA or sampling buffer, whereas empty spots are just blank. The detailed usage of these controls is discussed in the "data analysis" section.

2. Normalization and Data Analysis

Compared with traditional protein quantitation approaches, the protein array, especially high-throughput florescent chips, has the advantage of providing a huge abundance of data by precisely measuring thousands or even tens of thousands of proteins simultaneously with low sample volume consumption. How to manipulate these data effectively to reveal the biological mechanism behind the tiny sparkles has become a significant concern from the bioinformatics point of view. In this section, we focus primarily on a practical data analysis strategy for high-throughput florescent protein chips. Then the analysis of low-to-medium-throughput membrane arrays, which are manipulated purely by hand, is introduced briefly.

I. Quality Control

Prior to performing data analysis, the image and quality of each chip should be checked visually. It is important to make sure there are no apparent "clouds" across the spotting area. Each spot should have clear margins and no "tails." Generally speaking, positive controls should have relatively strong hybridization signals. On the other hand, no significant florescent should be detected in the negative control area or in the empty spots. The protein samples applied to the chip should be in an appropriate and equal concentration, as an extremely low concentration would exaggerate the system error and an extremely high concentration would over-saturate the antibodies and diminish potential statistical differences between samples.

II. Data Analysis

The data analysis can be divided into three steps: florescence intensity measurement, background subtraction, and normalization (Figure 2). Although the whole process is relatively straightforward and easy to understand, there are many details involved in each step, which might impact the final results.

Florescence intensity measurement: For the high-throughput florescent chip, the florescence intensity can be determined by appropriate software, generally that provided by the manufacturer of an array scanner. There should be a downloadable ".gal" file provided by the chip supplier that contains the information on each spot, such as position, antibody name, NCBI ID of each protein, etc. Before scanning slides, such a ".gal" file should be loaded into the scanner software. Then the florescence intensity of each spot will be recorded in an Excel file.

For the low-to-medium throughput array, especially membrane ones, the signal strength may not be generated automatically. In this case, an image-processing program is needed, such as Image J (http://rsbweb.nih.gov/ij). These are widely used in analyzing exposure X-ray films commonly used in Western blotting analysis.

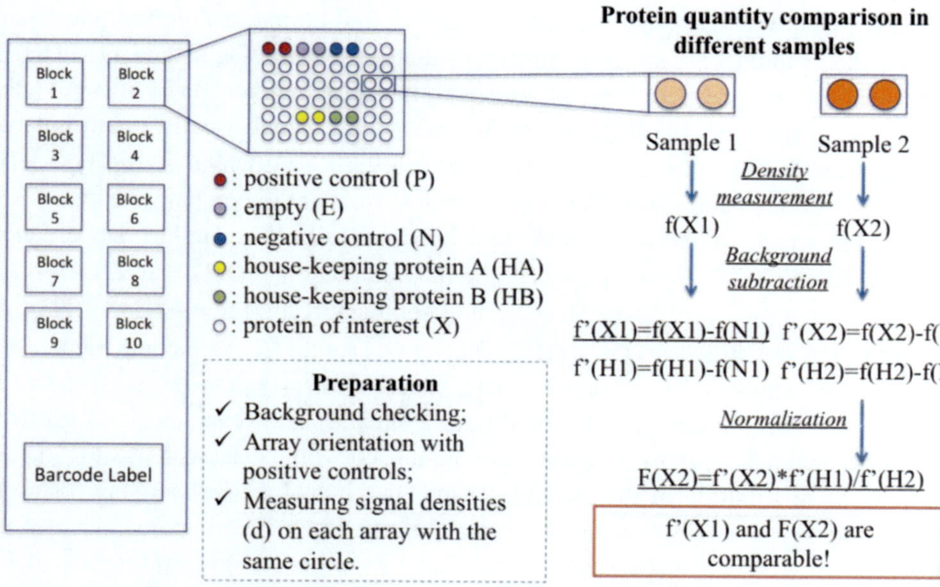

Figure 2. A fluorescent protein array usually contains a barcode area for its identification and one or more blocks holding antibodies. The block area contains different types of controls for qualification and data analysis, including a positive control, a negative control, a housekeeping control, and empty spots. Before performing any statistical analysis, it is important to check the array quality visually and quantitate each spot properly. The fluorescent intensity cannot be compared directly between different samples or experimental groups. The original data should be subjected to background subtraction and normalization to make them comparable across arrays.

By using these programs, the intensity of each spot can be measured manually by marking each spot with a circle. The circle should be roughly the same size as the largest spot on the array. Every spot should be quantified with the same circle, and the florescence signal intensity of each target spot can be recorded as f(X) (f = fluorescence).

Background subtraction: There are multiple possible causes of the appearance of "background," such as insufficient washing or unspecific binding of proteins or dyes to the chip platform. Even the chips with high signal-to-noise ratio can show "backgrounds." To quantify this "noise," several negative controls and/or empty spots are included on each array. Negative controls are usually BSA or the buffer used for antibody dilution. Thus, the negative controls can be used to represent the background. If there is no such negative control on the chip, some empty spots or blank areas can be used as the background signal. Let us record the negative control signal strength as f(N) (N = negative control). Then the florescence intensity of each spot after background subtraction is $f'(X) = f(X) - f(N)$.

Normalization: There are many factors, such as protein loading volume, protein concentration, dye concentration, or background, that may influence the signal intensity of each array.

Because of these potential variations among different arrays, it would be less meaningful if one compares these data directly across different arrays. Thus, normalization is a necessary step in diminishing the bias generated by the sample volume, dye concentration, or other factors, which makes data from different arrays comparable.

To do so, one needs to set one group as the reference (usually the control group) to which other arrays or samples are normalized. Let us assume that signal intensity after background subtraction is f′(X1) = f(X1) - f(N1). Next, we need to find the internal control on the array. Some arrays have multiple internal controls available, such as β-actin and GAPDH. It is essential to choose the one with appropriate signal intensity: neither too low to be detected nor too high to distinguish the difference (extremely high signal usually is attributable to over-saturated antibodies). Let us record this intensity as f(H1), which is f′(H1) = f(H1) - f(N1) after background subtraction. For other samples, we want to compare the data with the reference group, which can be recorded as the signal intensity in a similar manner, such as f(X2) and its background, f(N2). After background subtraction, f′(X2) = f(X2) - f(N2). Its internal control signal is f(H2). After background subtraction, f′(H2) = f(H2) - f(N2). After normalization: F(X2) = f′(X2) * f′(H1)/f′(H2). Then f′(X1) and F(X2) are comparable.

After normalization to the internal controls, usually housekeeping proteins, the relative expression levels become comparable for different samples. However, prior to identifying proteins changed significantly by a treatment or associated with a disease state, it is important to determine the distribution of the signal intensity. If the data are in Gaussian (or normal) distribution, the signal intensity value after normalization can be compared directly using Student's *t*-test or ANOVA, depending on how many experimental groups one has. If not, a logarithm transformation of all data points is needed prior to any further statistical analysis.

3. Bioinformatics Analysis from Single Protein to Pathway

High-throughput techniques such as microarray and protein array make it possible to determine a large number of genes or proteins in one experiment. Because the genes/proteins can be physically or functionally related to each other, the array techniques provide an opportunity to categorize these molecules on a higher level.

For the genes detected on microarray, researchers first try to group them according to their biological functions in a "gene ontology" (GO) manner. With the information provided by GO databases, each gene can be assigned to different cell activities, biological status, intracellular location, or molecular function. Researchers can use p values to rank GO "branches" based on the gene expression changes on the microarray and filter out the top ones as subjects for further study. A similar strategy can be applied to protein array data. However, given that most of the protein arrays contain only hundreds or a few thousand probes, which are more likely selected by researchers from certain pathways or diseases of interest, it may be less meaningful to do GO analysis in many cases. By taking this one step further from the GO analysis, which is in a gene-centric technique, pathway analysis can be carried out to sort the molecules significantly changed on the array in a biochemical manner (Thomas et al. 2007). Such an approach employs the accumulated knowledge from the literature and helps to visualize the molecules changed in a particular pathway. Because the up- or down-regulated molecules are highlighted in different colors by most pathway-related bioinformatics tools, the whole chain of events is more intuitive for researchers. Compared with RNA microarray, which detects the extent of expression of RNA transcripts, protein array with the capability of detecting protein expression and protein phosphorylation provides a more biologically meaningful way to examine those changes at the pathway level.

Obviously, microarray results show expression changes at the mRNA level, which could not directly reflect the protein–protein interaction or kinase-dependent signaling transduction.

In contrast, by using protein array researchers do not need to worry about post-transcriptional modifications of transcripts or the poor correlation between RNA and protein, as this technique is measuring the final products and functional effects directly inside the cells.

So far, a number of bioinformatics tools for pathway analysis have been developed for integrating, analyzing and displaying this high-throughput data, some of which are shown in Table 3. Almost all these tools are highly dependent on the pathway databases built from the accumulated literature, including Alliance for Cellular Signaling (Gilman et al. 2002), BioCarta (http://www.biocarta.com), EcoCyc (Karp et al. 2000), and the Kyoto Encyclopedia of Genes and Genomes (KEGG) (Kanehisa and Goto 2000). By using these tools, researchers can rank the pathways most associated with a treatment or disease of interest. One of the tools developed for this purpose is the GenMAPP (http://www.genmapp.org) (Dahlquist et al. 2002). With this application, researchers can perform pathway analysis with the option of constructing and modifying it on the basis of information provided through several major databases such as Ensembl (http://www.ensembl.org), Entrez Gene (http://www.ncbi.nlm.nih.gov/entrez), and Gene Ontology (Ashburner et al. 2000). The updated version GenMAPP 2 (Salomonis et al. 2007) with an expanded pathway pool can annotate each molecule in multiple colors with the goals of indicating different time points or expression forms (RNA and protein), which makes the output figure more informative. Further, it can analyze various types of data, such as microarray, protein array, and genotyping. On the other hand, there are some commercial bioinformatics tools available with similar features, which include Pathway Studio (Ariadne), GeneGo (MetaCore), and Ingenuity Pathway Analysis (Ingenuity). These tools are backed by pathway databases built with deeper literature research through commercial efforts (Werner 2008). By using these tools, researchers can easily and flexibly analyze the high-throughput data on the pathway level.

Although pathway analysis is used mostly for analyzing expression data, which is the major focus of this chapter, it has been applied to genomics data as well (Zhong et al. 2010). For example, Zhong et al. selected SNPs associated with gene expression (eSNPs) for type 2 diabetes from a public database and analyzed 110 KEGG pathways. They found 16 of them are enriched with eSNPs, many of which are biologically relevant to the disease of interest (Zhong et al. 2010).

Table 3. Pathway analysis websites

Tools	Websites
ArrayXPath	http://www.snubi.org/software/ArrayXPath
CRSD	http://140.120.203.40:8080/crsd
ErmineJ	http://www.chibi.ubc.ca/ermineJ
GenMAPP 2	http://www.genmapp.org/default.html
G:PROFILER	http://biit.cs.ut.ee/gprofiler/welcome.cgi
GSEA	http://www.broadinstitute.org/gsea/index.jsp
KOBAS 2.0	http://kobas.cbi.pku.edu.cn/home.do
MEGU	http://megu.iab.keio.ac.jp
Onto-Express	http://vortex.cs.wayne.edu/projects.htm
PathExpress	http://bioinfoserver.rsbs.anu.edu.au/utils/PathExpress
VisANT 4.0	http://visant.bu.edu

Application of Protein Arrays to Nicotine Research

As a stepwise demonstration of the application of protein array and how to make scientific sense of the results, we used a real dataset from one of our experiments (Cui et al. 2013). This experiment consisted of the following parts: 1) experiment design; 2) original data from protein array; 3) determination of differentially expressed proteins; 4) pathway analysis; and 5) data verification.

Nicotine is the major addictive component of tobacco smoke (Benowitz 1992). It has also been reported to have an immuno-modulatory effect (Wang et al. 2003). It is well known that nicotine suppresses the inflammation induced by TLR4 (Wang et al. 2003), the receptor for gram-negative bacterial lipopolysaccharide (LPS) (Wittebole et al. 2010), and multiple pathways are involved in this anti-inflammatory process (Sugano et al. 1998; Wang et al. 2004; de Jonge et al. 2005; Arredondo et al. 2006; Park et al. 2008; Cui and Li 2010). However, although nicotine also suppresses virus-induced inflammation (Razani-Boroujerdi et al. 2004), the underlying mechanism is unknown. Poly(I:C) is a synthetic double-stranded RNA (dsRNA), which activates Toll-like receptor 3 (TLR3) and is widely used in mimicking viral dsRNA-induced inflammation (Ranjith-Kumar et al. 2007).

To determine the pathways influenced by nicotine during poly(I:C)-induced inflammation, we treated mouse macrophage cells with nicotine, then stimulated them with poly(I:C), and finally used a protein array to examine the phosphorylation changes and analyzed the data on the pathway level. Briefly, this experiment had the following three groups:

1. Control: RAW264.7 cells treated with PBS buffer;
2. Poly(I:C): RAW264.7 cells treated with poly(I:C) 20 μg/ml for 30 min;
3. Nic+Poly(I:C): RAW264.7 cells treated with nicotine (5 μM) and then incubated with poly(I:C) for 30 min.

We used the Signaling Explorer Antibody Array provided by Full Moon Biosystems, Inc. (Sunnyvale, CA) to determine nicotine's influence on protein phosphorylation, which contains 1,318 antibodies primarily focusing on protein phosphorylation sites and is designed in a "direct labeling" manner. The array was processed strictly according to the manufacturer's protocol. The following is a brief summary of the procedure used:

1 Tagging samples: Incubated 25 μl of purified cell lysate from each sample (~7 μg/μl, as suggested by the manufacturer) with biotin/DMF; Array binding: Hybridized the biotin-labeled protein sample with the protein array;

3 Dye conjugation: After application of blocking buffer, the array was incubated with Cy3-streptavidin solution;

4 Quantification: Thoroughly washed the hybridized array and scanned it with the ScanArray Gx scanner. The fluorescent intensity of each spot was then quantified using the ScanArray Express microarray analysis system (PerkinElmer, Waltham, CA) with the ".gal" file provided by the manufacturer.

Table 3A shows the output results of three phosphor-proteins on the array as examples. It contains the coordinate information, the index, ID, protein name, phospho-site, and fluorescent intensity of one sample from each group. The coordinate information including "array row," "array column," "spot row," "spot column" indicates spot location, which helps users check them visually when necessary. The index number indicates the sequence of the spot on the array. ID is used to look up the detailed information of the protein in the "Dictionary" provided by the manufacturer. The last two rows of this table provide the information on negative controls and housekeeping genes on the same chip used to do data analysis. After obtaining this information, we first subtracted the background with the negative control florescence intensity (Table 3B). By taking protein MAPKAPK-2 (abbreviated "M-2" in the formula) as an example, we have:

$f_{Control}(M\text{-}2) = 631$; $f_{Poly(I:C)}(M\text{-}2) = 1841$; $f_{Nic+Poly(I:C)}(M\text{-}2) = 706$

$f'_{Control}(M\text{-}2) = f_{Control}(M\text{-}2) - f_{Control}(\text{negative control}) = 631\text{-}15 = 616$

$f'_{Poly(I:C)}(M\text{-}2) = f_{Poly(I:C)}(M\text{-}2) - f_{Poly(I:C)}(\text{negative control}) = 1841\text{-}0 = 1841$

$f'_{Nic+Poly(I:C)}(M\text{-}2) = f_{Nic+Poly(I:C)}(M\text{-}2) - f_{Nic+Poly(I:C)}(\text{negative control}) = 631\text{-}15=616$

Next, we calculated the fluorescence intensity after background subtraction for β-actin in the same manner:

$f'_{Control}(\beta\text{-actin}) = 4078\text{-}15 = 4063$

$f'_{Poly(I:C)}(\beta\text{-actin}) =4953\text{-}0= 4953$

$f'_{Nic+Poly(I:C)}(\beta\text{-actin}) = 3916\text{-}107 = 3809$

Then we used the control sample as reference and normalize others with the housekeeping protein's intensity (Table 3C):

$F_{Control}(M\text{-}2) = f'_{Control}(M\text{-}2) = 616$

$F_{Poly(I:C)}(M\text{-}2) = f'\ Poly_{(I:C)}(M\text{-}2)^{*}\ f'_{Control}(\beta\text{-actin})/f'_{Poly(I:C)}(\beta\text{-actin}) = 1510$

$F_{Nic+Poly(I:C)}(M\text{-}2) = f'_{Nic+Poly(I:C)}(\beta\text{-actin})^{*}\ f'_{Control}(\beta\text{-actin})/\ f'_{Nic+Poly(I:C)}(\beta\text{-actin}) = 639$

After a series of normalizations and statistical analysis, we obtained a list of changed proteins that are associated with each experimental group (Table 3D). As shown in this table, the fluorescence intensity for poly(I:C) group and Nic + poly(I:C) group are divided by the control group, so the control group would be "1" (not shown).

Any protein with a p value of < 0.05 is considered to be significant. It is clear that the p values for the three proteins are < 0.05, indicating nicotine significantly changed their phosphorylation during poly(I:C) stimulation. This is shown in a more intuitive fashion in Figure 3.

After identifying differentially expressed proteins on the array, we performed pathway analysis with the goal of determining nicotine's effects at the pathway level. For this experiment, our analysis revealed that 82 phospho-sites are significantly changed by nicotine treatment. And 27 of them, including the abovementioned three proteins, are involved in calcium signaling.

By using Ingenuity Pathway Analysis (http://www.ingenuity.com), we constructed a calcium pathway containing all the changed proteins on the array. Figure 4 shows one branch of it containing the three proteins in abovementioned experiment where up-regulated proteins are red highlighted and the down-regulated proteins are green highlighted.

(A). Detected fluorescence intensity of three representative proteins

Index	Array Row	Array Column	Spot Row	Spot Column	Name	ID	Control	Poly (I:C)	Nic+Poly (I:C)
1371	5	1	5	3	MAPKAPK-2 (Phospho-Thr222)	722	631	1841	706
325	2	1	2	2	Pyk2 (Phospho-Tyr402)	170	333	1679	456
1376	5	1	5	8	mTOR (Phospho-Ser2481)	727	859	3609	524
2481	8	1	9	11	Internal control (β-actin)	A	4078	4953	3916
2486	8	1	9	16	Negative control	N	15	0	107

(B). After background subtraction

Index	Array Row	Array Column	Spot Row	Spot Column	Name	ID	Control	Poly (I:C)	Nic+Poly (I:C)
1371	5	1	5	3	MAPKAPK-2 (Phospho-Thr222)	722	616	1841	599
325	2	1	2	2	Pyk2 (Phospho-Tyr402)	170	318	1679	349
1376	5	1	5	8	mTOR (Phospho-Ser2481)	727	844	3609	417
2481	8	1	9	11	Internal control (β-actin)	A	4063	4953	3809

(C). After normalization with β-actin

Index	Array Row	Array Column	Spot Row	Spot Column	Name	ID	Control	Poly (I:C)	Nic+Poly (I:C)
1371	5	1	5	3	MAPKAPK-2 (Phospho-Thr222)	722	616	1510	639
325	2	1	2	2	Pyk2 (Phospho-Tyr402)	170	318	1377	372
1376	5	1	5	8	mTOR (Phospho-Ser2481)	727	844	2961	445

(D). Final statistical analysis results for 3 representative proteins

Name	Phosphorylation site	Mean ± SEM		Ratio	p-value
		Poly (I:C)	Nic + Poly (I:C)		
MAPKAPK-2	Thr222	2.11 ± 0.17	1.18 ± 0.32	0.56	0.005
Pyk2	Tyr402	2.95 ± 0.62	1.41 ± 0.28	0.48	0.005
mTOR	Ser2481	2.71 ± 0.82	0.72 ± 0.15	0.26	0.001

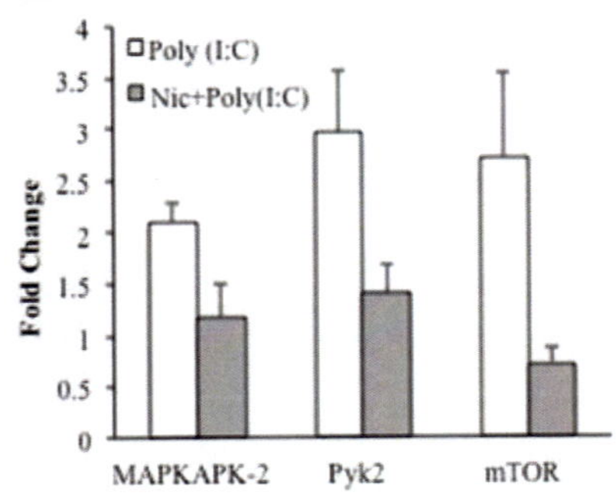

Figure 3. Data analysis procedure. (A) The original data from the array scanner usually include the protein index, location (array row, array column, spot row, and spot column), protein name, ID, and fluorescence intensity of the corresponding spot. (B) The original data are subject to background subtraction by deducing the fluorescence intensity of the negative control or empty spot. (C) The signals from internal controls are typically used to normalize the target proteins. In this step, the control group is set as the reference group, and the poly(I:C) and poly(I:C)+Nicotine groups are normalized to it. The fluorescence intensity values after normalization are comparable in different samples or groups. (D) If the fluorescence intensity of all the spots is in a Gaussian distribution, array data can be analyzed directly with at least three independent experiments. If not, array data should be logarithm-transformed. In this example, one-way ANOVA was applied to the data from the three proteins from three independent experiments. The bar chart shows the fold change of poly(I:C) and Nicotine+poly(I:C) group compared with the control group. The error bar represents the SEM of each group (n = 3).

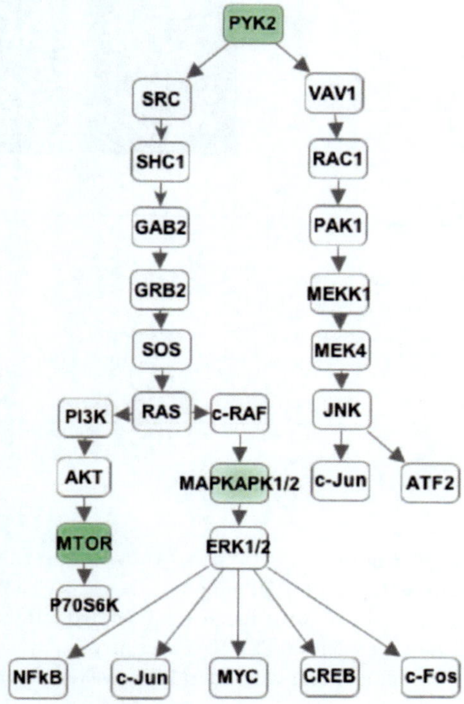

Figure 4. Pathway analysis. The calcium signaling pathway is found to be enriched with the nicotine-modulated proteins. By using the Ingenuity Pathway Analysis program, we constructed the calcium pathway based on the literature. In this figure, the three proteins mentioned in the example; i.e., MAPKAPK-2, PYK2 and MTOR, are marked in green, indicating they are down-regulated by nicotine.

According to the protein array results, we conclude that nicotine suppresses the calcium signaling pathway by decreasing a panel of protein phosphorylation, including CaMKIIα (data not shown). Because this kinase plays a key role in mediating TLR3-induced inflammation (Liu et al. 2008), we measured its phosphorylation level using Western blotting analysis to verify the array results. As shown in Figure 5A, nicotine significantly suppresses poly(I:C)-induced phosphorylation of CaMKIIα, which is consistent with what we found on the array. Furthermore, to prove calcium signaling is really altered by nicotine treatment during poly(I:C) stimulation, we performed calcium imaging to measure the intracellular calcium concentration changes directly in real time. As shown in Figure 5B, poly(I:C) significantly promotes calcium release inside the cell, which is totally blocked by prior nicotine treatment. This experiment further proved nicotine's suppressive effect on calcium signaling and verified the hypothesis we drew from the pathway analysis.

Although cigarette smoking has a significant impact on virus infection (Kark and Lebiush 1981; Kark et al. 1982; Cohen et al. 1993; Arcavi and Benowitz 2004) and nicotine suppresses influenza A-induced inflammatory responses while promoting virus infection (Liu et al. 2008), the molecular mechanism is largely unknown. Previously studies of the anti-inflammatory effect of nicotine focused predominantly on bacteria- or lipopolysaccharide-induced inflammation, which is mediated by TLR4. Although TLR3 and TLR4 are both members of the TLR family, their downstream signaling differs. The results from the TLR4 study thus cannot perfectly explain the observations in the TLR3 system.

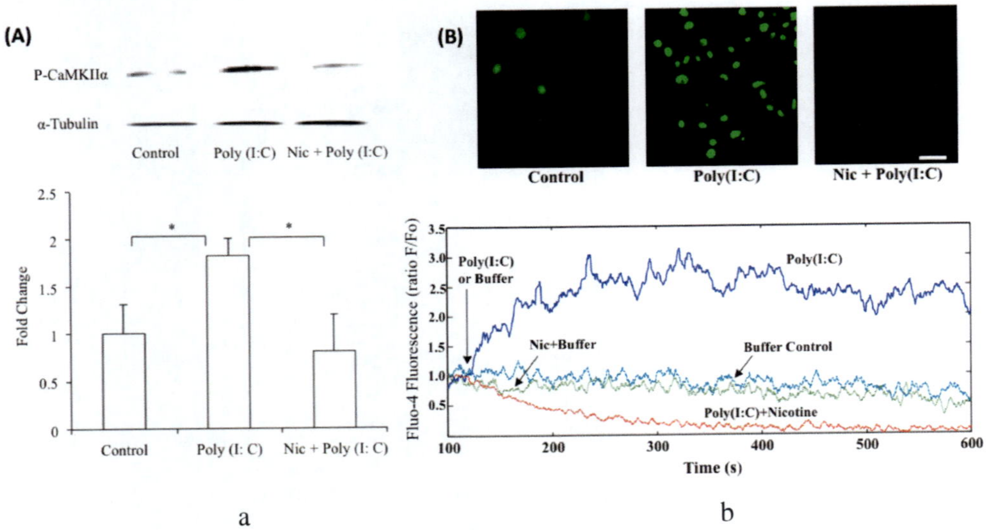

Figure 5. Protein array results verification. (a) RAW264.7 cells were treated with nicotine for 10 minutes and then stimulated with poly(I:C) for 30 minutes. Phospho-CaMKIIα was measured by Western blotting analysis. Data are shown as the mean ± S.E.M. from three independent experiments under identical experimental conditions (*$p < 0.05$). (b) After nicotine treatment, RAW264.7 cells were incubated with poly(I:C), and the intracellular calcium release was recorded by calcium imaging approach.

In this study, we used protein array technology to do parallel measurement of the phosphorylation of more than one thousand proteins. Based on the array data, we did pathway analysis and found nicotine suppresses calcium signaling during poly(I:C) stimulation, which is further proved by Western blotting analysis and calcium imaging experiments.

Moreover, most studies on the immuno-modulatory effect of nicotine rely heavily on the previous knowledge of the signaling molecules involved into the signal transduction of nicotine or immune pathways, such as NF-κB, IκB, PI3K, Jak2, and Stat3 (de Jonge et al. 2005; Blanchet et al. 2006; Cheng et al. 2007; Yoshikawa et al. 2006). By using the largest protein array available at that time, this study provides an unbiased search of the possible pathways and opens a new door to both academic research and clinical applications of abused drugs.

Nicotinic receptor functions as calcium channels in neuronal cells. Interestingly, the pathway of TLR3, the major receptor of poly(I:C), is mediated by calcium signaling (Liu et al. 2008). This implies that nicotine suppresses TLR-mediated inflammation through the calcium pathway.

Conclusion and Prospective

Development of array technology enabled researchers to do parallel large-scale gene and protein detection. Based on the concept of microarray chip, which detects mRNA expression, the protein array takes this one step further by directly measuring the effecter molecules in cells. However, there are several major hurdles faced by the practice of this emerging technique: 1) Proteins are less stable than cDNAs. In the artificial environment of an experiment, it is relatively hard to keep the naïvely folded conformation on which antibody affinity is highly dependent (Talapatra et al. 2002); 2) To profile the global protein expression in a cell, at least 25,000 antibodies are needed. Considering the post-translational modified forms, this number could be even larger, but the available antibodies are only a small proportion of them; 3) When printing a high density of antibodies on an array, avoiding cross-reactivity remains a serious issue; and 4) The dynamic range of proteins is much larger than that of mRNA. When the highly expressed proteins saturate the relevant antibodies, the signal of rarely expressed proteins may be undetectable.

Nevertheless, the protein array is still an attractive technique. It bridges the gap between genomics and proteomics and makes it possible to study the effecter molecules in different forms directly in a high-throughput manner. It can generate a significant amount of data, profiling not only the expression, but also the activation of different pathways and disease states, making it a promising tool in both academic research and clinical applications.

As an example of its potential, we used a protein array to examine how proteins and pathways are affected by nicotine in macrophages. Our results revealed that during viral infection mimicked by poly(I:C), nicotine significantly suppresses calcium signaling, which plays key role in mediating TLR-induced inflammation. This result is built on an unbiased pathway search using the largest protein array available at that time and provides a better understanding of the effect of nicotine on virus infection, contributing to the safe clinical usage of abused drugs.

References

Angenendt, P. (2005) Progress in protein and antibody microarray technology. *Drug discovery today* 10(7):503-511.

Arcavi, L., Benowitz, N. L. (2004) Cigarette smoking and infection. *Arch. Intern. Med.* 164 (20):2206-2216.

Arredondo, J., Chernyavsky, A. I., Jolkovsky, D. L., Pinkerton, K. E., Grando, S. A. (2006) Receptor-mediated tobacco toxicity: cooperation of the Ras/Raf-1/MEK1/ERK and JAK-2/STAT-3 pathways downstream of alpha7 nicotinic receptor in oral keratinocytes. *FASEB journal : official publication of the Federation of American Societies for Experimental Biology* 20(12):2093-2101.

Ashburner, M., Ball, C. A., Blake, J. A., Botstein, D., Butler, H., Cherry, J. M., Davis, A. P., Dolinski, K., Dwight, S. S., Eppig, J. T., Harris, M. A., Hill, D. P., Issel-Tarver, L., Kasarskis, A., Lewis, S., Matese, J. C., Richardson, J. E., Ringwald, M., Rubin, G. M., Sherlock, G. (2000) Gene ontology: tool for the unification of biology. The Gene Ontology Consortium. *Nat. Genet.* 25(1):25-29.

Benowitz, N. L. (1992) Cigarette smoking and nicotine addiction. *The Medical clinics of North America* 76(2):415-437.

Blanchet, M. R., Israel-Assayag, E., Daleau, P., Beaulieu, M. J., Cormier, Y. (2006) Dimethyphenylpiperazinium, a nicotinic receptor agonist, downregulates inflammation in monocytes/macrophages through PI3K and PLC chronic activation. *Am. J. Physiol. Lung Cell. Mol. Physiol.* 291(4):L757-763.

Cheng, P. Y., Lee, Y. M., Law, K. K., Lin, C. W., Yen, M. H. (2007) The involvement of AMP-activated protein kinases in the anti-inflammatory effect of nicotine in vivo and in vitro. *Biochem. Pharmacol.* 74(12):1758-1765.

Cohen, S., Tyrrell, D. A., Russell, M. A., Jarvis, M. J., Smith, A. P. (1993) Smoking, alcohol consumption, and susceptibility to the common cold. *Am. J. Public Health* 83(9):1277-1283.

Cui, W. Y., Li, M. D. (2010) Nicotinic modulation of innate immune pathways via alpha7 nicotinic acetylcholine receptor. *Journal of neuroimmune pharmacology : the official journal of the Society on NeuroImmune Pharmacology* 5(4):479-488.

Cui, W. Y., Zhao, S., Polanowska-Grabowska, R., Wang, J., Wei, J., Dash, B., Chang, S. L., Saucerman, J. J., Gu, J., Li, M. D. (2013) Identification and characterization of poly(I:C)-induced molecular responses attenuated by nicotine in mouse macrophages. *Mol. Pharmacol.* 83(1):61-72.

Dahlquist, K. D., Salomonis, N., Vranizan, K., Lawlor, S. C., Conklin, B. R. (2002) GenMAPP, a new tool for viewing and analyzing microarray data on biological pathways. *Nat. Genet.* 31(1):19-20.

De Jonge, W. J., van der Zanden, E. P., The, F. O., Bijlsma, M. F., van Westerloo, D. J., Bennink, R. J., Berthoud, H. R., Uematsu, S., Akira, S., van den Wijngaard, R. M., Boeckxstaens, G. E. (2005) Stimulation of the vagus nerve attenuates macrophage activation by activating the Jak2-STAT3 signaling pathway. *Nature immunology* 6(8): 844-851.

Ekins, R. P. (1998) Ligand assays: from electrophoresis to miniaturized microarrays. *Clinical chemistry* 44(9):2015-2030.

Elshal, M. F., McCoy, J. P. (2006) Multiplex bead array assays: performance evaluation and comparison of sensitivity to ELISA. *Methods* 38(4):317-323.

Fields, S., Song, O. (1989) A novel genetic system to detect protein-protein interactions. *Nature* 340(6230):245-246.

Gilman, A. G., Simon, M. I., Bourne, H. R., Harris, B. A., Long, R., Ross, E. M., Stull, J. T., Taussig, R., Bourne, H. R., Arkin, A. P., Cobb, M. H., Cyster, J. G., Devreotes, P. N., Ferrell, J. E., Fruman, D., Gold, M., Weiss, A., Stull, J. T., Berridge, M. J., Cantley, L. C., Catterall, W. A., Coughlin, S. R., Olson, E. N., Smith, T. F., Brugge, J. S., Botstein, D., Dixon, J. E., Hunter, T., Lefkowitz, R. J., Pawson, A. J., Sternberg, P. W., Varmus, H., Subramaniam, S., Sinkovits, R. S., Li, J., Mock, D., Ning, Y., Saunders, B., Sternweis, P. C., Hilgemann, D., Scheuermann, R. H., DeCamp, D., Hsueh, R., Lin, K. M., Ni, Y., Seaman, W. E., Simpson, P. C., O'Connell, T. D., Roach, T., Simon, M. I., Choi, S., Eversole-Cire, P., Fraser, I., Mumby, M. C., Zhao, Y., Brekken, D., Shu, H., Meyer, T., Chandy, G., Heo, W. D., Liou, J., O'Rourke, N., Verghese, M., Mumby, S. M., Han, H., Brown, H. A., Forrester, J. S., Ivanova, P., Milne, S. B., Casey, P. J., Harden, T. K., Arkin, A. P., Doyle, J., Gray, M. L., Meyer, T., Michnick, S., Schmidt, M. A., Toner, M., Tsien, R. Y., Natarajan, M., Ranganathan, R., Sambrano, G. R. (2002) Overview of the Alliance for Cellular Signaling. *Nature* 420(6916):703-706.

Gygi, S. P., Rochon, Y., Franza, B. R., Aebersold, R. (1999) Correlation between protein and mRNA abundance in yeast. *Molecular and cellular biology* 19(3):1720-1730.

Joos, T. O., Stoll, D., Templin, M. F. (2002) Miniaturised multiplexed immunoassays. *Current opinion in chemical biology* 6(1):76-80.

Kanehisa, M., Goto, S. (2000) KEGG: kyoto encyclopedia of genes and genomes. *Nucleic Acids Res.* 28(1):27-30.

Kark, J. D., Lebiush, M. (1981) Smoking and epidemic influenza-like illness in female military recruits: a brief survey. *Am. J. Public Health* 71(5):530-532.

Kark, J. D., Lebiush, M., Rannon, L. (1982) Cigarette smoking as a risk factor for epidemic a(h1n1) influenza in young men. *N Engl. J. Med.* 307(17):1042-1046.

Karp, P. D., Riley, M., Saier, M., Paulsen, I. T., Paley, S. M., Pellegrini-Toole, A. (2000) The EcoCyc and MetaCyc databases. *Nucleic Acids Res.* 28(1):56-59.

Kricka, L. J., Joos, T., Fortina, P. (2003) Protein microarrays: a literature survey. *Clinical chemistry* 49(12):2109.

Liu, X., Yao, M., Li, N., Wang, C., Zheng, Y., Cao, X. (2008) CaMKII promotes TLR-triggered proinflammatory cytokine and type I interferon production by directly binding and activating TAK1 and IRF3 in macrophages. *Blood* 112(13):4961-4970.

MacBeath, G. (2002) Protein microarrays and proteomics. *Nat. Genet.* 32 Suppl. 526-532.

Min, D. H., Mrksich, M. (2004) Peptide arrays: towards routine implementation. *Current opinion in chemical biology* 8(5):554-558.

Morgan, E., Varro, R., Sepulveda, H., Ember, J. A., Apgar, J., Wilson, J., Lowe, L., Chen, R., Shivraj, L., Agadir, A., Campos, R., Ernst, D., Gaur, A. (2004) Cytometric bead array: a multiplexed assay platform with applications in various areas of biology. *Clin. Immunol.* 110(3):252-266.

Nolan, J. P., Sklar, L. A. (2002) Suspension array technology: evolution of the flat-array paradigm. *Trends in biotechnology* 20(1):9-12.

Park, S. Y., Baik, Y. H., Cho, J. H., Kim, S., Lee, K. S., Han, J. S. (2008) Inhibition of lipopolysaccharide-induced nitric oxide synthesis by nicotine through S6K1-p42/44 MAPK pathway and STAT3 (Ser 727) phosphorylation in Raw 264.7 cells. *Cytokine* 44 (1):126-134.

Ranjith-Kumar, C. T., Miller, W., Xiong, J., Russell, W. K., Lamb, R., Santos, J., Duffy, K. E., Cleveland, L., Park, M., Bhardwaj, K., Wu, Z., Russell, D. H., Sarisky, R. T., Mbow, M. L., Kao, C. C. (2007) Biochemical and functional analyses of the human Toll-like receptor 3 ectodomain. *J. Biol. Chem.* 282(10):7668-7678.

Razani-Boroujerdi, S., Singh, S. P., Knall, C., Hahn, F. F., Pena-Philippides, J. C., Kalra, R., Langley, R. J., Sopori, M. L. (2004) Chronic nicotine inhibits inflammation and promotes influenza infection. *Cellular immunology* 230(1):1-9.

Salomonis, N., Hanspers, K., Zambon, A. C., Vranizan, K., Lawlor, S. C., Dahlquist, K. D., Doniger, S. W., Stuart, J., Conklin, B. R., Pico, A. R. (2007) GenMAPP 2: new features and resources for pathway analysis. *BMC bioinformatics* 8217.

Schena, M., Shalon, D., Davis, R. W., Brown, P. O. (1995) Quantitative monitoring of gene expression patterns with a complementary DNA microarray. *Science* 270(5235):467-470.

Service, R. F. (2000) Biochemistry. Protein arrays step out of DNA's shadow. *Science* 289 (5485):1673.

Service, R. F. (2001) Proteomics. Searching for recipes for protein chips. *Science* 294 (5549): 2080-2082.

Stoll, D., Templin, M. F., Bachmann, J., Joos, T. O. (2005) Protein microarrays: applications and future challenges. *Current opinion in drug discovery and development* 8(2):239-252.

Sugano, N., Shimada, K., Ito, K., Murai, S (1998) Nicotine inhibits the production of inflammatory mediators in U937 cells through modulation of nuclear factor-kappaB activation. *Biochem. Biophys. Res. Commun.* 252(1):25-28.

Talapatra, A., Rouse, R., Hardiman, G. (2002) Protein microarrays: challenges and promises. *Pharmacogenomics* 3(4):527-536.

Templin, M. F., Stoll, D., Schwenk, J. M., Potz, O., Kramer, S., Joos, T. O. (2003) Protein microarrays: promising tools for proteomic research. *Proteomics* 3(11):2155-2166.

Thomas, P. D., Mi, H., Lewis, S. (2007) Ontology annotation: mapping genomic regions to biological function. *Current opinion in chemical biology* 11(1):4-11.

Tyers, M., Mann, M. (2003) From genomics to proteomics. *Nature* 422(6928):193-197.

Walter, G., Bussow, K., Cahill, D., Lueking, A., Lehrach, H. (2000) Protein arrays for gene expression and molecular interaction screening. *Current opinion in microbiology* 3(3): 298-302.

Wang, H., Liao, H., Ochani, M., Justiniani, M., Lin, X., Yang, L., Al-Abed, Y., Wang, H., Metz, C., Miller, E. J., Tracey, K. J., Ulloa, L. (2004) Cholinergic agonists inhibit HMGB1 release and improve survival in experimental sepsis. *Nature medicine* 10(11): 1216-1221.

Wang, H., Yu, M., Ochani, M., Amella, C. A., Tanovic, M., Susarla, S., Li, J. H., Wang, H., Yang, H., Ulloa, L., Al-Abed, Y., Czura, C. J., Tracey, K. J. (2003) Nicotinic acetylcholine receptor alpha7 subunit is an essential regulator of inflammation. *Nature* 421(6921):384-388.

Werner, T. (2008) Bioinformatics applications for pathway analysis of microarray data. *Current opinion in biotechnology* 19(1):50-54.

Wittebole, X., Castanares-Zapatero, D., Laterre, P. F. (2010) Toll-like receptor 4 modulation as a strategy to treat sepsis. *Mediators of inflammation* 2010568396.

Yoshikawa, H., Kurokawa, M., Ozaki, N., Nara, K., Atou, K., Takada, E., Kamochi, H., Suzuki, N. (2006) Nicotine inhibits the production of proinflammatory mediators in human monocytes by suppression of I-kappaB phosphorylation and nuclear factor-kappaB transcriptional activity through nicotinic acetylcholine receptor alpha7. *Clin. Exp. Immunol.* 146(1):116-123.

Zhong, H., Yang, X., Kaplan, L. M., Molony, C., Schadt, E. E. (2010) Integrating pathway analysis and genetics of gene expression for genome-wide association studies. *Am. J. Hum. Genet.* 86(4):581-591.

Zhu, H., Snyder, M. (2003) Protein chip technology. *Current opinion in chemical biology* 7 (1):55-63.

In: Recent Advances in Proteomics Research
Editor: Ming D. Li
ISBN: 978-1-62948-218-7

Chapter 9

Quantitative Proteomic Approaches for Environmental Toxicology

***John M. Prins*[1], *Yinsheng Wang*[1] *and D. I. Lurie*[2,*]**
[1]Department of Chemistry, University of California, Riverside, CA, US
[2]Department of Biomedical and Pharmaceutical Sciences, College of Health Professions and Biomedical Sciences, The University of Montana, Missoula, MT, US

Abstract

Environmental toxicology is an interdisciplinary scientific field focusing on investigating the harmful effects of environmental toxicants on biological systems, including humans and animals. In order to identify critical proteins and cellular pathways that are primarily affected by exposure to environmental toxicants, it is often helpful to measure changes in global protein expression levels. For this reason, mass spectrometry-based quantitative proteomics has emerged as a powerful technique for toxicological studies since it can be used to quantitate a large number of proteins simultaneously in a single sample preparation. Over the past several decades a variety of quantitative proteomic techniques have been developed and used for measuring changes in protein expression, both in vivo and in vitro, following exposure to a wide variety of toxic chemicals and heavy metals. This review will focus on two commonly used mass spectrometry-based quantitative proteomic techniques for measuring relative changes in protein expression in biological samples following exposure to environmental toxicants: (i) Two-dimensional gel electrophoresis-based proteomics; and (ii) Stable Isotope Labeling of Amino acids in Cell culture (SILAC). In the first technique, proteins are separated according to their isoelectric point (pI) (first dimension) followed by separation based on their molecular weight using sodium dodecyl sulfate polyacrylamide gel electrophoresis (SDS PAGE) (second dimension). The separated proteins are stained and relative protein expression changes are determined by comparing the staining density of individual protein spots between the experimental and control samples. Those proteins

* Corresponding author: Dr. Diana I. Lurie, Dept. of Biomedical and Pharmaceutical Sciences, College of Health Professions and Biomedical Sciences, Skaggs Building Room 383, The University of Montana, Missoula, MT 59812-1552. Phone: 406-243-2103; Fax: 406-243-5228; -E-mail: Diana.lurie@umontana.edu.

displaying significant changes in spot density are excised from the gels and digested with trypsin or other suitable enzyme into peptides.

Peptide samples are then analyzed by mass spectrometry and protein databases are used for protein identification. The second approach relies on metabolic labeling for protein quantitation. Briefly, when cells are grown in culture they must acquire essential amino acids from the culture media in order to grow and divide. SILAC takes advantage of this requirement to incorporate a label into proteins for mass spectrometry based quantitative proteomics. When cells are grown in culture media containing a given 'light' or 'heavy' form of essential amino acids, they become incorporated into proteins as the cells proliferate. Generally, one or two specific amino acids are labeled, inducing a predictable mass shift in the peptide which can be measured by LC-MS/MS. The corresponding mass shift enables peptide pairs to be easily identified and allows for the accurate estimation of protein expression ratios. Here we describe the use of both of these approaches to assess perturbations in protein expression levels following exposure to various environmental toxicants, including heavy metals and tobacco-specific *N*-nitrosamines in several different biological settings.

Keywords: Quantitative proteomics, Environmental toxicology, Mass Spectrometry, Two-dimensional gel electrophoresis (2-D SDS PAGE), Stable Isotope Labeling of Amino acids in Cell culture (SILAC), Tobacco Specific Nitrosamines, Heavy Metals

Introduction

Toxicants are present throughout the environment and can be found in air, dust, soil, food, and water. Therefore, humans are unavoidably exposed to a wide variety of environmental toxicants (xenobiotics) such as polychlorinated biphenyls (PCBs), pesticides, and heavy metals that are associated with an increased risk for a number of diseases, including cancer, Parkinson's disease, and Alzheimer's disease (Dik et al. 2012; Mates et al. 2010; Mostafalou and Abdollahi 2013; Fitzmaurice et al. 2013; Cocco et al. 2008). Environmental toxicology incorporates many scientific disciplines including chemistry, cell biology, and pharmacology to investigate the harmful effects of man-made and naturally occurring environmental toxicants on biological systems, including animals and humans. Research in environmental toxicology is often aimed at elucidating the mechanisms of action of toxicants by identifying the cellular pathways and proteins targeted by specific xenobiotics. The results from these studies are extremely important for public health risk assessment, establishing environmental guidelines and policies for toxicants, as well as diagnosis and treatment of human disease (Ge et al. 2007).

Identifying the molecular mechanisms, cellular pathways, and adverse effects of a specific xenobiotic on living systems can be extremely challenging. Ideally, for each toxicant, data needs to be collected for several different biologically relevant doses and durations of exposure. In addition, experiments often need to be conducted for several different biologically relevant model systems including species and organ systems. Therefore, studies aimed at identifying unknown mechanisms of action for specific toxicants can quickly become quite large and complicated. When choosing an experimental technique, high-throughput technologies with discovery potential that allow for the investigation into how a biological system responds as a whole to environmental toxicants are very desirable and extremely beneficial for identifying critical proteins and cellular pathways targeted by specific xenobiotics.

In many cases, organisms respond to exposure to environmental toxicants through changes in the expression of genes and proteins allowing it to adapt to changes in the environment. Identifying the proteins that demonstrate up-regulated or down-regulated expression levels following exposure to a specific toxicant can provide insight into the physiological mechanisms of action. For this reason, the field of mass spectrometry based-quantitative proteomics has emerged as a powerful technique that can be employed in environmental toxicology studies (Wetmore and Merrick 2004). Quantitative proteomic experiments are well suited for environmental toxicology studies since they allow for the comparison of two or more experimental conditions; facilitate the identification of thousands of proteins in a single sample preparation; and provide quantitative information about proteins and/or peptides. Therefore, this approach provides a means for high-throughput large-scale analysis of protein changes induced by exposure to environmental toxicants (George et al. 2010).

In general, quantitative proteomics techniques can be divided into three main categories: 1) Expression proteomics, which involves the quantitation of global changes in protein and peptide expression; 2) Interaction proteomics, which focuses on characterizing specific and functional protein-protein interactions (Blagoev et al. 2003); and 3) Protein modification proteomics, where post-translational modifications (e.g., phosphorylation, glycosylation, etc.) across sample species (control vs. treated) are identified and quantitated (Jensen 2004). Although all three strategies are important for toxicological assessment, this chapter will focus on expression proteomics and two very common mass spectrometry (MS)-based quantitative proteomic techniques used for environmental toxicology studies: (i) Two-dimensional gel electrophoresis-based proteomics (2-DE); and (ii) Stable Isotope Labeling by Amino acids in Cell culture (SILAC). Here we describe the use of both approaches to assess perturbations in global protein expression levels following exposure to various environmental toxicants, including heavy metals and tobacco-specific *N*-nitrosamines (TSNAs) in several different biological settings.

1. Overview of Proteomic Techniques

A variety of quantitative proteomic techniques have recently been developed for assessing protein expression levels (Monteoliva and Albar 2004). In general, there are two main strategies for expression proteomic studies, gel-based approaches which primarily utilize two-dimensional gel electrophoresis (2-DE) to assess protein expression levels and non-gel based methods that rely on MS for assessment of protein expression. 2-DE methods for expression proteomic studies rely on protein staining density to determine the relative abundance of proteins across sample species prior to protein identification (Sa-Correia and Teixeira 2010). Once significantly altered protein spots have been identified by 2-DE, the individual protein spots are excised from the gel and identified with mass spectrometry. Although 2-DE is still widely used, these classical approaches have largely been replaced with high-throughput non-gel-based liquid chromatography coupled with tandem mass spectrometry (LC-MS/MS) methods for large-scale expression proteomic studies.

A popular approach for non-gel based methods relies on introducing an isotopic label into proteins and peptides for differential protein expression measurements.

Isotopic labeling induces a predictable mass shift allowing for the generation of two or more unique sample populations which can be differentiated by the mass spectrometer (Monteoliva and Albar 2004). The corresponding mass shift can be used to easily identify differentially expressed proteins across different samples and allows for the accurate estimation of protein expression ratios.

1.1. Protein Identification by Mass Spectrometry

An extensive review of mass-spectrometry techniques and instrumentation for quantitative proteomics is beyond the scope of this chapter; however, several excellent reviews can be found in the literature that cover this topic in detail (Cottrell 2011; Bantscheff et al. 2007; Bantscheff et al. 2012; Ong et al. 2003; Lanucara and Eyers 2011). In this chapter, we will briefly discuss two widely used techniques for protein identification commonly associated with 2-DE- and SILAC-based quantitative proteomic studies, peptide mass fingerprinting and peptide sequencing.

Peptide mass fingerprinting is a traditional technique for protein identification usually associated with 2-DE proteomics and relies upon a comparison of the experimentally determined MS peak mass values of individual peptides. Peptide samples generated by proteolytic digestion are usually analyzed by matrix-assisted laser desorption-ionization time-of-flight mass spectrometry (MALDI-TOF) to determine peptide mass values. Proteins are identified using a protein database (Mascot, SwissProt, etc.) to compare the experimentally determined peptide mass values with the theoretically predicted molecular mass values. Statistical confidence for protein identification increases with increasing peptide mass matches to a particular protein and increasing sequence coverage.

Similarly, statistical confidence decreases with any unmatched masses appearing in the spectra, which makes it difficult to identify more than one protein at a time with peptide mass fingerprinting. For this reason, peptide mass mapping is primarily applicable to 2-DE approaches since two-dimensional gel separation is capable of resolving hundreds or even thousands of protein spots. In theory, every protein spot represents an individual protein; therefore, these proteins can be excised from the gel and analyzed individually (Thiede et al. 2005).

Peptide sequencing is based on the fragmentation pattern of individual peptides generated by collision-induced dissociation (CID) from LC-MS/MS and well-known patterns of peptide fragmentation (Dancik et al. 1999).

For this strategy, comparisons are made between the molecular mass values for the experimentally generated fragment ions and all theoretically predicted peptide fragments of the appropriate mass.

Theoretically, protein identification can be achieved by matching one unique peptide, although statistical confidence for protein identification is increased by matching multiple peptides and increasing sequence coverage. Peptide sequencing can be used for both 2-DE and SILAC approaches. However, this technique is particularly well suited for SILAC experiments. In SILAC experiments, protein quantification is achieved by comparing the abundance of 'light' and 'heavy' peptides in a given sample by mass spectrometry, eliminating the need to completely resolve individual proteins as in 2-DE.

For SILAC experiments, proteins only need to be partially purified through 1D-SDS PAGE or HPLC which helps simplify the proteome prior to LC-MS/MS analysis. CID and peptide sequencing are then capable of effectively identifying a large number of the proteins in the simplified mixture (Ong and Mann 2005).

1.2. Comparison of Protein Expression Using 2-DE

Although 2-DE is largely being replaced by LC-MS/MS-based approaches to assess relative changes in protein expression, it continues to be used and remains a valuable tool for environmental toxicology studies. The 2-DE approach is particularly valuable when examining protein changes in response to a toxicant in animal models, where proteins in individual tissues and organs can be quantified and analyzed. An experimental outline for a typical 2-DE proteomic experiment is shown in Figure 1.

Briefly, for 2-DE experiments, isoelectric focusing (IEF) is used to separate sample proteins according to their net charge or isoelectric point (pI) (first dimension) followed by separation based on the protein's molecular weight using SDS PAGE electrophoresis (second dimension). Following separation, proteins are usually detected by staining with colloidal Coomassie blue, silver stain, or fluorescent dyes (e.g., SYPRO Ruby, Pro-Q Diamond, Pro-Q Emerald, etc.).

After protein separation and staining, additional computer-based analysis is necessary to digitally image and align protein maps between control and treatment groups. Gels are then analyzed to detect relative protein expression changes by comparing the staining density of individual protein spots between the experimental samples and control samples. Those proteins displaying significant changes in spot density are excised from the gels and digested with trypsin or other suitable enzymes into peptides. Peptide sample analysis by mass spectrometry is then used to generate peptide mass maps of the individual proteins which can be compared to theoretical peptide masses of peptides generated from in-silico digestion of protein in databases (e.g., Mascot, SwissProt) for protein identification (Carrette et al. 2006).

The use of 2-DE for assessing relative changes in protein expression across sample species has proven to be a useful strategy for the identification of differentially expressed proteins following exposure to environmental toxicants.

In our lab, we were able to use this approach to investigate subtle changes in protein expression in auditory neurons of the brainstem following exposure to lead (Pb) (Prins et al. 2010). Pb is a risk factor for neurological and behavioral dysfunction and exposure during development is associated with auditory temporal processing deficits in both humans and animals. Auditory temporal processing is the processing of sound in terms of both time and space that occurs in the brain once the neural signal enters the brainstem.

This auditory processing allows us to resolve complex sounds and to hear conversation in a noise background. Pb disrupts these auditory processing functions, and deficits in auditory temporal processing are associated with behavioral syndromes such as Attention Deficit Hyperactivity Disorder (ADHD) (Lidsky and Schneider 2003; Braun et al. 2006). But it was completely unknown how Pb altered the proteome in the auditory brainstem and how these protein changes might be associated with these behavioral syndromes.

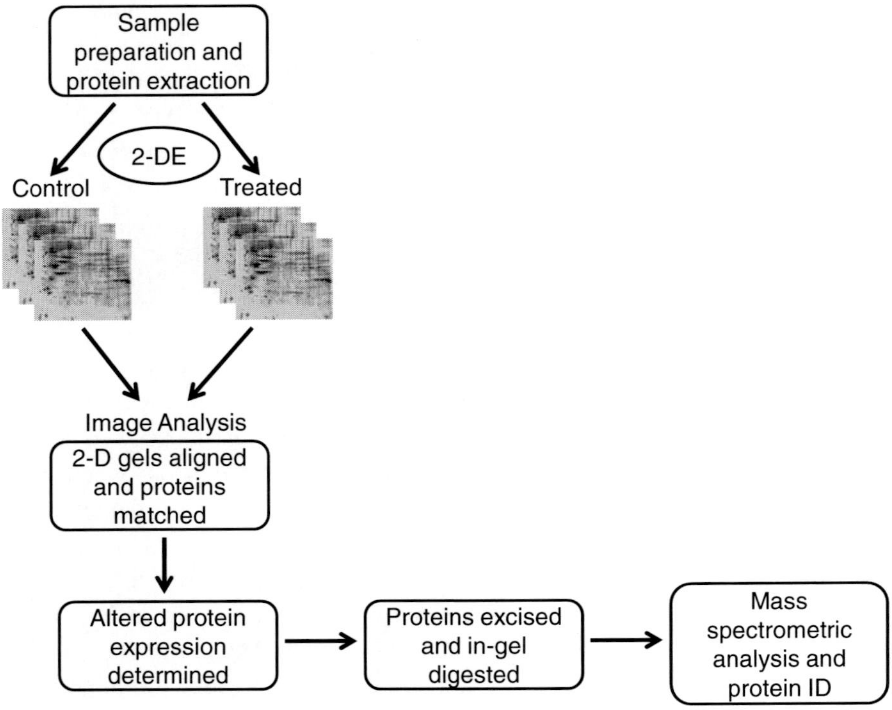

Figure 1. Experimental workflow for a classical 2-DE experiment. Control and treated samples are initially resolved by 2-DE in replicate gels and stained with a suitable protein stain (e.g., Coomassie Blue, Silver-stain, Sypro Ruby, etc.). Two dimensional gel images are generated and analyzed by specific software and differentially expressed protein spots are identified. Protein spots are excised from the gel, digested into peptides and analyzed by mass spectrometry for protein identification.

In our studies, CBA/CaJ mice (n=3–5/group) were exposed to 0.01, or 0.1 mM lead acetate during development via drinking water. At postnatal day (PND) 21, the auditory regions (ventral brainstem region (VBS)) were isolated from the rest of the brainstem (dorsal brainstem region (DBS)). Samples of the VBS and DBS regions were collected for control mice, as well as, 0.01 mM and 0.1 mM Pb-treated mice. A sub-cellular protein extraction method was used to isolate and enrich proteins based on their cellular localization so that cytosolic, membrane, and nuclear fractions could be obtained. 2-DE-based quantitative proteomics was used to assess relative changes in protein expression in each fraction and for each treatment group in both DBS and VBS regions of the brainstem. With this approach, we were able to identify a total of 187 total proteins among which 36 were significantly altered in response to Pb exposure. Most of the proteins that changed expression following Pb exposure were specific to the VBS (auditory brainstem) fraction and did not change within the DBS fraction, suggesting that Pb has a greater effect on the auditory region of the brainstem. The majority of the significantly altered proteins were associated with cellular energy metabolism, including the voltage-dependant ion channel (VDAC) and creatine kinase (CK) as well as several proteins of the glycolytic pathway, the phosphocreatine circuit, and oxidative phosphorylation (Prins et al. 2010). Cellular energy metabolism and efficient energy production are crucial for auditory neurons to maintain high rates of synaptic activity; therefore, subtle changes in cellular energy producing systems in brainstem auditory neurons may have a detrimental effect on auditory temporal processing.

The proteomic approach was essential in identifying VDAC as a novel target of Pb, and demonstrated for the first time that this key protein, involved in mitochondrial energy production, is modulated by Pb toxicity. Previously, no studies had associated Pb exposure with changes in VDAC and thus the proteomic analysis yielded a new target for Pb.

The 2-DE proteomic approach has also been utilized to examine protein changes induced by other environmental toxicants including dioxins, bisphenol A, cigarette smoke, and pesticides in animal models. Chen et al performed a proteomic analysis of the rat ovary following chronic exposure to the endocrine disrupter 2,3,7,8-tetrachlorodibenzo-*p*-dioxin (TCDD). TCDD is a very potent toxin and is released into the environment primarily through the combustion of fossil fuels where it is ubiquitously distributed. TCDD is considered a carcinogen and an endocrine disrupter, and is associated with immunotoxicity and hepatotoxicity. Previous proteomic studies had found that TCDD exposure resulted in changes in immune function proteins as well as cellular homeostasis and cell survival (Chen et al. 2009). This study examined the proteome of the ovary following chronic TCDD exposure in order to elucidate how disruption in endocrine function could be related to TCDD-induced protein changes in the ovary. Using the 2-DE approach, this group found that TCDD exposure led to up- and down-regulation of 5 and 2 proteins, respectively. One of the upregulated proteins was Selenium-binding protein 2 (SBP2) which specifically binds selenium and APAP (*N*-acetyl-*p*-aminophenol). The function of SBP2 in the ovary is not known, but it may be related to the reproductive senescence produced by TCDD (Chen et al. 2009). Once again, the proteomic analysis revealed a novel target of TCDD and found increases in SBP2, a protein not usually associated with ovarian function. Other proteins modulated by TCDD were associated with oxidative stress, transcription of the estrogen receptor, chaperones, and endothelial exocytosis.

In a similar study, a proteomic analysis was performed on another endocrine disrupter, bisphenol A (BPA) (Betancourt et al. 2012). BPA is used to manufacture polycarbonate plastics and resins, and is found in cans and plastic beverage bottles. Early pre-pubertal exposure to BPA *increases* the susceptibility to chemically induced mammary cancer in a rat model. Interestingly, this study also examined the isoflavone genistein, a phytoestrogen found in many soy-based food products. Genistein *decreases* the susceptibility of the mammary gland to carcinogenesis in the rat when given pre-pubertally. Both BPA and genistein have estrogenic properties and the goal of this study was to identify proteomic changes in the rat mammary gland induced by either BPA or genistein that might account for their differential physiological effects on carcinogenesis. Using 2-DE, 18 proteins were found to be altered by BPA and 16 proteins were altered by genistein (Betancourt et al. 2012). Both compounds induced protein changes associated with binding and catalytic activity, antioxidant activity, and regulation of enzyme activity. They also identified one protein, annexin A2 that was increased by BPA and decreased by genistein. Annexin A2 is a calcium-dependent, phospholipid-binding protein that is overexpressed in several cancers, including breast cancer. It is involved in cellular pathways that are related to tumorogenesis and thus it was significant that BPA increased annexin A2 levels by 73% and genistein decreased annexin A2 levels by 67%. Interestingly, selenium binding protein was also decreased by BPA in this study, and selenium binding protein is associated with carcinogenesis, where its expression is reduced in multiple tumor types. Thus, the proteomic analysis again revealed several novel proteins modulated by toxicants in a specific tissue.

Another good example of the power of 2-DE proteomics for analyzing the effects of toxicants on in vivo systems is a study examining the effect of the pesticide cypermethrin on the brain (Singh et al. 2011). Cypermethrin crosses the blood-brain barrier where it induces free radical generation leading to oxidative damage in the dopaminergic neurons of the nigrostriatal pathway. This damage leads to a Parkinson's phenotype in animal models. In this study, rats were treated with cypermethrin from postnatal day 5 through 4, 8, and 12 weeks. A proteomic analysis was performed of the striatum and substantial nigra. They observed significant changes in 36 protein spots induced by the cypermethrin treatment and identified 15 proteins by MS. These proteins included Hsp-60 and70, MAPK-activated kinase 5, SNAP-25, and phosphatidyl-ethanolamine-binding protein-1 (PEBP-1). They then compared the proteome to cypermethrin treated animals that had also been treated with minocycline, an inhibitor of microglial activation, in order to assess the role of microglial activation in cypermethrin mediated neurotoxicity. They found that minocycline inhibited the microglial activation, restored dopamine levels, and improved behavioral function in this model. The proteomic analysis revealed that minocycline treatment restored the expression levels of seven proteins including SNAP-25, and Hsp-60. Thus, the proteomic analysis also enabled the identification of proteins that were changed by cypermethrin, and then restored with minocycline treatment, and they could associate these proteins with the cellular and behavioral deficits induced by cypermethrin through microglial activation.

Finally, prenatal exposure to cigarette smoke is associated with low birth weight and reductions in kidney mass. Jagadapillai et al., 2012, examined the effect of prenatal cigarette smoke exposure (CSE) on the kidney proteome with an emphasis on metabolic proteins in order to determine whether they could identify cellular pathways that might account for the higher risk of diabetes, hypertension, and cardiovascular disease associated with CSE. Pregnant dams and their subsequent litters were exposed to cigarette smoke for 6 h per day, 7 days a week from gestation through PD21. Kidney weights were decreased in the smoke-exposed animals compared to controls, and approximately 45 proteins were found to be differentially expressed between the smoke exposure and sham controls. Proteins with decreased expression upon smoke exposure included those that were part of the lipid metabolism network and the small molecule biochemistry network. In addition, several proteins involved in inflammation, nucleic acid metabolism, and respiratory diseases were also decreased. Several cell-signaling proteins were increased (Jagadapillai et al. 2012). The authors concluded that cigarette smoke exposure during development altered several metabolic pathways in the kidney, including lipid metabolism, inflammation, and respiratory disease pathways. Thus, they were able to identify several metabolic pathways in the kidney that were modulated by smoke exposure and could influence the development of diabetes, hypertension, and cardiovascular disease.

The 2-DE proteomic approach has been successfully applied to in vivo models to determine cellular pathways and proteins that are modulated by various toxicants. An advantage of this approach is that it can be used to analyze the proteome in any tissue or organ system. Animal models can therefore be used to assess the effects of toxicants, and entire physiological networks can be analyzed. Unfortunately, the detection sensitivity and linearity of 2-DE is relatively low; therefore, relatively few proteins (usually under a 100 proteins) are usually identified as being altered following treatment. As a consequence, these studies likely underestimate the number of proteins and cellular pathways that are changing following exposure.

That being said, this approach allows for direct comparisons between in vivo and in vitro studies since it is applicable to both in vivo and in vitro studies, making 2-DE a valuable tool for studies on the mechanisms of toxicity.

In 2004, Lau et al., (Lau et al. 2004) used 2-DE based proteomics to investigate the effects of arsenite exposure on protein expression in cultured lung epithelial cells (LECs). For these studies, LECs were generally treated with 40 μM of $NaAsO_2$ via the culture media and harvested at 0, 2, 4, 6, 8, 12 and 24 h time points. 2-DE allowed for the separation and visualization of over 1000 protein spots for the LECs. Of these 1000 protein spots, they were able to identify seven proteins with significantly altered expression levels in response to arsenite treatment. The identified up-regulated proteins were heat shock protein 70 (HSP70), aldose reductase, heme oxygenase-1, heat shock protein 27 (HSP27), ferritin light chain and αB-crystallin. In addition, the glycolytic enzyme glyceraldehyde-3-phosphate dehydrogenase was down-regulated following arsenite treatment. Several follow-up studies were then used to demonstrate that pretreatment with the thiol antioxidants glutathione or *N*-acetylcysteine could effectively block the arsenite-induced up-regulation of these proteins and sustained cell viability. The results from this study suggested that high levels of arsenite exposure primarily induced apoptosis through an oxidative stress mechanism in LECs.

The 2-DE approach was also successfully utilized to examine the effect of arsenic in the lungs of mice (Lantz et al. 2007). Twenty-one day old mice were exposed to arsenic through the drinking water for a total of four weeks. At the end of the exposure, a bronchoaveolar lavage fluid (BALF) was collected and a proteomic analysis performed. This group identified a total of 25 spots and found that arsenic decreased glutathione-S-transferase (GST)-omega 1, contraspin, apolipoprotein A-1 and A-IV, peroxiredoxin-6, enolase 1, and receptor for advanced glycation end products (RAGE). RAGE is associated with chronic inflammatory diseases and normal levels in the lung are fairly high. However, there is an association between decreased plasma levels of RAGE and coronary artery disease and rheumatoid arthritis, suggesting that decreased RAGE results in deficient control of inflammation. The authors then examined RAGE expression in human sputum and compared this with urinary arsenic levels. They found that there was a significant negative association between sputum RAGE concentration and total urinary inorganic arsenic (Lantz et al. 2007). Thus, the in vivo animal proteomic study led to the identification of RAGE as a protein of interest that was modulated by arsenic. RAGE was then found to be reduced in people with elevated urinary arsenic levels. RAGE had not been previously identified as a biomarker for arsenic exposure in lung lining fluid, and these in vivo studies found a unique target for arsenic. Taken together, the in vivo and in vitro studies identified similar routes of toxicity for arsenic on the lung, namely, that arsenic induces apoptosis in lung cells primarily through oxidative stress, and that arsenic modulates the lung inflammatory response in both mice and humans. These studies provide a more complete picture of how arsenic modulates cellular function in the lung.

Gel-to-gel variation is a major drawback for 2-DE-based quantitative proteomics and can greatly limit the ability to detect subtle changes in protein expression. In order to reduce some of the inherent errors associated with 2-DE, Amersham Biosciences developed a hybrid 2-DE technique called differential in-gel electrophoresis (DIGE). The DIGE approach takes advantage of spectrally resolvable fluorescent dyes (Cy2, Cy3, and Cy5). Protein samples from different treatment groups are differentially labeled with these dyes and co-resolved within the same 2-D gel for direct comparison and quantification of proteins across sample species (Tannu and Hemby 2006).

In a typical DIGE experiment, an internal standard is prepared by mixing equal amounts of protein from both the control and treatment group. The control, treatment group, and internal standard are all labeled with a different fluorescent dye, mixed together at equal protein concentration ratios and separated within a single gel. The internal standard is used to normalize the gel and eliminate any errors associated with gel-to-gel variation (Tannu and Hemby 2006). A recent study by Kodavanti et al., (Kodavanti et al. 2011) utilized the 2-D DIGE approach to investigate the effects of the PCB Aroclor 1254, a developmental neurotoxicant, on protein networks in rat cerebellum and hippocampus. For this study, pregnant (Long Evans) rats were given 0.0 or 6.0 mg/kg/day of Aroclor-1254 from gestation day 6 through PND 21. At PND 14, the cerebellum and hippocampus was isolated and analyzed to assess any relative changes in protein expression induced by Aroclor 1254 treatment. The results from the quantitative proteomics experiments identified two proteins in the cerebellum and 18 proteins in the hippocampus that were differentially expressed following PCB exposure. Several mitochondrial proteins were differentially expressed following Aroclor 1254 treatment, including ATP synthase, subunit β (ATP5B), creatine kinase, malate dehydrogenase, voltage-dependent anion-selective channel protein 1 (VDAC1) and ryanodine receptor type II (RyR2). In addition, Aroclor 1254 treatment altered the expression of dihydropyrimidinase-related protein 4 (DPYSL4) and valosin-containing protein (VCP) which are proteins associated with nervous system development. The results from this study suggest that Aroclor 1254 may induce developmental neurotoxicity by disrupting energy metabolism and intracellular signaling in the cerebellum and hippocampus.

Some of the major drawbacks of using 2-DE for protein expression analysis include limited protein loading capacity, low detection sensitivity and linearity, poor solubility of membrane proteins, and low gel reproducibility. In addition, 2-DE is largely limited to proteins with molecular weights between 10-120 kDa (Monteoliva and Albar 2004). Although some hybrid 2-DE methods, such as DIGE, have been developed to minimize some of these errors, this technique is largely being replaced by non-gel based methods. One contributing factor is that 2-DE is both labor-intensive and time-consuming, which greatly limits the amount of throughput that can be achieved with this method. Several automated instrumental platforms have been developed for spot identification, spot excision, and digestion; however, they are relatively expensive and require experienced technicians to operate. For this reason, non-gel based methods are becoming increasingly popular for quantitative proteomic studies since the majority of the steps are easily automated.

1.3. Comparison of Protein Expression Using Stable Isotope Labeling by Amino Acids in Cell Culture (SILAC)

SILAC is a metabolic labeling method for incorporating an isotopic label into proteins for quantitative proteomics and allows for the complete labeling of all proteins within a given proteome (Ong et al. 2002). This technique is becoming increasingly popular for toxicological studies. When cells are grown in culture they must acquire essential amino acids from the culture media in order to proliferate. SILAC studies take advantage of this requirement to introduce a non-radioactive isotope label into proteins for quantitative proteomic experiments.

Isotopic labeling is introduced through chemical or metabolic incorporation, with the labeling strategy determined by the type of sample being analyzed and the mass spectrometer being used for analysis. When the cells are grown in culture media containing a given 'light' or 'heavy' form of essential amino acids they become incorporated into the proteins, which allows for the generation of two unique sample populations that can be differentiated by a mass spectrometer (Ong et al. 2002).

Generally, one or two specific amino acids are labeled inducing a predictable mass shift in the peptide which can be measured by LC-MS/MS. The corresponding mass shift enables peptide pairs to be easily identified and allows for the accurate estimation of protein expression ratios (Ong and Mann 2006).

Generally, cells need to be cultured in SILAC culture media for approximately 10-14 days, allowing for the complete metabolic incorporation of the 'light' or 'heavy' form of the essential amino acid. Since the 'light' and 'heavy' forms of the amino acids only differ by their molecular weight, the two cell populations should be functionally identical. This allows for one cell population to be used for a toxicant treatment group and the other as a control group. The relative changes in protein expression between the control and treatment groups represents how the cell responds to exposure to the xenobiotic tested.

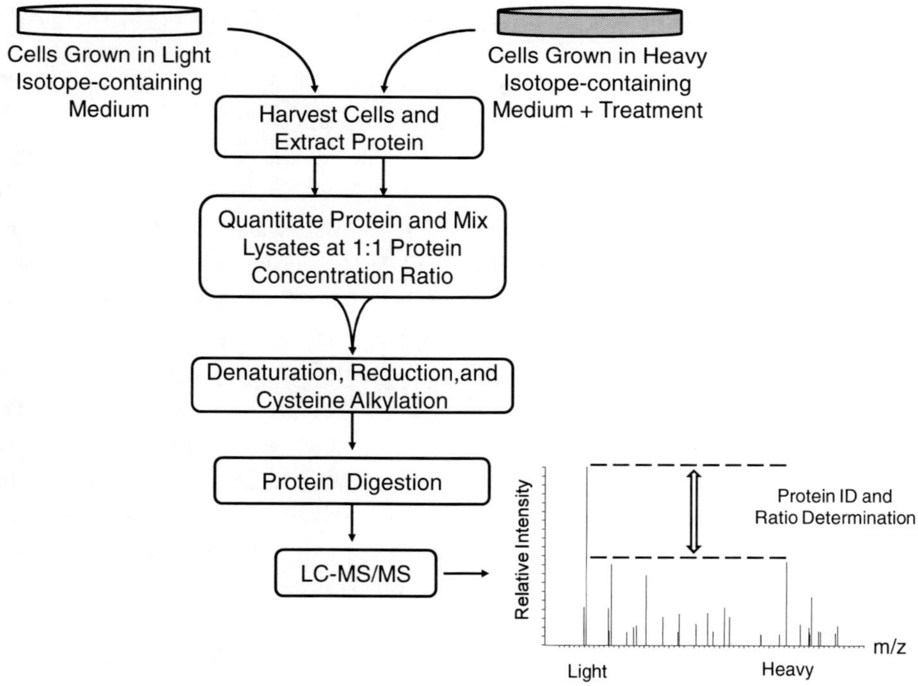

Figure 2. Experimental workflow for a typical SILAC-based quantitative proteomic study. Cell samples are cultured in the presence of 'light' or 'heavy' isotopically labeled amino acids in order to generate two unique cell populations. One cell population is used for control samples and the other is used for a treatment group. Cell samples are harvested, lysed and protein concentration determined for each sample group. Protein samples are then mixed at a 1:1 protein concentration ratio and digested with trypsin or other suitable enzyme. Subsequently, peptide samples are subjected to LC-MS/MS to obtain relative protein abundance and protein identification by database searching.

The cells from each population are then simultaneously harvested, proteins extracted, and protein concentrations determined. Since SILAC allows for the complete labeling of proteins, the protein extracts from the control and treatment groups are mixed together at equal protein concentration ratios and further purified using chromatography or SDS-PAGE. After the purification step, proteolyic enzymes are used to digest protein samples into peptides and samples are subjected to LC-MS/MS for analysis (Figure 2). The resultant mass spectra are then used to identify proteins via protein database search (Mascot, Swiss Prot, etc.). Proteins are quantified by the ratio of intensities of the 'light' and 'heavy' labeled peptide peaks in a given mass spectrum, which gives a relative ratio of abundance of the two samples under comparison. Identified proteins displaying significant increases or decreases in response to exposure can be further analyzed using a variety of software programs to construct protein networking pathways and their associations with cellular signaling pathways and disease processes. SILAC is considered a relatively high-throughput technique, since the majority of the steps in a typical SILAC experiment can be automated. The major drawback of this approach is that SILAC labeling can only be achieved in live proliferating cells and is generally not applicable to tissue samples; although, some in vivo studies are becoming possible with the development of the SILAC mouse (Kruger et al. 2008) and SILAC fly (Sury et al. 2010). Despite this limitation, SILAC is becoming increasingly popular for *in vitro* toxicological studies, mainly due to the increased throughput and the increased number of proteins that can be identified and quantified with this technique.

Recently, we utilized a SILAC labeling approach along with LC-MS/MS to investigate the effects of exposure to the TSNA 4-(methylnitrosamino)-1-(3-pyridinyl)-1-butanone (NNK) on protein expression levels in cultured human cell lines. NNK is a well-known carcinogen and present at relatively high levels in tobacco and its smoke. Although the ability of the metabolically activated form of NNK to generate DNA adducts is well established, little is known about the cellular pathways perturbed by NNK in its native state. In this study, we utilized SILAC-based quantitative proteomics to assess the perturbation of protein expression in human skin fibroblast cells (GM00637 cells) following exposure to NNK. With this approach, we were able to quantify 1412 proteins and 137 of them showed significantly altered expression following NNK exposure, with 38 being down-regulated and 99 up-regulated. The significantly altered proteins were associated with a variety of molecular and cellular functions, including mitochondrial proteins, cytoskeleton-associated proteins, and proteins involved in glycolysis and gluconeogenesis. A major finding from this study was the up-regulation of essentially all subunits of the 20S proteasome core complex. The up-regulation of the 20S core complex is accompanied with a significant increase in 20S proteasome activities in GM00637, IMR90 (human lung fibroblast), and MCF-7 (human breast cancer cell line) cells upon NNK treatment. Furthermore, the β-adrenergic receptor (β-AR) antagonist propranolol inhibited the NNK-induced increase in proteasome activity in all the three cell lines, suggesting that up-regulation of the 20S proteasome may be mediated through the β-AR. Results from this study provided novel insights into the cellular mechanisms targeted by NNK (Prins and Wang 2012).

Non-gel based techniques have benefitted greatly over the last 10 years through improvements in biological mass spectrometry, chromatography, and database search engines. Additionally, non-gel based methods are free of some of the major drawbacks associated with 2-DE, including limited gel loading capacity and low detection sensitivity.

These methods can provide quantitative information for high and low molecular weight proteins (proteins greater than 120 kDa or less than 10 kDa). Taken together, these advancements have led to marked increases in the number of proteins that can be identified and quantified with these techniques.

A recent study by Xiong and Wang (Xiong and Wang 2010), used SILAC together with LC-MS/MS to investigate the effects of arsenite treatment on protein expression levels in HL-60 cells (human promyelocytic leukemia cells). For this study, HL-60 cells were SILAC labeled and treated with 5 μM arsenite via culture media. After 24 hours of arsenite treatment, the cells were harvested and analyzed using the SILAC-based quantitative proteomics approach described above. This technique led to the identification and quantification of 1067 proteins in the HL-60 cells, 56 had significantly altered levels of expression induced by arsenite treatment. These included the up-regulation of core histones, neutrophil elastase, α-mannosidase as well as the down-regulation of fatty acid synthase and protein phosphatase 1α. The SILAC labeling approach, when coupled to LC-MS/MS, offers a very sensitive and robust platform for identifying relative changes in protein expression levels in cultured cells. For comparison, consider the arsenite study performed by Lau et al., 2004 described above (Lau et al. 2004).

This latter study utilized a 2-DE based quantitative proteomics to investigate the effects of arsenite exposure on protein expression in cultured lung epithelial cells (LECs). In this study 2-DE allowed for the separation and visualization of 1000 protein spots and the identification of only seven proteins with significantly altered expression levels in response to arsenite treatment.

The SILAC approach has also been utilized to examine protein phosphorylation changes induced by the dioxin TCDD (Schulz et al. 2013). Recall that Chen et al, 2009 examined the effect of TCDD on the ovary, and found that TCDD exposure resulted in a total of five proteins that were up-regulated and two proteins that were down-regulated. However, when phosphorylation changes in rat hepatoma cells were quantified using SILAC proteomics, more than 6500 phosphopeptides were identified and approximately 32 phosphopeptides displayed phosphorylation changes induced by TCDD, suggesting that phosphorylation changes were associated with only a few signaling pathways (Schulz et al. 2013). The results from this study demonstrated that TCDD exposure primarily altered the phosphorylation of transcriptional regulators, with the second highest group being regulators of small GTPases. These results were not entirely unexpected, since TCDD is known as a modulator of transcriptional regulation. However, the fact that TCDD modulated regulators of small GTPases was a novel finding, and provides valuable insight into the toxic mechanisms of TCDD.

A similar study used SILAC to investigate the phosphoproteome of a macrophage cell line following exposure to the common food contaminant, trichothecene mycotoxin deoxynivalenol (DON). DON targets the innate immune system and stimulates proinflammatory gene expression via a ribotoxic stress response (Pan et al. 2013). A total of 188 proteins whose phosphorylation state was significantly up- or down-regulated by DON were identified. The results from the SILAC analysis suggested that the DON-induced RSR in macrophages was primarily elicited through a multifaceted signaling network composed of MAPK, NF-κB, AKT, and AMPK pathways and transcriptional regulation was also impacted (Pan et al. 2013).

A recent study ~~done~~ by Zhang et al., (2013), utilized SILAC labeling coupled with 2-DE to investigate methamphetamine (METH) neurotoxicity and protein tyrosine nitration mediated through the DDAH/ADMA/NOS (dimethylarginine dimethylaminohydrolase I/ asymmetric dimethylated L-arginine/nitric oxide synthase) pathway (Zhang et al, 2013). Elevated levels of nitric oxide (NO) are associated with a variety of neurodegenerative diseases, including Alzheimer's and Parkinson's diseases.

In addition, elevated levels of NO have been observed in METH addiction. The cellular mechanisms involved in METH-induced neurotoxicity remain largely unresolved; therefore, this study was aimed at investigating the possible relationship between protein nitration and METH-induced neurotoxicity.

In this study, the effects of METH exposure on nitroprotein expression levels in PC-12 cells and the striatum of Sprague-Dawley rats were assessed by SILAC- and 2-DE-based quantitative proteomics, respectively. The results from these studies indicated the METH treatment resulted in the elevated expressions of three nitroproteins in rat striatum and the up-regulated expression of 27 nitroproteins were observed in PC12 cells. These nitrated proteins are involved in a variety of cellular functions, including Cdk5 activation, cytoskeletal structure, and ribosomal function. Overall, the results from this study indicated that protein nitration plays a significant role in acute METH induced neurotoxicity. This is an interesting study, since it incorporates both 2-DE and SILAC approaches in order to compare the effects of METH exposure in two different model systems and provides a more detailed description of protein nitration in METH-induced neurotoxicity.

SILAC-based quantitative proteomics studies can be utilized to reveal very complex networks of signaling pathways that are modulated by toxicants, both at the protein and post-translational levels. We have described above two such examples of how posttranslational modifications can be altered by a toxicant and the utility of SILAC in studying the effects to toxicant exposure on protein signaling pathways. Thus, one of the strengths of SILAC-based quantitative proteomics is the ability to determine how toxicants can perturb networks of proteins at the posttranslational as well as the translational level. This gives us insight into the ability of toxicants to impact a more subtle, regulatory layer of cellular function.

Conclusion and Prospective

In order to obtain an accurate understanding of how environmental toxicants affect physiology, it is necessary to identify the cellular pathways, molecular mechanisms, and adverse effects of individual toxicants on living systems.

Most toxicants target more than one cellular pathway and perturb a variety of molecular mechanisms. Elucidating the toxic mechanisms of action for a specific toxicant can be an extremely daunting task if approached with the traditional biological methods used for analyzing expression changes for individual proteins. The ability to assess global changes in protein expression provides valuable insight into how a biological system is responding as a whole to exposure to specific toxicants. For this reason, quantitative proteomics studies have become increasingly popular for environmental toxicology studies over the last several decades since they provide a means for high-throughput, large-scale analysis of protein expression changes.

Classical 2-DE approaches provide the ability to resolve and visualize thousands of proteins in complex protein samples through gel staining followed by protein identification by mass spectrometry and database searching. Although, the labor-intensive nature of 2-DE, along with the analytical challenges presented by this method, limits the level of throughput that can be achieved with this technique, it is still very useful and has produced many contributions to toxicology studies. One particularly valuable aspect of the 2-DE approach is that it can be used in animal models and will successfully identify and quantify protein changes in individual tissues and organs. As we have seen in this chapter, this approach has identified novel proteins that are altered by various toxicants in animal models. For example, VDAC was modulated in the auditory brainstem following low-level Pb exposure (Prins et al. 2010). VDAC is a membrane protein involved in mitochondrial energy production and this was the first time Pb had been shown to affect VDAC expression in any system. In addition, no previous studies had examined the role of VDAC in the central auditory system, where auditory neurons are very metabolically active. So this proteomic study was able to not only identify a novel target for Pb, but also identified a potentially important novel protein, VDAC, that might influence central auditory function. Similarly, TCDD was found to up-regulate SBP2 in the rat ovary (Chen et al. 2009), but the function of SBP2 in the ovary is not known. Yet this study suggests that SBP2 might play a role in the normal functioning of the ovary, and should be a target for future studies.

In addition to identifying novel proteins, 2-DE allows for the visualization of protein isoforms and potential posttranslational modifications, using physical parameters such as the molecular weight and isoelectric point of individual protein spots. Taken together, these two aspects present 2-DE-based quantitative proteomics as an extremely informative technique for toxicology studies and also can provide a greater understanding of the basic biology of a tissue or organ system.

However, traditional 2-DE-based quantitative proteomic techniques are largely being replaced by non-gel based techniques which are more amenable to automation and capable of high-throughput analysis. SILAC-based quantitative proteomics is particularly well suited for in vitro toxicological studies, mainly due to the increased throughput and the increased number of proteins that can be identified and quantified with this technique. The SILAC approach has been used to identify and obtain quantitative information for thousands of proteins in a single sample preparation, allowing for the identification of specific protein networking pathways and their associations with cellular signaling pathways. SILAC is particularly well suited for the examination of post-translational modifications such as phosphorylation.

For example, the 2-DE approach identified a total of five proteins that were changed by TCDD in the rat ovary (Chen et al. 2009), but using SILAC to examine protein phosphorylation changes induced by TCDD in rat heaptoma cells, more than 6500 phosphopeptides were identified and approximately 32 phosphopeptides were altered by this toxicant (Schulz et al. 2013).

A major limitation of SILAC-based quantitative proteomics has been that it is primarily limited to live proliferating cell models and is generally not applicable to tissue samples; however, with the development of SILAC animal models some *in vivo* studies are becoming possible. A recent study by Huang et al., (2012) utilized a SILAC mouse model to study the effects of deleting the ribonuclease III enzyme Dicer in adult mice, which plays a role in the maturation and biogenesis of miRNAs and siRNAs.

In this study, Dicer knockout mice demonstrated an increase in lipid accumulation in the small intestine and SILAC mice were used for the proteomic profiling of the small intestine in order to elucidate the underlying mechanisms. A total of 646 proteins were identified and 155 proteins were significantly altered as a result of Dicer deletion, including the down-regulation of several enzymes in lipid metabolic pathways (microsomal triglyceride transfer protein, long-chain fatty acyl-CoA ligase 5, fatty acid binding protein, and very-long-chain fatty acyl-CoA dehydrogenase) (Huang et al. 2012). The results from this study demonstrate the utility of the SILAC mouse model for *in vivo* assessment of protein expression changes in complex biological systems.

Protein labeling strategies for non-gel based techniques such as ICAT (isotope coded affinity tags), and iTRAQ (isobaric tags for relative and absolute quantitation) are rapidly being developed for the chemical labeling of proteins and peptides *in vitro*, respectively (Ong and Mann 2006; Gygi et al. 2002; Boersema et al. 2009; Unwin et al. 2010; Wu et al. 2006). For ICAT and iTRAQ analysis, proteins are extracted from cell or tissue samples prior to labeling, allowing these strategies to be used with *in vivo* model systems. Briefly, for ICAT analysis, proteins are extracted from samples and chemically labeled at cysteine residues with 'light' and 'heavy' versions of the ICAT reagent. Labeled proteins are mixed together at an equal concentration ratio, digested into peptides, and then analyzed by mass spectrometry (Gygi et al. 2002). A primary limitation to the ICAT labeling approach is that proteins containing few or no cysteine residues will not be identified and information for post-translational modifications may be lost. In contrast, iTRAQ analysis relies on the isobaric tagging of peptides for relative and absolute quantification of proteins; therefore, protein samples are generally digested into peptides prior to labeling (Unwin et al. 2010). A major advantage to this approach is that iTRAQ reagents can be used to label all peptides within a sample through covalent labeling at the N-terminus and the lysine side chains; therefore, this approach is well suited for the quantification of posttranslational modifications (e.g., phosphorylation, glycosylation, etc.). ICAT and iTRAQ labeling strategies offer the advantage of coupling non-gel based quantitative proteomic approaches to animal model systems; although, the efficiency of labeling and co-elution of chemically similar peptides are limitations to their application (Wu et al. 2006). In addition to the labeling methods, several robust and cost-effective mass spectrometry-based label-free quantitative proteomic strategies have been reported (Smaczniak et al. 2012; Bluemlein and Ralser 2011). Label-free methods rely on peak area ratios and spectral counting to quantify peptides; therefore, samples must be analyzed individually. For this reason, reproducible label-free quantification requires consistent sample handling and chromatography conditions between samples; therefore this technique is not very useful for reliably assessing small changes in protein expression between samples and is not very practical for a large number of samples. These labeling strategies and label-free techniques along with advances in techniques for peptide and protein sample preparation and enrichment has allowed for the development of several non-gel based techniques, both *in vitro* and *in vivo*, for assessing relative changes in protein expression, posttranslational modifications (phosphorylation, glycosylation, etc.) and cellular signaling networks.

Advances in instrument sensitivity, automation, and mass spectrometry-based proteomic methods has led to the rapid growth of quantitative proteomics in the field of toxicology. Gel- and non-gel-based approaches are closely related, and each technique is useful in providing distinct information about proteins.

As we have seen in this chapter, these techniques are particularly effective when used together for the identification of cellular pathways and proteins targeted by specific xenobiotics. The use of quantitative proteomics for toxicology studies is a powerful technique due to the discovery potential and the capability for high-throughput analysis. Thus, the field will likely see increased use of proteomics for elucidating pathways and proteins modulated by environmental toxicants. As these technologies continue to develop and mature, they will deliver increasingly meaningful biological insight into the effects of environmental toxicants on physiological systems.

References

Bantscheff, M., Lemeer, S., Savitski, M. M., Kuster, B. (2012) Quantitative mass spectrometry in proteomics: critical review update from 2007 to the present. *Anal. Bioanal. Chem.* 404(4):939-965.

Bantscheff, M., Schirle, M., Sweetman, G., Rick, J., Kuster, B. (2007) Quantitative mass spectrometry in proteomics: a critical review. *Anal. Bioanal. Chem.* 389(4):1017-1031.

Betancourt, A. M., Wang, J., Jenkins, S., Mobley, J., Russo, J., Lamartiniere, C. A. (2012) Altered carcinogenesis and proteome in mammary glands of rats after prepubertal exposures to the hormonally active chemicals bisphenol a and genistein. *J. Nutr.* 142 (7): 1382S-1388S.

Blagoev, B., Kratchmarova, I., Ong, S. E., Nielsen, M., Foster, L. J., Mann, M. (2003) A proteomics strategy to elucidate functional protein-protein interactions applied to EGF signaling. *Nat. Biotechnol.* 21(3):315-318.

Bluemlein, K., Ralser, M. (2011) Monitoring protein expression in whole-cell extracts by targeted label- and standard-free LC-MS/MS. *Nat. Protoc.* 6(6):859-869.

Boersema, P. J., Raijmakers, R., Lemeer, S., Mohammed, S., Heck, A. J. (2009) Multiplex peptide stable isotope dimethyl labeling for quantitative proteomics. *Nat. Protoc.* 4(4): 484-494.

Braun, J. M., Kahn, R. S., Froehlich, T., Auinger, P., Lanphear, B. P. (2006) Exposures to environmental toxicants and attention deficit hyperactivity disorder in US children. *Environ. Health Perspect.* 114(12):1904-1909.

Carrette, O., Burkhard, P. R., Sanchez, J. C., Hochstrasser, D. F. (2006) State-of-the-art two-dimensional gel electrophoresis: a key tool of proteomics research. *Nat. Protoc.* 1(2):812-823.

Chen, X., Ma, X. M., Ma, S. W., Coenraads, P. J., Zhang, C. M., Liu, J., Zhao, L. J., Sun, M., Tang, N. J. (2009) Proteomic analysis of the rat ovary following chronic low-dose exposure to 2,3,7,8-tetrachlorodibenzo-p-dioxin (TCDD). *J. Toxicol. Environ. Health A* 72(11-12):717-726.

Cocco, P., Brennan, P., Ibba, A., de Sanjose Llongueras, S., Maynadie, M., Nieters, A., Becker, N., Ennas, M. G., Tocco, M. G., Boffetta, P. (2008) Plasma polychlorobiphenyl and organochlorine pesticide level and risk of major lymphoma subtypes. *Occup. Environ. Med.* 65(2):132-140.

Cottrell, J. S. (2011) Protein identification using MS/MS data. *J. Proteomics* 74(10):1842-1851.

Dancik, V., Addona, T. A., Clauser, K. R., Vath, J. E., Pevzner, P. A. (1999) De novo peptide sequencing via tandem mass spectrometry. *J. Comput. Biol.* 6(3-4):327-342.

Dik, S., Scheepers, P. T., Godderis, L. (2012) Effects of environmental stressors on histone modifications and their relevance to carcinogenesis: a systematic review. *Crit. Rev. Toxicol.* 42(6):491-500.

Fitzmaurice, A. G., Rhodes, S. L., Lulla, A., Murphy, N. P., Lam, H. A., O'Donnell, K. C., Barnhill, L., Casida, J. E., Cockburn, M., Sagasti, A., Stahl, M. C., Maidment, N. T., Ritz, B., Bronstein, J. M. (2013) Aldehyde dehydrogenase inhibition as a pathogenic mechanism in Parkinson disease. *Proc. Natl. Acad. Sci. US* 110(2):636-641.

Ge, Y., Preston, R. J., Owen, R. D. (2007) Toxicoproteomics and its application to human health risk assessment. *Proteomics Clin. Appl.* 1(12):1613-1624.

George, J., Singh, R., Mahmood, Z., Shukla, Y. (2010) Toxicoproteomics: new paradigms in toxicology research. *Toxicol. Mech. Methods* 20(7):415-423.

Gygi, S. P., Rist, B., Griffin, T. J., Eng, J., Aebersold, R. (2002) Proteome analysis of low-abundance proteins using multidimensional chromatography and isotope-coded affinity tags. *Journal of proteome research* 1(1):47-54.

Huang, T. C., Sahasrabuddhe, N. A., Kim, M. S., Getnet, D., Yang, Y., Peterson, J. M., Ghosh, B., Chaerkady, R., Leach, S. D., Marchionni, L., Wong, G. W., Pandey, A. (2012) Regulation of lipid metabolism by Dicer revealed through SILAC mice. *Journal of proteome research* 11(4):2193-2205.

Jagadapillai, R., Chen, J., Canales, L., Birtles, T., Pisano, M. M., Neal, R. E. (2012) Developmental cigarette smoke exposure: kidney proteome profile alterations in low birth weight pups. *Toxicology* 299(2-3):80-89.

Jensen, O. N. (2004) Modification-specific proteomics: characterization of post-translational modifications by mass spectrometry. *Curr. Opin. Chem. Biol.* 8(1):33-41.

Kodavanti, P. R., Osorio, C., Royland, J. E., Ramabhadran, R., Alzate, O. (2011) Aroclor 1254, a developmental neurotoxicant, alters energy metabolism- and intracellular signaling-associated protein networks in rat cerebellum and hippocampus. *Toxicol. Appl. Pharmacol.* 256(3):290-299.

Kruger, M., Moser, M., Ussar, S., Thievessen, I., Luber, C. A., Forner, F., Schmidt, S., Zanivan, S., Fassler, R., Mann, M. (2008) SILAC mouse for quantitative proteomics uncovers kindlin-3 as an essential factor for red blood cell function. *Cell* 134(2):353-364.

Lantz, R. C., Lynch, B. J., Boitano, S., Poplin, G. S., Littau, S., Tsaprailis, G., Burgess, J. L. (2007) Pulmonary biomarkers based on alterations in protein expression after exposure to arsenic. *Environ. Health Perspect.* 115(4):586-591.

Lanucara, F., Eyers, C. E. (2011) Mass spectrometric-based quantitative proteomics using SILAC. *Methods Enzymol.* 500133-150.

Lau, A. T., He, Q. Y., Chiu, J. F. (2004) A proteome analysis of the arsenite response in cultured lung cells: evidence for in vitro oxidative stress-induced apoptosis. *Biochem. J.* 382(Pt 2):641-650.

Lidsky, T. I., Schneider, J. S. (2003) Lead neurotoxicity in children: basic mechanisms and clinical correlates. *Brain* 126(Pt 1):5-19.

Mates, J. M., Segura, J. A., Alonso, F. J., Marquez, J. (2010) Roles of dioxins and heavy metals in cancer and neurological diseases using ROS-mediated mechanisms. *Free Radic. Biol. Med.* 49(9):1328-1341.

Monteoliva, L., Albar, J. P. (2004) Differential proteomics: an overview of gel and non-gel based approaches. *Brief Funct. Genomic Proteomic* 3(3):220-239.

Mostafalou, S., Abdollahi, M. (2013) Pesticides and human chronic diseases: Evidences, mechanisms, and perspectives. *Toxicol. Appl. Pharmacol.* 268(2):157-177.

Ong, S. E., Blagoev, B., Kratchmarova, I., Kristensen, D. B., Steen, H., Pandey, A., Mann, M. (2002) Stable isotope labeling by amino acids in cell culture, SILAC, as a simple and accurate approach to expression proteomics. *Mol. Cell. Proteomics* 1(5):376-386.

Ong, S. E., Foster, L. J., Mann, M. (2003) Mass spectrometric-based approaches in quantitative proteomics. *Methods* 29(2):124-130.

Ong, S. E., Mann, M. (2005) Mass spectrometry-based proteomics turns quantitative. *Nat. Chem. Biol.* 1(5):252-262.

Ong, S. E., Mann, M. (2006) A practical recipe for stable isotope labeling by amino acids in cell culture (SILAC). *Nat. Protoc.* 1(6):2650-2660.

Pan, X., Whitten, D. A., Wu, M., Chan, C., Wilkerson, C. G., Pestka, J. J. (2013) Global protein phosphorylation dynamics during deoxynivalenol-induced ribotoxic stress response in the macrophage. *Toxicol. Appl. Pharmacol.* 268(2):201-211.

Prins, J. M., Brooks, D. M., Thompson, C. M., Lurie, D. I. (2010) Chronic low-level Pb exposure during development decreases the expression of the voltage-dependent anion channel in auditory neurons of the brainstem. *Neurotoxicology* 31(6):662-673.

Prins, J. M., Wang, Y. (2012) Quantitative proteomic analysis revealed 4-(methylnitrosamino)-1-(3-pyridinyl)-1-butanone-induced up-regulation of 20S proteasome in cultured human fibroblast cells. *Journal of proteome research* 11(4):2347-2354.

Sa-Correia, I., Teixeira, M. C. (2010) 2D electrophoresis-based expression proteomics: a microbiologist's perspective. *Expert Rev. Proteomics* 7(6):943-953

Schulz, M., Brandner, S., Eberhagen, C., Eckardt-Schupp, F., Larsen, M. R., Andrae, U. (2013) Quantitative phosphoproteomic analysis of early alterations in protein phosphorylation by 2,3,7,8-tetrachlorodibenzo-p-dioxin. *Journal of proteome research* 12 (2):866-882.

Singh, A. K., Tiwari, M. N., Dixit, A., Upadhyay, G., Patel, D. K., Singh, D., Prakash, O., Singh, M. P. (2011) Nigrostriatal proteomics of cypermethrin-induced dopaminergic neurodegeneration: microglial activation-dependent and -independent regulations. *Toxicol. Sci.* 122(2):526-538.

Smaczniak, C., Li, N., Boeren, S., America, T., van Dongen, W., Goerdayal, S. S., de Vries, S., Angenent, G. C., Kaufmann, K. (2012) Proteomics-based identification of low-abundance signaling and regulatory protein complexes in native plant tissues. *Nat. Protoc.* 7(12):2144-2158.

Sury, M. D., Chen, J. X., Selbach, M. (2010) The SILAC fly allows for accurate protein quantification in vivo. *Mol. Cell Proteomics* 9(10):2173-2183.

Tannu, N. S., Hemby, S. E. (2006) Two-dimensional fluorescence difference gel electrophoresis for comparative proteomics profiling. *Nat. Protoc.* 1(4):1732-1742.

Thiede, B., Hohenwarter, W., Krah, A., Mattow, J., Schmid, M., Schmidt, F., Jungblut, P. R. (2005) Peptide mass fingerprinting. *Methods* 35(3):237-247.

Unwin, R. D., Griffiths, J. R., Whetton, A. D. (2010) Simultaneous analysis of relative protein expression levels across multiple samples using iTRAQ isobaric tags with 2D nano LC-MS/MS. *Nat. Protoc.* 5(9):1574-1582.

Wetmore, B. A., Merrick, B. A. (2004) Toxicoproteomics: proteomics applied to toxicology and pathology. *Toxicol. Pathol.* 32(6):619-642.

Wu, W. W., Wang, G., Baek, S. J., Shen, R. F. (2006) Comparative study of three proteomic quantitative methods, DIGE, cICAT, and iTRAQ, using 2D gel- or LC-MALDI TOF/TOF. *Journal of proteome research* 5(3):651-658

Xiong, L., Wang, Y. (2010) Quantitative proteomic a.nalysis reveals the perturbation of multiple cellular pathways in HL-60 cells induced by arsenite treatment. *Journal of proteome research* 9(2):1129-1137.

In: Recent Advances in Proteomics Research
Editor: Ming D. Li

ISBN: 978-1-62948-218-7

Chapter 10

Toxin Proteomics for the Identification and Differentiation of Botulinum Neurotoxins

***Suzanne R. Kalb** and John R. Barr*†**

Centers for Disease Control and Prevention, National Center for Environmental Health, Division of Laboratory Sciences, Atlanta, GA, US

Abstract

Botulinum neurotoxins (BoNTs) cause the disease botulism, which can be lethal if untreated. There are seven known serotypes of BoNT, A-G, defined by their response to antisera. Many serotypes have toxin subtypes based on amino acid sequence and immunogenic properties, and some subtypes are further differentiated into toxin variants which have minor amino acid changes. Toxin characterization is important as different types and subtypes of BoNT can respond differently to medical countermeasures for botulism, and detailed characterization of the toxin can aid in epidemiologic and forensic investigations. Proteomic techniques have been established to determine the serotype, subtype, or toxin variant of BoNT. These techniques involve serotype specific antibody extraction of the toxin, digestion with proteolytic enzymes, MS/MS analysis of the resulting peptides, and database searching to identify the BoNT protein. These techniques demonstrate the capability to characterize BoNT and their neurotoxin-associated proteins and have been capable of differentiating toxins, some of which have been up to 99.9% identical. This differentiation can be accomplished from toxins present in a complex matrix such as stool, food, or bacterial cultures and no DNA is required.

Keywords: Mass Spectrometry, Proteomics, Botulinum Neurotoxin, Botulism

* Suzanne R. Kalb: Centers for Disease Control and Prevention, National Center for Environmental Health, Division of Laboratory Sciences, 4770 Buford Hwy, NE, MS-F50, Atlanta, GA 30341. E-mail: skalb@cdc.gov.
† John R. Barr. E-mail: jbarr@cdc.gov.

Abbreviations

BoNT	Botulinum neurotoxin
SNAP	synaptosomal-associated protein
NAP	neurotoxin-associated protein
mAb	monoclonal antibody
MALDI	matrix-assissted laser desorption/ionization
PSD	post-source decay
PMF	peptide mass fingerprinting
HA	hemagglutinin
NTNH	nontoxic-nonhemagglutinin
ESI	electrospray ionization

Introduction

Botulinum neurotoxins (BoNTs) are a group of highly toxic protein neurotoxins secreted by *Clostridum botulinum*, *C. butyricum*, *C. baratii*, and *C. argentinense*. Exposure to BoNTs can lead to death through paralysis and causes the disease known as botulism. BoNT consists of a 150 kDa heavy chain, which binds to receptors on the neuron, and a 50 kDa light chain with proteolytic activity. The BoNT light chain causes paralysis by cleaving proteins necessary for nerve signal transmission. Botulinum neurotoxins are currently classified into seven serotypes labeled A-G, based on their response to antisera. BoNT/A, /C, and /E cleave SNAP-25 (synaptosomal-associated protein) (Binz et al. 1994; Blasi et al. 1993; Foran et al. 1996; Schiavo et al. 1993; Schiavo et al. 1993; Williamson et al. 1996), and BoNT/B, /D, /F, and /G cleave synaptobrevin-2 (also known as VAMP-2) (Kalb et al. 2012; Schiavo et al. 1992; Schiavo et al. 1994; Schiavo et al. 1993; Yamasaki et al. 1994; Yamasaki et al. 1994) as seen in Figure 1. BoNT is secreted as part of a protein complex combined with neurotoxin-associated proteins (NAPs).

The seven different serotypes of BoNT are differentiated according to their capability to be neutralized by an antiserum produced using a specific BoNT type. As each serotype is neutralized by a different antiserum, identification of the serotype of BoNT is important. BoNTs can also be differentiated below the serotype level, and this is known as subtype differentiation. Subtypes can be defined as nucleotide or amino acid sequence variation within BoNTs in strains of a serotype. However, historically subtypes have been defined by cultural/biochemical characteristics (Wagner et al. 1925), functional differences (Kozaki et al. 1998), or differential binding of monoclonal antibodies (mAbs) (Gibson et al. 1988; Kozaki et al. 1995; Smith et al. 2005). Although amino acid variance within a serotype is typically less than 10%, it can be as much as 30% in the case of BoNT/F1 and /F5 (Raphael et al. 2010). Production of the new toxin variant (neurotoxin protein) could be due to a new strain (*Clostridium* organism), and some of the neurotoxins of some strains have as few as a single amino acid difference, or 0.08% difference.

Differentiation or identification of the subtype of BoNT is important. Different subtypes of toxin within a serotype show different binding characteristics to panels of monoclonal antibodies and in some cases even to polyclonal antibodies (Kozaki et al. 1998; Umeda et al. 2009).

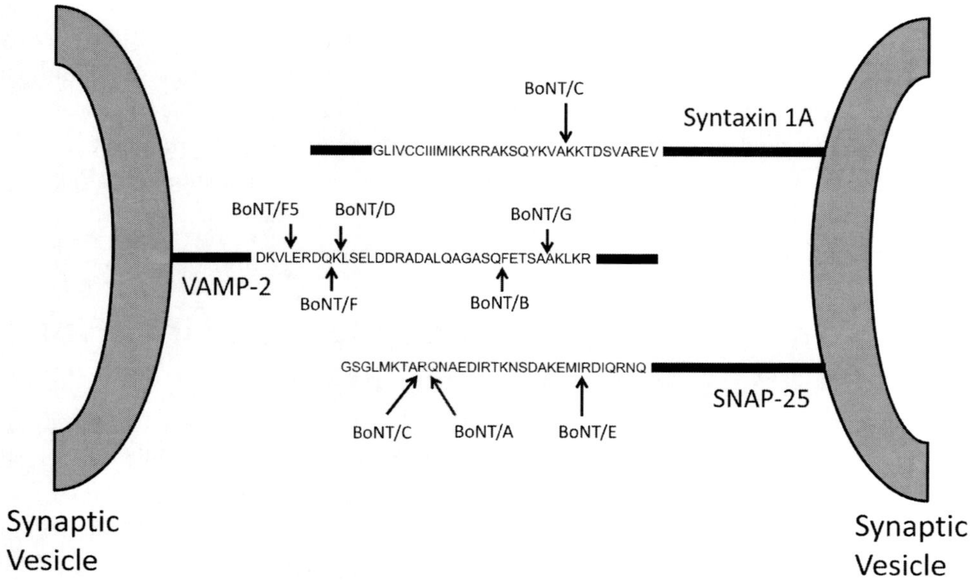

Figure 1. Schematic of the SNARE complex proteins necessary for nerve transmission via acetylcholine release. The cleavage sites of the proteins by individual BoNT serotypes are depicted.

As researchers search for alternative treatments to the equine immunoglobulin currently in use to treat botulism, assessment of differential binding becomes important. Some mAb are under investigation (Marks 2004) to treat botulism and differential binding of these antibodies to different subtypes would be a critical component for treatment efficacy from botulism caused by different subtypes. Additionally, differentiation of the BoNT subtype is important to epidemiologic and forensic investigations attempting to determine the origin of the toxin, its spread in a botulism incident, and commonality/differences in concurrent botulism outbreaks.

Pulsed field gel electrophoresis (Lin and Johnson 1995) has been a common technique to indicate genetic differences between strains in botulism outbreaks. Toxin gene sequencing has traditionally been the approach to identify subtypes but is time consuming and labor intensive. Other techniques to determine the toxin subtype include randomly amplified polymorphic DNA analysis (Hielm et al. 1999), amplified fragment-length polymorphism analysis (Keto-Timonen et al. 2005), *fla*A variable-region sequencing (Paul et al. 2007), multilocus sequence typing (Jacobson et al. 2008), multiple locus variable-number tandem repeat analysis (Macdonald et al. 2008), and a comparative genomic hybridization microarray (Raphael et al.).

One major limitation of all these methods is that they rely on the presence of bacterial DNA in the sample material. Botulism is a toxin mediated disease and is caused by intoxication with BoNT. The disease does not require the presence of the organism, and in some botulism cases, the bacterium is not present while the toxin is present. In such a circumstance, toxin subtype identification or strain characterization using DNA-based methods is challenging, but could be possible (Macdonald et al. 2008; De Medici et al. 2009) if samples still contain small amounts of DNA. In cases where no DNA is found, then DNA-based methods cannot be utilized for subtype identification/differentiation. In this situation, an alternate method which does not rely upon the presence of DNA would be the only option for subtype/toxin variant identification.

Because BoNT is present, a proteomic technique could be utilized to identify/differentiate the amino acid composition of the neurotoxin in question. Also, proteomic techniques are very rapid and have been used to promptly answer critical questions in the necessary time frame of a botulism outbreak investigation. In this review, we detail proteomic methods which yield information on the amino acid sequence of the toxin in question and thus reveal its serotype or subtype/toxin variant.

Botulinum Neurotoxin Serotype Identification by Mass Spectrometry

The seven known serotypes of BoNT are 64% or less identical at the amino acid sequence level. The toxins from different serotypes contain hundreds of amino acid differences and can be differentiated from each other through determination of their amino acid sequences as seen in Figure 2.The toxins are first digested with an enzyme such as trypsin which breaks the protein into many smaller peptides. The peptides are then fragmented inside the mass spectrometer to produce MS/MS which are specific for the amino acid sequence of the peptide. By knowing the amino acid sequence of a peptide originating from the toxin, the amino acid sequence and hence the identity of the toxin can be determined.One example of this is the difference between BoNT/A and /B as depicted in Figure 3. Residues which differ between BoNT/A and /B are marked in bold and underlined. It is apparent from the large number of amino acid differences that BoNT/B is very dissimilar from BoNT/A, and knowledge of these differences would lead to the ability to differentiate these serotypes.

Mass spectrometry is a powerful tool for identification of the amino acid sequence of a protein. Ben van Baar et al. reported on mass spectral analysis of BoNT/A1 and /B1 (van Baar et al. 2002). They digested purified BoNT/A and /B protein complexes with trypsin and analyzed the mix of peptides by MALDI-MS in order to obtain a peptide mass fingerprinting (PMF) map. In some selected cases, peptides were sequenced using MALDI-PSD.

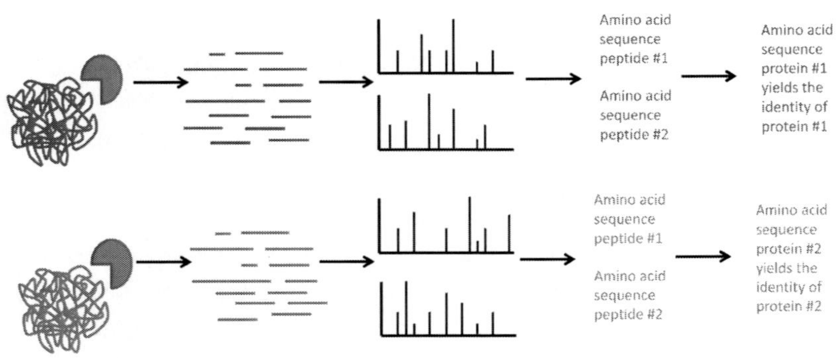

Figure 2. Graphical representation of the use of proteomics techniques to determine the identity of BoNT.

PFVNKQFNYK DPVNGVDIAY IKIPNAGQMQ PVKAFKIHNK IWVIPERDTF TNPEEGDLNP
PPEAKQVPVS YYDSTYLSTD NEKDNYLKGV TKLFERIYST DLGRMLLTSI VRGIPFWGGS
TIDTELKVID TNCINVIQPD GSYRSEELNL VIIGPSADII QFECKSFGHE VLNLTRNGYG
STQYIRFSPD FTFGFEESLE VDTNPLLGAG KFATDPAVTL AHELIHAGHR LYGIAINPNR
VFKVNTNAYY EMSGLEVSFE ELRTFGGHDA KFIDSLQENE FRLYYYNKFK DIASTLNKAK
SIVGTTASLQ YMKNVFKEKY LLSEDTSGKF SVDKLKFDKL YKMLTEIYTE DNFVKFFKVL
NRKTYLNFDK AVFKINIVPK VNYTIYDGFN LRNTNLAANF NGQNTEINNM NFTKLKNFTG
LFEFYKLLCV RGIITSKTKS LDKGYNKALN DLCIKVNNWD LFFSPSEDNF TNDLNKGEEI
TSDTNIEAAE ENISLDLIQQ YYLTFNFDNE PENISIENLS SDIIGQLELM PNIERFPNGK
KYELDKYTMF HYLRAQEFEH GKSRIALTNS VNEALLNPSR VYTFFSSDYV KKVNKATEAA
MFLGWVEQLV YDFTDETSEV STTDKIADIT IIIPYIGPAL NIGNMLYKDD FVGALIFSGA
VILLEFIPEI AIPVLGTFAL VSYIANKVLT VQTIDNALSK RNEKWDEVYK YIVTNWLAKV
NTQIDLIRKK MKEALENQAE ATKAIINYQY NQYTEEEKNN INFNIDDLSS KLNESINKAM
ININKFLNQC SVSYLMNSMI PYGVKRLEDF DASLKDALLK YIYDNRGTLI GQVDRLKDKV
NNTLSTDIPF QLSKYVDNQR LLSTFTEYIK NIINTSILNL RYESNHLIDL SRYASKINIG
SKVNFDPIDK NQIQLFNLES SKIEVILKNA IVYNSMYENF STSFWIRIPK YFNSISLNNE
YTIINCMENN SGWKVSLNYG EIIWTLQDTQ EIKQRVVFKY SQMINISDYI NRWIFVTITN
NRLNNSKIYI NGRLIDQKPI SNLGNIHASN NIMFKLDGCR DTHRYIWIKY FNLFDKELNE
KEIKDLYDNQ SNSGILKDFW GDYLQYDKPY YMLNLYDPNK YVDVNNVGIR GYMYLKGPRG
SVMTTNIYLN SSLYRGTKFI IKKYASGNKD NIVRNNDRVY INVVVKNKEY RLATNASQAG
VEKILSALEI PDVGNLSQVV VMKSKNDQGI TNKCKMNLQD NNGNDIGFFI GFHQFNNIAK
LVASNWYNRQ IERSSRTLGC SWEFIPVDDG WGERPL

Figure 3. The amino acid sequence of BoNT/A. Residues in bold and underlined differ between BoNT/A and /B.

Additionally, the tryptic digests of BoNT/A and /B were further analyzed by LC- ESI-MS/MS on a QTOF instrument. PMF and LC-MS/MS data were searched against the NCBI database containing BoNT protein sequences. In this manner, van Baar reported sequence coverage of greater than 35% for BoNT/A1 and 25% for BoNT/B1. These sequence coverages were sufficient to identify the protein as BoNT and to identify the serotype of the toxin, as BoNT/A or /B.

In 2004, van Baar et al. extended their previous work to incorporate serotypes /C, /D, /E, and /F (van Baar et al. 2004). Using the same procedures as detailed in their 2002 publication [29], they reported identification of the amino acid sequence for 11% of BoNT/C, 49% of BoNT/D, 30% of BoNT/E3, and 53% of BoNT/F1. Because these proteins are 36% or more dissimilar, these sequence coverages were enough to identify the protein as BoNT and to differentiate the serotype of BoNT. Detection of BoNT and differentiation of the serotype of BoNT by mass spectrometry is possible with other techniques [31, 32] that do not examine the amino acid composition of the neurotoxin.

These methods detect and identify BoNT by detecting the specific enzymatic activity of the neurotoxin. Each serotype of toxin has a different enzymatic target or cleavage site on synaptobrevin-2 or SNAP-25, so mass spectrometric detection of the cleavage of synaptobrevin-2 or SNAP-25 at different cleavage sites will identify the presence of BoNT and differentiate the serotype.

These methods are suitable for detecting the presence of active BoNT and differentiating the serotype with far greater sensitivity than the proteomic approaches; however, such a method is deficient in differentiating beyond the serotype levels as all BoNT subtypes within a serotype cleave synaptobrevin-2 or SNAP-25 in the same location (Kalb et al. 2008), excluding BoNT/F5 (Kalb et al. 2012).

Neurotoxin-Associated Proteins Identification with Mass Spectrometry

BoNTs are produced by *Clostridium botulinum* and other closely-related organisms and are generated as part of a protein complex also known as the progenitor toxin, comprising the neurotoxin and neurotoxin-associated proteins (NAPs). The composition and structure of the associated proteins of the progenitor toxin often differs between serotypes. It is generally believed that the NAPs shield the neurotoxin from the harsh conditions found in the stomach and digestive tract, especially low pH and digestive enzymes (Sugii et al. 1977) but the full role of these proteins is not completely understood. Additionally, it has been suggested that the NAPs aid translocation of the neurotoxin across the intestinal epithelium (Fujinaga et al. 2000), and the NAPs may add to the immunogenicity of BoNT/A (Lee et al. 2005). Identification and characterization of the NAPs might support BoNT serotype identification as the NAPs often differ between serotypes.

Mass spectral identification of the NAPs associated with BoNT/A-G was first reported in 2005 (Hines et al. 2005). The protein complexes of purified BoNT/A-G were digested with trypsin and the resulting peptides were then separated by LC and both MS and MS/MS spectra were recorded on a QTOF mass spectrometer. The mass spectral data were searched against the NCBI database for protein identification. In this fashion, the neurotoxins were identified and differentiated from each other with sequence coverage of 18%, 23.5%, 33.3%, 18.6%, 36.5%, 21.9%, and 16.1% for BoNT/A-G respectively. These sequence coverages are comparable to what was previously reported by van Baar (van Baar et al. 2002; van Baar et al. 2004). However, this work also reported for the first time, mass spectral identification of the NAPs, and reported that the NAPs varied between the serotypes. BoNT/A was identified in a complex with nontoxic-nonhemagglutinin and hemagglutinin proteins; specifically, NTNH, HA-70, HA-33, HA-17, and flagellin (Hines et al. 2005), while BoNT/B was found complexed with NTNH, HA-70, HA-33, HA-17 and flagellin. In all cases other than HA-70, the NAPs from BoNT/A and /B were found to have different amino acid sequences, making it possible to distinguish serotypes (Hines et al. 2005). BoNT/C complex is composed of NTNH, HA-70, HA-33, and HA-17; BoNT/D complex has NTNH, HA-70, HA-33, and flagellin; BoNT/E complexes NTNH, ORF-X1, and NBP; BoNT/F with NTNH only; and BoNT/G was reported in a complex with NTNH, HA-70, and HA-17. The serotype of toxin complex could be discerned based on at least one of the associated proteins as most of the NAPs have different amino acid sequences (Hines et al. 2005).

Sequence coverage of 10-50% has been found to be adequate for BoNT serotype identification, but improved sequence coverage substantiates differentiation. The work summarized previously in this section (van Baar et al. 2002; van Baar et al. 2004; Hines et al. 2005) was all performed on purified toxin complexes, which is a mixture of proteins. Assessment of protein mixtures often causes decreases in the sequence coverage for any single protein. In 2011, Moura et al. published on the mass spectral identification of purified BoNT/C and BoNT/D complexes (Moura et al. 2011). In this work, they separated the individual proteins of the progenitor complexes by gel and then digested each gel band containing a single protein. This work confirmed the previous studies (Hines et al. 2005), but adding gels to the front end of the LC-MS/MS analysis yielded higher sequence coverage of most proteins despite the fact that less toxin was used for this experiment (Moura et al. 2011).

Also, this work introduced a label-free quantitation method to report that 31% of the BoNT/C complex consisted of the neurotoxin although only 22% of the BoNT/D complex consisted of the neurotoxin (Moura et al. 2011). Many of these same procedures were used by Terilli et al in 2011, reporting on the proteomic analysis of BoNT/G. Sequence coverages were reported as 66% and higher for BoNT/G, NTNH, HA-70, and HA-17 (Terilli et al. 2011), allowing for simple discrimination of the serotype of toxin complex. The label-free quantitation method reported a molar ratio of BoNT:NTNH:HA-70:HA-17 as 1:1:2:1. Additionally, work described by Moura and Terrili used a 3 minute in-solution digestion of the BoNT/C, /D, and /G complexes (Moura et al. 2011; Terilli et al. 2011). Reducing the digestion time from a standard overnight digestion to a 3 minute process allows for faster BoNT serotype identification because the overnight digestion step is the longest step in the process.

Serotype Detection of Bont in Complex Matrices

BoNTs have been purified and are commercially available for research purposes as purified neurotoxins or purified progenitor complexes. Toxins from botulism outbreaks can sometimes be grown in bacterial cultures, or be present in associated foods, and animal/human clinical samples. The clinical samples include serum, gastric, and stool. Matrices such as these generally contain relatively low toxin levels with high levels of matrix proteins. Therefore, for botulism outbreak investigations, it is critical to identify and distinguish the serotype of BoNT present in a complex matrix.

Klaubert et al. published in 2009 on a method to detect BoNT/A, /B, /E, and /F (the four serotypes which are associated with human cases of botulism) in bacterial cultures (Klaubert et al. 2009). The method used pepsin digestion of the culture at pH 2. BoNT is shielded by the NAPs at low pH in the digestive track and thus was partially shielded from digestion by the pepsin while the unwanted matrix proteins were digested to small peptides. Peptides were then separated from proteins using molecular weight cut-off filters. The retained proteins (consisting of mostly BoNT and its NAPs) were digested with trypsin and separated by 2D nano-LC using ion exchange chromatography and reversed phase chromatography. The peptides were analyzed by nano LC-ESI MS/MS and the data were searched against the NCBI database. Although sequence coverage was somewhat low (less than 10%), toxin from bacterial cultures of four serotypes could be detected and differentiated (Klaubert et al. 2009).

An alternate method to detect BoNT in the presence of a complex matrix is immunoaffinity purification of the toxin from matrix proteins. Antibodies to the toxin allows for enrichment of the toxin from food (Kalb et al. 2005; Kull et al. 2010; Wang et al. 2012) or bacterial culture (Kull et al. 2010; Kalb et al. 2012). Through this process, both BoNT/A (Kalb et al. 2005; Kull et al. 2010; Wang et al. 2012) and BoNT/B (Kull et al. 2010; Kalb et al. 2012) were identified while in complex matrices. Interestingly, this technique reported exquisitely high sequence coverage—between 65 and 98%. Such high sequence coverage ensures correct identification and differentiation of these toxins. These techniques use antibodies which are serotype-specific for toxin extraction; however, an antibody has been described to bind all BoNT/A, /B, /E, and /F, the four serotypes associated with human botulism (Kalb et al. 2010).

Theoretically, this antibody could be utilized for purification of any BoNT associated with human cases present in a complex matrix.

Mass Spectrometry-Based Differentiation of Bont at the Subtype/Toxin Variant Level

Differentiation of BoNT below the serotype level to the subtype/toxin variant level can be important for forensic or epidemiological purposes and for the assessment of potential antitoxin medical countermeasures.Although there are many assays to detect BoNT and distinguish it at the serotype level, few methods exist to differentiate the toxin below the serotype level, and the majority of these methods depend upon the analysis of DNA for that purpose. The serotypes of BoNT are easier to differentiate because they are only 34-64% identical at the amino acid level. At the subtype level the amino acid identity increases to between 70-97.5% and further increases to greater than 97.5% at the toxin variant level. Thus, it is critical to have high sequence coverage to correctly identify a neurotoxin below the serotype level. As an example, Figure 4 is the amino acid sequence of BoNT/A1. Residues which differ between BoNT/A1 and /A2 are bolded and underlined. These two protein sequences are approximately 90% identical.

Differentiation of BoNT below the serotype level was first reported in 2005 (Kalb et al. 2005). In this work, BoNT/A1 and /A2 were added to milk and extracted using serotype-specific antibodies to BoNT/A. Both toxins were identified as BoNT/A via analysis of the enzymatic activity of their light chains upon a peptide substrate. The toxins were then digested with trypsin and analyzed by LC-MS/MS.

```
PFVNKQFNYKDPVNGVDIAYIKIPNAGQMQPVKAFKIHNKIWVIPERDTFTNPEEGDLNP
PPEAKQVPVSYYDSTYLSTDNEKDNYLKGVTKLFERIYSTDLGRMLLTSIVRGIPFWGGS
TIDTELKVIDTNCINVIQPDGSYRSEELNLVIIGPSADIIQFECKSFGHEVLNLTRNGYG
STQYIRFSPDFTFGFEESLEVDTNPLLGAGKFATDPAVTLAHELIHAGHRLYGIAINPNR
VFKVNTNAYYEMSGLEVSFEELRTFGGHDAKFIDSLQENEFRLYYYNKFKDIASTLNKAK
SIVGTTASLQYMKNVFKEKYLLSEDTSGKFSVDKLKFDKLYKMLTEIYTEDNFVKFFKVL
NRKTYLNFDKAVFKINIVPKVNYTIYDGFNLRNTNLAANFNGQNTEINNMNFTKLKNFTG
LFEFYKLLCVRGIITSKTKSLDKGYNKALNDLCIKVNNWDLFFSPSEDNFTNDLNKGEEI
TSDTNIEAAEENISLDLIQQYYLTFNFDNEPENISIENLSSDIIGQLELMPNIERFPNGK
KYELDKYTMFHYLRAQEFEHGKSRIALTNSVNEALLNPSRVYTFFSSDYVKKVNKATEAA
MFLGWVEQLVYDFTDETSEVSTTDKIADITIIIPYIGPALNIGNMLYKDDFVGALIFSGA
VILLEFIPEIAIPVLGTFALVSYIANKVLTVQTIDNALSKRNEKWDEVYKYIVTNWLAKV
NTQIDLIRKKMKEALENQAEATKAIINYQYNQYTEEEKNNINFNIDDLSSKLNESINKAM
ININKFLNQCSVSYLMNSMIPYGVKRLEDFDASLKDALLKYIYDNRGTLIGQVDRLKDKV
NNTLSTDIPFQLSKYVDNQRLLSTFTEYIKNIINTSILNLRYESNHLIDLSRYASKINIG
SKVNFDPIDKNQIQLFNLESSKIEVILKNAIVYNSMYENFSTSFWIRIPKYFNSISLNNE
YTIINCMENNSGWKVSLNYGEIIWTLQDTQEIKQRVVFKYSQMINISDYINRWIFVTITN
NRLNNSKIYINGRLIDQKPISNLGNIHASNNIMFKLDGCRDTHRYIWIKYFNLFDKELNE
KEIKDLYDNQSNSGILKDFWGDYLQYDKPYYMLNLYDPNKYVDVNNVGIRGYMYLKGPRG
SVMTTNIYLNSSLYRGTKFIIKKYASGNKDNIVRNNDRVYINVVVKNKEYRLATNASQAG
VEKILSALEIPDVGNLSQVVVMKSKNDQGITNKCKMNLQDNNGNDIGFIGFHQFNNIAK
LVASNWYNRQIERSSRTLGCSWEFIPVDDGWGERPL
```

Figure 4. The amino acid sequence of BoNT/A1. Residues in bold and underlined differ between BoNT/A1 and /A2.

By comparing the obtained data against protein sequences in the NCBI database, the toxins were differentiated as either BoNT/A1 or /A2 with sequence coverages of 65-70% (Kalb et al. 2005). As an example, Figure 5A is a MS/MS spectrum of the peptide SFGHEVLNLTR from the tryptic digest of BoNT/A1 and Figure 5B is SFGHDVLNLTR from BoNT/A2. Although the peptides are similar, they differ in the fifth residue. This difference results in a measurable mass difference of the intact peptide and a different fragmentation pattern.

In 2010, differentiation of two subtypes of BoNT/B; BoNT/B1 and /B4, was reported. The toxins were extracted from culture supernatants using immunoaffinity capture and then digested with trypsin followed by MS/MS analysis by MALDI-TOF (Kull et al. 2010). Using similar techniques, differentiation of BoNT/B1-/B5 extracted from culture supernatants was reported in 2012 (Kalb et al. 2012). High sequence coverages of 66-77% allowed for differentiation of these five different toxins at the subtype level in spite of similarities as high as 98% in the case of BoNT/B2 and /B3 (Kalb et al. 2012).

Recently, Wang et al. reported of a new technique designed to enhance sequence coverage of BoNT proteins (Wang et al. 2012). The toxin complexes were subjected to SDS-PAGE to separate the components of the complex, followed by multiple-enzyme digestion of each gel band in a sequential fashion. Peptides from the gel band comprising the neurotoxin were then analyzed by LC-MS/MS followed by database searching. Analysis in this fashion yielded sequence coverage of 90% and greater for all serotypes of BoNT (Wang et al. 2012).

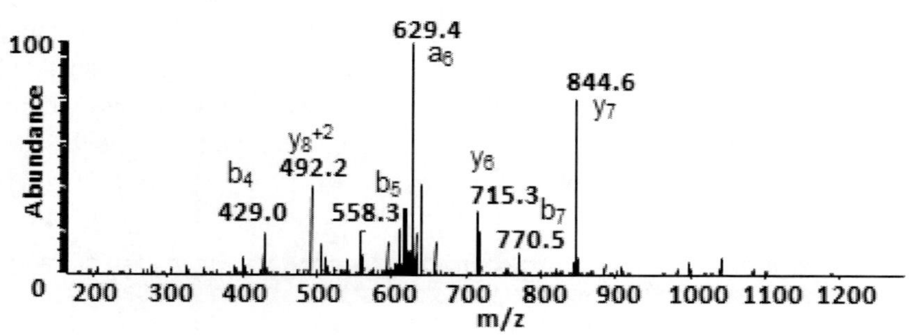

a

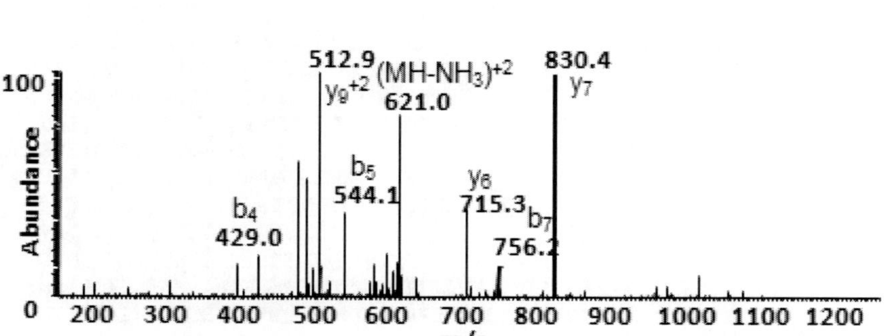

b

Figure 5. MS/MS spectra of the peptides SFGHEVLNLTR (5A) and SFGHDVLNLTR (5B) from the tryptic digests of BoNT/A1 and /A2 respectively.

Additionally, this technique was able to confidently identify the toxin present in a toxin-contaminated carrot juice sample associated with a botulism outbreak. For this analysis, the toxin was first extracted from the carrot juice using immunoaffinity capture prior to separation of the extracted proteins by SDS-PAGE. The sequence coverage of the toxin was 98.6%, securing a determination of the toxin as BoNT/A1 (GenBank accession no. ABY56330) (Wang et al. 2012).

In fact, this extremely high level of sequence coverage allowed for determination of the toxin beyond the subtype level. Toxin variants exist in some subtypes of BoNT, and the neurotoxins produced by different strains of *Clostridium botulinum* within the same subtype can be up to 99.9% identical at the amino acid level.As an example, Figure 6 is the amino acid sequence of BoNT/A1 produced by the Hall strain of *Clostridium botulinum*. Residues bolded and underlined vary between A1 Hall and A1(B) CDC 1744. The amino acid sequences of these proteins are 99.8% identical with only 2 amino acids which differ in a protein of 1290 amino acids. A sequence coverage of 98.6% as reported in the above work allowed for the determination of the toxin variant in the carrot juice outbreak as there are two similar toxin variants which are 99.9% and 99.6% identical to the toxin variant identified in this sample.

These proteomic methods (*vide supra*) rely upon fitting the mass spectral data from a protein of interest to amino acid sequences of known proteins within a database. This technique can result in an incorrect identification if the true sequence of the protein is not yet known and in the database. To alleviate this problem, in 2012, a method was reported to identify novel subtypes or toxin variants of BoNT/B which have not yet been sequenced and input into the database (Kalb et al. 2012). This method used an amino acid substitution database in which each amino acid within the sequence of BoNT/B1 was individually altered to the other 19 possibilities. The MS/MS data were then searched against this database of possible mutations for a potential match.

PFVNKQFNYKDPVNGVDIAYIKIPN**A**GQMQPVKAFKIHNKIWVIPERDTFTNPEEGDLNP
PPEAKQVPVSYYDSTYLSTDNEKDNYLKGVTKLFERIYSTDLGRMLLTSIVRGIPFWGGS
TIDTELKVIDTNCINVIQPDGSYRSEELNLVIIGPSADIIQFECKSFGHEVLNLTRNGYG
STQYIRFSPDFTFGFEESLEVDTNPLLGAGKFATDPAVTLAHELIHAGHRLYGIAINPNR
VFKVNTNAYYEMSGLEVSFEELRTFGGHDAKFIDSLQENEFRLYYYNKFKDIASTLNKAK
SIVGTTASLQYMKNVFKEKYLLSEDTSGKFSVDKLKFDKLYKMLTEIYTEDNFVKFFKVL
NRKTYLNFDKAVFKINIVPKVNYTIYDGFNLRNTNLAANFNGQNTEINNMNFTKLKNFTG
LFEFYKLLCVRGIITSKTKSLDKGYNKALNDLCIKVNNWDLFFSPSEDNFTNDLNKGEEI
TSDTNIEAAEENISLDLIQQYYLTFNFDNEPENISIENLSSDIIGQLELMPNIERFPNGK
KYELDKYTMFHYLRAQEFEHGKSRIALTNSVNEALLNPSRVYTFFSSDYVKKVNKATEAA
MFLGWVEQLVYDFTDETSEVSTTDKIADITIIIPYIGPALNIGNMLYKDDFVGALIFSGA
VILLEFIPEIAIPVLGTFALVSYIANKVLTVQTIDNALSKRNEKWDEVYKYIVTNWLAKV
NTQIDLIRKKMKEALENQAEATKAIINYQYNQYTEEEKNNINFNIDDLSSKLNESINKAM
ININKFLNQCSVSYLMNSMIPYGVKRLEDFDASLKDALLKYIYDNRGTLIGQVDRLKDKV
NNTLSTDIPFQLSKYVDNQRLLSTFTEYIKNIINTSILNLRYESNHLIDLSRYASKINIG
SKVNFDPIDKNQIQLFNLESSKIEVILKNAIVYNSMYENFSTSFWIRIPKYFNSISLNNE
YTIINCMENNSGWKVSLNYGEIIWTLQDTQEIKQRVVFKYSQMINISDYINRWIFVTITN
NRLNNSKIYINGRLIDQKPISNLGNIHASNNIMFKLDGCRDTHRYIWIKYFNLFDKELNE
KEIKDLYDNQSNSGILKDFWGDYLQYDKPYYMLNLYDPNKYVDVNNVGIRGYMYLKGPRG
SVMTTNIYLNSSLYRGTKFIIKKYASGNKDNIVRNNDRVYINVVVKNKEYRLATNASQAG
VEKILSALEIPDVGNLSQVVVMKSKNDQGITNKCKMNLQDNNGNDIGFIGFHQFNNIAK
LVASNWYNRQIERSSRTLGCSWEFIPVDDGWGERPL

Figure 6. The amino acid sequence of BoNT/A1, Hall strain. Residues in bold and underlined differ between /A1 Hall and /A1(B), CDC 1744.

The use of this amino acid substituted database was used for analysis of MS/MS data obtained from a toxin extracted from a culture supernatant of BoNT/B by immunoaffinity purification and resulted in the identification of 5 novel amino acid differences and the identification of a new subtype of BoNT/B, BoNT/B7 (Kalb et al. 2012). DNA sequencing confirmed the mass spectrometry results, and when the correct amino acid sequence was added to the database, sequence coverage of the toxin was 68% (Kalb et al. 2012).

Conclusion and Future Directions

Mass spectral proteomics has proven to be a powerful technique for the analysis of botulinum neurotoxins and has yielded identifications of the toxin at the serotype, subtype, and toxin variant level. This type of work has the challenge of obtaining unusually high sequence coverage of the toxin with relatively low levels of protein. These analyses have been successfully performed by ESI and MALDI mass spectrometers and can include identification of the neurotoxin-associated proteins as well as the neurotoxin, as seen in Table 1. BoNTs have been identified and differentiated within complex matrices, including food and bacterial culture.

De novo identification of novel BoNTs has been successful through mass spectrometric techniques and importantly, these determinations can be achieved in the absence of DNA.

Table 1. Summary of proteomics information presented in this chapter

Reference	Method used	Sequence coverage on toxin(s)	Complex matrix?	NAPs?
2002 van Baar	MALDI-PSD/ LC-ESI-MS/MS	35% BoNT/A; 25% BoNT/B	No	No
2004 van Baar	MALDI-PSD/ LC-ESI-MS/MS	11% BoNT/C; 49% BoNT/D, 30% BoNT/E, 53% BoNT/F	No	No
2005 Hines	LC-ESI-MS/MS	18% BoNT/A; 23.5% BoNT/B; 33.3% BoNT/C; 18.6% BoNT/D; 36.5% BoNT/E; 21.9% BoNT/F; 16.1% BoNT/G	No	Yes
2011 Moura	LC-ESI-MS/MS	44% BoNT/C; 18.4% BoNT/DC	No	Yes
2011 Terilli	LC-ESI-MS/MS	66% BoNT/G	No	Yes
2009 Klaubert	LC-ESI-MS/MS	6% BoNT/B, not mentioned for BoNT/A, /E, or /F	Culture supernatant	Yes
2005 Kalb	LC-ESI-MS/MS	65-70% BoNT/A1 and /A2	Milk	No
2010 Kull	MALDI-MS	Not mentioned	Milk, apple juice, orange juice, ham	No
2012 Wang	Gel then LC-ESI-MS/MS	99% BoNT/A; 95% BoNT/B; 90% BoNT/C; 96% BoNT/D; 92% BoNT/E; 97% BoNT/F; 91% BoNT/G	Carrot juice	No
2012 Kalb	LC-ESI-MS/MS	76% BoNT/B1; 76% BoNT/B2; 66% BoNT/B3; 75% BoNT/B4; 74% BoNT/B5	Culture supernatant	No

Identification and differentiation of botulinum neurotoxins is an emerging field, so there are many possible new directions. Some possibilities include detection of the toxin at lower levels. This is important as toxin is often present in clinical samples at very low levels, as little as 10 pg (67 attomole) per mL of sample. Current techniques do not have sufficient sensitivity to obtain high sequence coverage of attomole toxins in a complex matrix such as serum or stool. Rapid subtyping of toxin can play an important role in timely epidemiologic or forensic investigations. Increased sequence coverage to permit more confident differentiation of the subtype or toxin variant would also aid in epidemiologic or forensic investigations and will continue to be explored. Techniques to obtain additional amino acid sequence information and therefore increased sequence coverage could be obtained through alternate mass spectrometric fragmentation techniques such as ETD or ECD or perhaps a combination of those techniques. The use of a more targeted analysis with multiple reaction monitoring on triple quadrupole mass spectrometers may allow for greater sensitivity and may be another new direction for this field. A decrease in the time required for these analyses would also be an important advancement. Many possibilities for improvement exist in this developing field to combat the deadly disease of botulism.

References

Binz, T., Blasi, J., Yamasaki, S., Baumeister, A., Link, E., Sudhof, T. C., Jahn, R., Niemann, H. (1994) Proteolysis of SNAP-25 by types E and A botulinal neurotoxins. *J. Biol. Chem.* 269(3):1617-1620.

Blasi, J., Chapman, E. R., Link, E., Binz, T., Yamasaki, S., De Camilli, P., Sudhof, T. C., Niemann, H., Jahn, R. (1993) Botulinum neurotoxin A selectively cleaves the synaptic protein SNAP-25. *Nature* 365(6442):160-163.

De Medici, D., Anniballi, F., Wyatt, G. M., Lindstrom, M., Messelhausser, U., Aldus, C. F., Delibato, E., Korkeala, H., Peck, M. W., Fenicia, L. (2009) Multiplex PCR for detection of botulinum neurotoxin-producing clostridia in clinical, food, and environmental samples. *Appl. Environ. Microbiol.* 75(20):6457-6461.

Foran, P., Lawrence, G. W., Shone, C. C., Foster, K. A., Dolly, J. O. (1996) Botulinum neurotoxin C1 cleaves both syntaxin and SNAP-25 in intact and permeabilized chromaffin cells: correlation with its blockade of catecholamine release. *Biochemistry* 35 (8):2630-2636.

Fujinaga, Y., Inoue, K., Nomura, T., Sasaki, J., Marvaud, J. C., Popoff, M. R., Kozaki, S., Oguma, K. (2000) Identification and characterization of functional subunits of Clostridium botulinum type A progenitor toxin involved in binding to intestinal microvilli and erythrocytes. *FEBS letters* 467(2-3):179-183.

Gibson, A. M., Modi, N. K., Roberts, T. A., Hambleton, P., Melling, J. (1988) Evaluation of a monoclonal antibody-based immunoassay for detecting type B Clostridium botulinum toxin produced in pure culture and an inoculated model cured meat system. *The Journal of applied bacteriology* 64(4):285-291.

Hielm, S., Bjorkroth, J., Hyytia, E., Korkeala, H. (1999) Ribotyping as an identification tool for Clostridium botulinum strains causing human botulism. *Int. J. Food Microbiol.* 47(1-2):121-131.

Hines, H. B., Lebeda, F., Hale, M., Brueggemann, E. E. (2005) Characterization of botulinum progenitor toxins by mass spectrometry. *Appl. Environ. Microbiol.* 71(8):4478-4486.

Jacobson, M. J., Lin, G., Whittam, T. S., Johnson, E. A. (2008) Phylogenetic analysis of Clostridium botulinum type A by multi-locus sequence typing. *Microbiology* 154(Pt 8): 2408-2415.

Kalb, S. R., Baudys, J., Rees, J. C., Smith, T. J., Smith, L. A., Helma, C. H., Hill, K., Kull, S., Kirchner, S., Dorner, M. B., Dorner, B. G., Pirkle, J. L., Barr, J. R. (2012) De novo subtype and strain identification of botulinum neurotoxin type B through toxin proteomics. *Analytical and bioanalytical chemistry* 403(1):215-226.

Kalb, S. R., Baudys, J., Webb, R. P., Wright, P., Smith, T. J., Smith, L. A., Fernandez, R., Raphael, B. H., Maslanka, S. E., Pirkle, J. L., Barr, J. R. (2012) Discovery of a novel enzymatic cleavage site for botulinum neurotoxin F5. *FEBS letters* 586(2):109-115.

Kalb, S. R., Garcia-Rodriguez, C., Lou, J., Baudys, J., Smith, T. J., Marks, J. D., Smith, L. A., Pirkle, J. L., Barr, J. R. (2010) Extraction of BoNT/A, /B, /E, and /F with a single, high affinity monoclonal antibody for detection of botulinum neurotoxin by Endopep-MS. *PLoS One* 5(8):e12237.

Kalb, S. R., Goodnough, M. C., Malizio, C. J., Pirkle, J. L., Barr, J. R. (2005) Detection of botulinum neurotoxin A in a spiked milk sample with subtype identification through toxin proteomics. *Anal. Chem.* 77(19):6140-6146.

Kalb, S. R., Smith, T. J., Moura, H., Hill, K., Lou, J., Geren, I. N., Garcia-Rodriguez, C., Marks, J. D., Smith, L. A., Pirkle, J. L., Barr, J. R. (2008) The use of Endopep-MS to detect multiple subtypes of botulinum neurotoxins A, B, E, and F. *International Journal of Mass Spectrometry* 278(2-3):101-108.

Keto-Timonen, R., Nevas, M., Korkeala, H. (2005) Efficient DNA fingerprinting of Clostridium botulinum types A, B, E, and F by amplified fragment length polymorphism analysis. *Appl. Environ. Microbiol.* 71(3):1148-1154.

Klaubert, B., Vujtovic-Ockenga, N., Wermter, R., Schad, K., von Meyer, L. (2009) Determination of botulinum toxins after peptic sample pre-treatment by multidimensional nanoscale liquid chromatography and nano-electrospray ion-trap mass spectrometry. *Journal of chromatography B, Analytical technologies in the biomedical and life sciences* 877(11-12):1084-1092.

Kozaki, S., Kamata, Y., Nishiki, T., Kakinuma, H., Maruyama, H., Takahashi, H., Karasawa, T., Yamakawa, K., Nakamura, S. (1998) Characterization of Clostridium botulinum type B neurotoxin associated with infant botulism in japan. *Infection and immunity* 66 (10): 4811-4816.

Kozaki, S., Nakaue, S., Kamata, Y. (1995) Immunological characterization of the neurotoxin produced by Clostridium botulinum type A associated with infant botulism in Japan. *Microbiology and immunology* 39(10):767-774.

Kull, S., Pauly, D., Stormann, B., Kirchner, S., Stammler, M., Dorner, M. B., Lasch, P., Naumann, D., Dorner, B. G. (2010) Multiplex detection of microbial and plant toxins by immunoaffinity enrichment and matrix-assisted laser desorption/ionization mass spectrometry. *Anal. Chem.* 82(7):2916-2924.

Lee, J. C., Yokota, K., Arimitsu, H., Hwang, H. J., Sakaguchi, Y., Cui, J., Takeshi, K., Watanabe, T., Ohyama, T., Oguma, K. (2005) Production of anti-neurotoxin antibody is enhanced by two subcomponents, HA1 and HA3b, of Clostridium botulinum type B 16S toxin-haemagglutinin. *Microbiology* 151(Pt 11):3739-3747.

Lin, W. J., Johnson, E. A. (1995) Genome analysis of Clostridium botulinum type A by pulsed-field gel electrophoresis. *Appl. Environ. Microbiol.* 61(12):4441-4447.

Macdonald, T. E., Helma, C. H., Ticknor, L. O., Jackson, P. J., Okinaka, R. T., Smith, L. A., Smith, T. J., Hill, K. K. (2008) Differentiation of Clostridium botulinum serotype A strains by multiple-locus variable-number tandem-repeat analysis. *Appl. Environ. Microbiol.* 74(3):875-882.

Marks, J. D. (2004) Deciphering antibody properties that lead to potent botulinum neurotoxin neutralization. *Mov. Disord.* 19 Suppl. 8S101-108.

Moura, H., Terilli, R. R., Woolfitt, A. R., Gallegos-Candela, M., McWilliams, L. G., Solano, M. I., Pirkle, J. L., Barr, J. R. (2011) Studies on botulinum neurotoxins type/C1 and mosaic/DC using Endopep-MS and proteomics. *FEMS Immunol. Med. Microbiol.* 61(3): 288-300.

Paul, C. J., Twine, S. M., Tam, K. J., Mullen, J. A., Kelly, J. F., Austin, J. W., Logan, S. M. (2007) Flagellin diversity in Clostridium botulinum groups I and II: a new strategy for strain identification. *Appl. Environ. Microbiol.* 73(9):2963-2975.

Raphael, B. H., Choudoir, M. J., Luquez, C., Fernandez, R., Maslanka, S. E. (2010) Sequence diversity of genes encoding botulinum neurotoxin type F. *Appl. Environ. Microbiol.* 76 (14):4805-4812.

Raphael, B. H., Joseph, L. A., McCroskey, L. M., Luquez, C., Maslanka, S. E. Detection and differentiation of Clostridium botulinum type A strains using a focused DNA microarray. *Molecular and cellular probes* 24(3):146-153.

Schiavo, G., Benfenati, F., Poulain, B., Rossetto, O., Polverino de Laureto, P., DasGupta, B. R., Montecucco, C. (1992) Tetanus and botulinum-B neurotoxins block neurotransmitter release by proteolytic cleavage of synaptobrevin. *Nature* 359(6398):832-835.

Schiavo, G., Malizio, C., Trimble, W. S., Polverino de Laureto, P., Milan, G., Sugiyama, H., Johnson, E. A., Montecucco, C. (1994) Botulinum G neurotoxin cleaves VAMP/ synaptobrevin at a single Ala-Ala peptide bond. *J. Biol. Chem.* 269(32):20213-20216.

Schiavo, G., Rossetto, O., Catsicas, S., Polverino de Laureto, P., DasGupta, B. R., Benfenati, F., Montecucco, C. (1993) Identification of the nerve terminal targets of botulinum neurotoxin serotypes A, D, and E. *J. Biol. Chem.* 268(32):23784-23787.

Schiavo, G., Santucci, A., Dasgupta, B. R., Mehta, P. P., Jontes, J., Benfenati, F., Wilson, M. C., Montecucco, C. (1993) Botulinum neurotoxins serotypes A and E cleave SNAP-25 at distinct COOH-terminal peptide bonds. *FEBS letters* 335(1):99-103.

Schiavo, G., Shone, C. C., Rossetto, O., Alexander, F. C., Montecucco, C. (1993) Botulinum neurotoxin serotype F is a zinc endopeptidase specific for VAMP/synaptobrevin. *J. Biol. Chem.* 268(16):11516-11519.

Smith, T. J., Lou, J., Geren, I. N., Forsyth, C. M., Tsai, R., Laporte, S. L., Tepp, W. H., Bradshaw, M., Johnson, E. A., Smith, L. A., Marks, J. D. (2005) Sequence variation within botulinum neurotoxin serotypes impacts antibody binding and neutralization. *Infection and immunity* 73(9):5450-5457.

Sugii, S., Ohishi, I., Sakaguchi, G. (1977) Correlation between oral toxicity and in vitro stability of Clostridium botulinum type A and B toxins of different molecular sizes. *Infection and immunity* 16(3):910-914.

Terilli, R. R., Moura, H., Woolfitt, A. R., Rees, J., Schieltz, D. M., Barr, J. R. (2011) A historical and proteomic analysis of botulinum neurotoxin type/G. *BMC microbiology* 11232.

Umeda, K., Seto, Y., Kohda, T., Mukamoto, M., Kozaki, S. (2009) Genetic characterization of Clostridium botulinum associated with type B infant botulism in Japan. *J. Clin. Microbiol.* 47(9):2720-2728.

Van Baar, B. L., Hulst, A. G., de Jong, A. L., Wils, E. R. (2002) Characterisation of botulinum toxins type A and B, by matrix-assisted laser desorption ionisation and electrospray mass spectrometry. *Journal of chromatography* 970(1-2):95-115.

Van Baar, B. L., Hulst, A. G., de Jong, A. L., Wils, E. R. (2004) Characterisation of botulinum toxins type C, D, E, and F by matrix-assisted laser desorption ionisation and electrospray mass spectrometry. *Journal of chromatography* 1035(1):97-114.

Wagner, E., Meyer, K. F., Dozier, C. C. (1925) Studies on the Metabolism of B. Botulinus in Various Media. Xxvi. *Journal of bacteriology* 10(4):321-412.

Wang, D., Baudys, J., Rees, J., Marshall, K. M., Kalb, S. R., Parks, B. A., Nowaczyk, L., 2nd, Pirkle, J. L., Barr, J. R. (2012) Subtyping botulinum neurotoxins by sequential multiple endoproteases in-gel digestion coupled with mass spectrometry. *Anal. Chem.* 84(11): 4652-4658.

Williamson, L. C., Halpern, J. L., Montecucco, C., Brown, J. E., Neale, E. A. (1996) Clostridial neurotoxins and substrate proteolysis in intact neurons: botulinum neurotoxin C acts on synaptosomal-associated protein of 25 kDa. *J. Biol. Chem.* 271(13):7694-7699.

Yamasaki, S., Baumeister, A., Binz, T., Blasi, J., Link, E., Cornille, F., Roques, B., Fykse, E. M., Sudhof, T. C., Jahn, R., et al. (1994) Cleavage of members of the synaptobrevin/ VAMP family by types D and F botulinal neurotoxins and tetanus toxin. *J. Biol. Chem.* 269(17):12764-12772.

Yamasaki, S., Binz, T., Hayashi, T., Szabo, E., Yamasaki, N., Eklund, M., Jahn, R., Niemann, H. (1994) Botulinum neurotoxin type G proteolyses the Ala81-Ala82 bond of rat synaptobrevin 2. *Biochemical and biophysical research communications* 200(2):829-835.

In: Recent Advances in Proteomics Research
Editor: Ming D. Li

ISBN: 978-1-62948-218-7

Chapter 11

Challenges of Mining Low to Medium Abundant Proteins for Proteomics

***Balamurugan Natarajan*[1], *Rahim Abdullah*[2], *Nazrul Islam*[2] *and Savithiry S. Natarajan*[3,*]**

[1]USDA-ARS, Invasive Insects Biocontrol & Behavior Laboratory, Beltsville, US

[2]University of Maryland, Department of Plant Science and Landscape Architecture, College Park, US

[3]USDA-ARS, Soybean Genomics and Improvement Laboratory, Beltsville, US

Abstract

Low to medium abundant proteins play an integral role in biological processes. Harnessing these proteins using proteomic techniques is paramount for an in-depth understanding of the different biological processes. Proteomics encompasses four principal applications, namely protein mining, protein expression profiling, protein-network mapping and mapping of protein modifications. The results in these applications depend mostly on the quality of protein extraction and sample preparation. Two major challenges of sample preparation in proteomics for low to medium abundant proteins are the removal of contaminants and depletion of the highly abundant proteins. In this chapter, methods of contaminant removal such as TCA/acetone, NH_4AC/methanol, and phenol/chloroform, are discussed along with methods of depletion of highly abundant proteins including precipitation/fractionation, antibody based affinity, and filter-aided separation. In addition, we provide some perspectives on sample preparation strategies for further research related to proteomics.

[*]Corresponding author: Dr. Savithiry S. Natarajan, Soybean Genomics and Improvement Laboratory, USDA-ARS, 10300 Baltimore Avenue, Beltsville, MD 20705, Phone Number: 301-504-5258, Fax Number: 301-504-5728, Email: savi.natarajan@ars.usda.gov.

Keywords: Proteome, contaminants, high abundant proteins, low abundant proteins, medium abundant proteins, recalcitrant plant/animal tissues, sample preparation

1. Introduction

The inception of genome sequencing projects over the past decade has generated a large set of genomic databases. While further development of these large databases is imminent, the challenge now is to unravel the complex biological systems encoded by these genomes. One way of gaining insight into these complex biological processes is to focus on the network of gene products (Heazlewood et al. 2003; Hirano et al. 2004; Pandey and Mann 2000) by course of proteome analyses. Recent advances in protein electrophoresis and mass spectrometry made the study of the proteome possible. Proteome analysis can be studied in three ways: 1) Structural proteomics (high throughput X-ray crystallography/modeling,high throughput NMR spectroscopy/modeling); 2) Expression or analytical proteomics (electrophoresis, protein chips, DNA chips, 2D-HPLC, mass spectrometry, microsequencing); 3) Functional or interaction proteomics (HT functional assays, ligand chips,yeast 2-hybrid, deletion analysis, motif analysis). Although global profiling of proteins is necessary to understand expression proteomics, a comprehensive analysis of medium to low abundance proteins, which plays a key role as cellular ligands and signaling molecules, could greatly facilitate the detection of informative biomarkers necessary to assess early events and therapeutic target discoveries in the biological processes (Millioni et al. 2011). However, presence of contaminants and highly abundant proteins makes the process more challenging. Therefore, removal of abundant proteins is crucial to the characterization of low abundance proteins (Figure 1). For example, investigations of human plasma proteomics are difficult due to the presence of large proportion of albumin (55%, in the order of 30 g/L) and the wide range of predominant glycoprotein. In addition, there are other abundant proteins also present in plasma proteome, among which immunoglobulins (Igs), fibrinogen, transferrin, haptoglobin, and lipoproteins are the main constituents. In general, the top ten most abundant plasma proteins account for approximately 90% of the total protein content (Liumbruno et al. 2010).

Isolation of low to medium abundance proteins without first removing the highly abundant proteins of human plasma is a difficult task. In this chapter, we discuss about how to remove contaminants and deplete highly abundant proteins before separation of low to medium abundance proteins either by gel-based method, two-dimensional electrophoresis (2-DE) or by gel-free/LC-based proteomic approaches.

The improvements in sample preparation, application of the ready made IPG strips, and the further advancement of bioinformatics tools led scientists to adopt 2-DE methodologies to analyze protein characterization. Analysis of complex mixtures of proteins by 2-DE are based on two-dimensional gel separation. In the first dimension, the proteins are separated on the basis of protein isoelectric point technically known as IEF (*pI*). In the second dimension, the focused proteins are further separated by electrophoresis based on the molecular weight (Mr). Although limitations are present regarding the reproducibility and sensitivity of 2-DE results in characterizing the proteins, this technique is still considered good method for separating and visualizing proteins from a complex mixture. 2-DE is widely used and it allows for the separation of proteins from various crops and is a useful tool in studying protein variation due to different stress stimuli, temperature effects, mineral toxicity, and introduction or silencing of genes (Iwahashi and Hosoda 2000; Luo et al. 2009; Salekdeh and Komatsu 2007; Klubicová et al. 2012; Herman et al. 2003; Danchenko et al. 2009; Ren et al. 2009; Agrawal et al. 2008). Rabilloud et al. (1999) extracted parasite-encoded membrane proteins and reported that technical difficulties arise from poor solubilization and/or focusing of these membrane proteins. Optimizing extraction techniques by altering conditions including the use of urea, thiourea, and new zwitterionic detergents in both extraction and isoelectric focusing, allowed the enrichment of more protein spots using 2-DE. The authors concluded that sample preparation greatly influences the protein extraction, isolation, and further isoelectric focusing (IEF) in the first dimensional protein separation (Xu et al. 2003; Sivaraman et al. 1997). Analysis becomes more complicated when proteins are extracted from recalcitrant plant tissues which contain cell wall associated polysaccharides and secondary metabolites that interfere in IEF and cause several artifacts on 2-DE gels (Neilson et al. 2010). Unfortunately, due to the diversity in sample sources, there is no single method that can be universally applied to plant proteins analyzed by 2-DE. Since plant cells are well protected by the cell walls, protein extraction starts with cell lysis and disruption of sub-cellular components. Cell lysis can be achieved by freeze thawing, detergent lysis, sonication, grinding with liquid nitrogen, using a high pressure French press, homogenization with glass beads/sands, or nitrogen cavitation in buffer-based media. Buffers commonly used for protein solubilization contain single/multiple chaotropes, nonionic detergent(s), zwitterionic detergent(s), reducing agent(s), and protease inhibitors to prevent protein degradation during separation. In this chapter, we have discussed different methodologies used to remove the contaminants after crushing the samples.

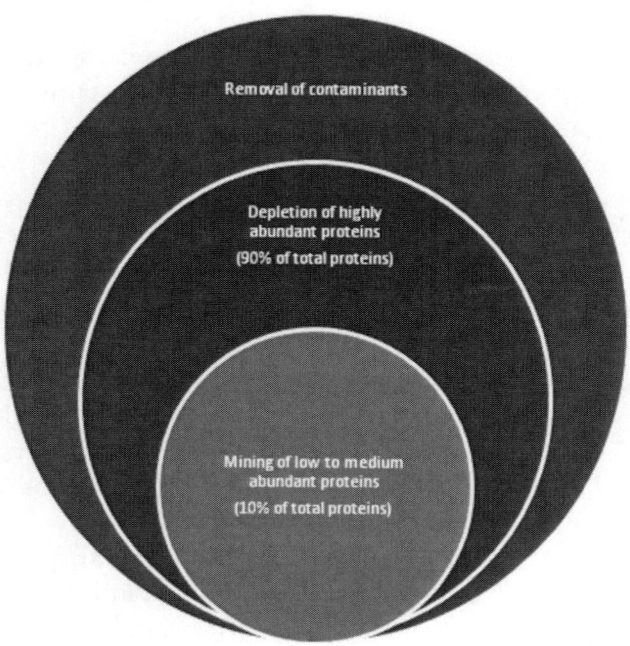

Figure 1. Schematic illustration of mining low to medium abundant proteins after removing contaminants and depleting highly abundant proteins.

Like 2-DE, a prerequisite for optimal proteins separation by LC-based proteomic is the availability of a sample of well-resolved proteins or peptides, devoid of interfering substances.LC-based proteomics has been further revolutionized with the introduction of "multidimensional protein identification technology (MudPIT)". In this technique, peptide separation is based on a strong cation exchange (SCX) followed by RP-HPLC before injecting the peptides into the mass spectrometry (Florens and Washburn 2006). As MudPIT relies on bi-phasic micro column for peptide separation, the presence of any contaminants and high abundant proteins might jeopardize the separation. Although different affinity based cleaning procedure have been designed for the LC-based separation, a pre-cleaning of the sample and LC compatible buffers are still a challenging task in LC-based proteomics.

Several methodologies have been used to remove impurities and contaminants (e.g., lipids, cellulose, nucleic acids, etc) from protein preparations, such as the implementation of detergents, reductants, chaotropic denaturing agents, and digestive enzymes. The use of detergents (SDS, CHAPS, cholate, and tween) is used to solubilize membrane proteins and aid in separation from lipids. Reductants (e.g., dithiothreital (DTT), β-mercaptoethanol) are generally necessary to reduce disulfide bonds and to prevent protein oxidation. Denaturing agents (e.g., urea thiourea, and acids) are frequently used to disrupt protein-protein interactions and the protein's secondary and tertiary structure by altering solution ionic strength. Digestive enzymes (e.g., DNAse, RNAse) may be utilized to digest contaminating nucleic acids, carbohydrates, and lipids. In this chapter, we focus on some common contaminants and highly abundant proteins present in plant/animal samples and discuss three methods for each with their pros and cons.

Common Contaminants in Recalcitrant Plant/Animal Tissues

Unlike microbial samples, extraction of proteins from some plant and animal samples are challenging. As mentioned earlier, plant cells contain many components including cell wall and vacuoles that account for the majority of the cell mass. In addition the presence of several proteases, oxidative enzymes, large quantities of polysaccharides, lipids, phenolics, and other secondary metabolites that may interfere with protein extraction and its separation add to the difficulty (Isaacson et al. 2006; Thelen and Peck 2007). For example, protein extraction from fruit tissues is difficult, due to the presence of low protein content and different interfering compounds including pigments, starch, polyphenols, polysaccharides, tannins, and organic acids that can lead to a high degree of protein denaturation and inactivation (Clements 1970). Similarly, plant bark, wood and the roots and aged leaves are also considered as recalcitrant tissues and protein extraction from these tissues are difficult due the abundance of secondary metabolites that strongly interfere with 2-DE, resulting in streaking, smearing, poor resolution, and reduced number of discrete protein spots (Hirano et al. 2004; Islam et al. 2004).

Similar to the challenges of recalcitrant plant tissues, protein extraction from mammalian tissues such as formalin-fixed paraffin-embedded tissues, cartilage/skeleton/bone, blood, muscle and other organs is less advanced because of technical challenges related to their biochemical properties. For example, cartilage is characterized by low cellular content and high cellular matrix (ECM). The ECM is composed of highly anionic macromolecules that are known to interfere with IEF (Wilson et al. 2008). Recently a unified method of protein extraction from cartilage has been reported (Desjardin et al. 2012). However, the removal of most of the contaminants is yet to be addressed.

Removal of Contaminants

Precipitation is generally employed to selectively separate proteins in the sample from contaminating species such as salts, detergents, nucleic acids, lipids, and pigments that can interfere with the 2-DE and other separation protocols. Several protein precipitation methods were implemented including the use of an *organic solvent* or addition of a *salt,* or *changing the pH* to alter the nature of the protein-solution interactions. In this section, some of the methods used for protein precipitation are discussed and summarized in Table 1.

a. Trichloroacetic acid (TCA)/acetone: TCA/acetoneprecipitation is a more traditionally used method for protein precipitation compared to other methods, allowing the removal of the non-precipitated contaminants (Méchin et al. 2007). Although the cleaning of proteins with TCA precipitation is a very common practice, surprisingly little is known of the actual mechanism of TCA precipitation. As noted by investigators, TCA precipitation follows a U-shaped curve with optimum precipitationat a TCA concentration of roughly 15% suggested that the mechanism of TCA protein precipitation is hydrophobic aggregation (Xu et al. 2003; Sivaraman et al. 1997). While TCA precipitation is frequently used to remove interfering compounds from proteins, TCA precipitated proteins are often difficult to solubilize and requires additional chaotropes or detergents. To address this problem, TCA precipitated proteins are pretreated with 0.2 M NaOH for less than 5 min and by the addition of a standard sample solubilization buffer (SSSB; Urea, CHAPS, DTT, and Ampholytes) (Nandakumar et al. 2003). Fic et al. (2010), reported protein recovery after protein precipitation with TCA/acetone followed by sonication and found that only 33% of total proteins were recovered. Another disadvantage of cleaning proteins with TCA is that the TCA adheres to the protein after washing with cold acetone. Ashoub et al. (2011) extracted wild type barley leaf proteins by using absolute ethanol containing 10 mM DTT followed by resuspension in a urea/thiourea and NP-40 containing solution. The resulting proteins extract was separated by 2-DE and was compared with protein extracts using established method of 10% TCA-Acetone /0.07% 2-mercaptoethanol precipitation. The authors tested the quality of both protein extracts by separating through 2-DE and concluded that more protein spots at the basic range with low molecular weight proteins in the range of 25-15 kDa were recovered with precipitation of ethanol containing 10mM DTT when compared to TCA/acetone extraction with mercaptoethanol (Ashoub et al. 2011). Natarajan et al. (2005) compared four different extraction methodologies (thiourea/urea, urea, modified TCA/acetone and phenol) to solubilize soybean seed proteins and further analyzed the quality of the proteins by 2-DE and MS. The authors reported that modified TCA/acetone with the addition of thiourea showed higher protein resolution and discrete protein spots in 2-DE gels especially in the lower molecular weight range when compared with urea only and phenol methods. Using the optimal trichloroacetic acid (TCA)/acetone precipitation followed by solubilization in urea, the authors were able to identify more protein spots from different classes of proteins (storage, allergen, and anti-nutritional) by MS analysis. In another report, the authors used the TCA/acetone method to extract soybean low abundant seed proteins. In this investigation, the authors identified twenty low abundant proteins that are involved in seed metabolism and compared the protein variations among16 soybean genotypes (Natarajan et al. 2012). Xu et al. (2007) used the same method and characterized the distribution of soybean major seed allergens such as Gly m Bd 60K, Gly m Bd 30K, and Gly m Bd 28K in wild, landrace, ancestral, and modern soybean genotypes.

b. Ammonium acetate in methanol followed by phenol extraction (NH_4AC/methanol): This technique has proven useful with plant samples containing high levels of interfering substances like phenolics, fats and carbohydrates. Removing interfering substances from proteins by precipitating with NH_4AC/methanol instead of TCA/acetone has not been commonly used. However, it appears to be promising when compared to the TCA/acetone for a number of reasons. As the pH of a NH_4AC/methanol solution (pH 7.5) is similar to that of typical protein extraction buffers (urea buffer, pH 8.5), we anticipate that any NH_4AC/methanol that could not be removed after washing with cold acetone will not significantly alter the dielectric constant of the medium. The dielectric constant of the media is considered an important criterion in determining protein solubility (Rabilloud et al. 1999). In addition, unlike TCA, the interference of residual NH_4AC/methanol with IEF has not been reported. Barbosa et al. (2013) reported seventeen different protein extraction procedures for sunflower seed embryos with different extraction times, temperatures, type of solvents and varying analytical procedures. For all procedures, the authors used a buffer that contains KCl, DTT, PMSF, and SDS and precipitate the extracted proteins using ammonium acetate in methanol. The authors reported that an efficient method involves the use of deionized water at 50◦ C for extraction, with manual maceration and showed 66 well resolved protein spots using 2D-PAGE.

c. Phenol extraction: A combination of phenol and sugar is an effective protein extraction method for recalcitrant tissues. However, phenol extraction is discouraged for a number of reasons, which include health risks. Rodrigues et al. (2009) compared four extraction methodologies to solubilize papaya leaf proteins such as phenol extraction/methanol-ammonium acetate precipitation, no precipitation fractionation, TCA/acetone precipitation, and TCA/acetone precipitation with fractionation. The authors reported that the papaya protein extract prepared using TCA and TCA with fractionation resulted in numerous protein bands in 1-DE (one dimensional electrophoresis) compared to other methods.The authors selected these two methods for further 2-DE and MS analysis and concluded that TCA method without fractionation proved to be more useful (Rodrigues et al. 2009). Other researchers reported that phenol extraction is effective for extracting protein from selective plant tissues (Rode et al. 2012). Zhang et al. (2012) initially evaluated three protein extraction methods for banana peels, namely TCA-acetone precipitation, phenol extraction, and TCA precipitation. The authors concluded that modified phenol method was efficient in solubilizing high quality proteins suitable for 2D-PAGE analysis. Boonmee et al. (2011) used phenol extraction method and subsequently followed by NH_4AC/methanol precipitation to isolate *Curcuma comosa* proteins. The authors tested the quality of the extracted protein by analyzing 2-DE separation and MS analysis and identified 183 well resolved proteins. Keiblinger et al. (2012) compared four methods, namely SDS extraction without phenol, NaOH with subsequesnt phenol extraction, SDS-phenol extraction, and SDS-phenol extraction with prior washing steps to extract protein from two different soil samples to study soil metaproteomics.

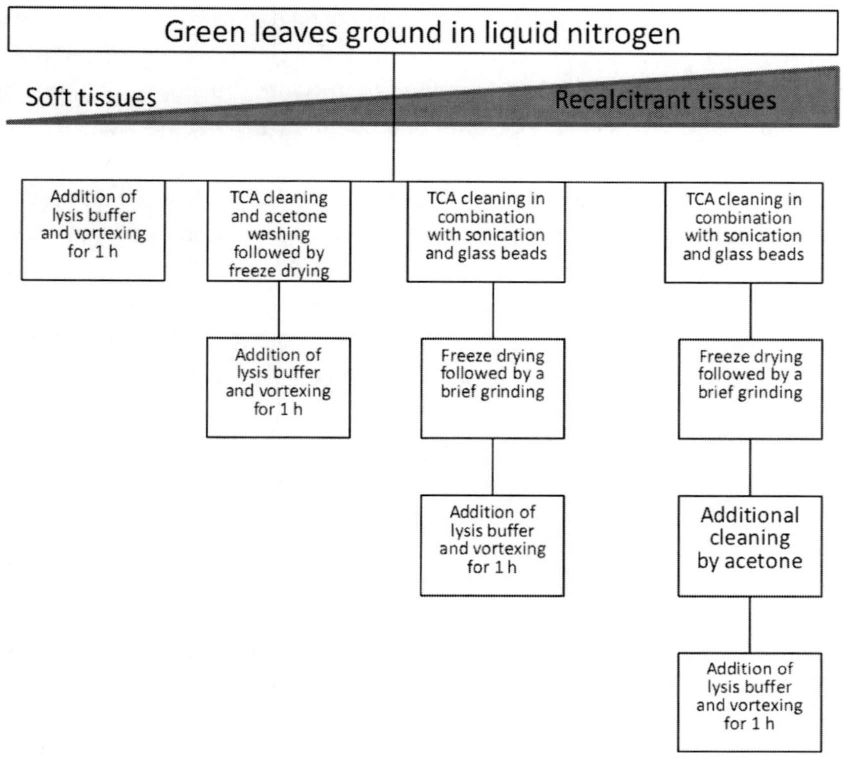

Figure 2. An outline of removing contaminants from plant tissue using Trichloro acetic acid (TCA) and other mechanical devices (Source: Adapted from Islam et al. (2004)).

The authors concluded that the SDS-phenol extraction method resulted a greater number of protein spots analyzed by 2-DE and LC-MS/MS compared to the other three methods. Rodrigues et al. (2012) and Chatterjee et al. (2012) initially compared different methods and finally selected phenol method as an efficient method to extract proteins from soybean and chickpea root tissues and further analyzed proteins using 2-DE and MS analysis. Further, the authors concluded that addition of SDS in phenol improved the efficiency of protein extraction. Protein extraction from roots is difficult compared to other plant tissues, since root tissueis highly vacuolated with low quantity of protein and high quantity of proteases, storage polysaccharides, phenolics, and secondary metabolites (Tsugita and Kamo 1999). Phenol with SDS produced well-resolved protein spots extracted from soybean roots (Rodrigues et al. 2012). Chatterjee et al. (2012) reported that in addition to phenol with SDS, pulsatory (sonication) treatments proved to be most suitable method for chickpea root protein extraction and characterization.Wang et al. (2006) used a universal and rapid protocol for protein extraction from recalcitrant plant tissues including the aged leaves of bamboo, grape, iris, olive, lemon, pine, redwood, sugarcane, and tobacco. The authors also used tissues of posidonia grass plants growing under water in the Mediterranean Sea as well as tissues from apple, pear, banana, Kiwi, grape, tomato and orange fruits. The authors reported that a TCA/acetone method combined with phenol and methanol washes is efficient for extracting proteins from recalcitrant plant tissues. The authors concluded that this method efficiently removes high levels of polyphenols (olive and pine leaves), and fruit samples which have high sugar (banana), acidity (grape and orange), and pigments (olive and tomato). In addition, the protein extract is compatible with silver staining post 1-DE or 2-DE analysis.Wang et al. (2007) compared four methods, TCA, E-TCA, phenol extraction, and a modified phenol extraction called the BPP (Borax/polyvinylpolypyrrolidone /phenol) method to extract *Salicornia europaea* proteins. The authors reported that the BPP method yielded more distinct and resolved protein spots on 2-DE gels and the quality of the protein was suitable for subsequent MS analysis. In addition, the authors concluded that BPP method allows simultaneous separation of salt and proteins in the halophytic samples by partitioning them into different phases resulting improvement in protein yield.

Although there is no universal procedure for removing contaminants from samples, a general outline suggesting different steps involved in protein extraction from recalcitrant tissue using an example of TCA/acetone followed by urea lysis buffer is shown in Figure 2.

Common Highly Abundant Proteins in Plant/Animal Tissues

Resolution of protein spots in 2-DE is greatly dependent on the isoelectric focusing (IEF) step. High resolution is often restricted by the presence of high abundance proteins. Even with better resolution, the existence of low abundant proteins is masked by the presence of highly abundant proteins in the second molecular size dimension. In plants, some of the most abundant proteins are Ribulose-1,5-bisphosphate carboxylase oxygenase (*RuBisCO)* in leaves, glycinin and beta-conglycinin storage proteins in soybean and other legumes seeds, patatin (glycoprotein) in potato tubers, glutens in wheat seeds, and bark storage protein in poplar bark. Like abundant RuBisCO in plants, the detection of low abundant proteins including red blood cell (RBC) membranes and cystolic proteinsin mammalian tissues is also difficult due to the presence of highly abundant hemoglobin. To overcome this limitation, researchers used several specific and efficient approaches for the removal of abundant hemoglobin including the application of strong cation exchange (Bhattacharya et al. 2007), gel filteration (Kakhniashvili et al. 2004), clear native electrophoresis (D'Amici et al. 2012), solution phase isoelectric focusing (Prabakaran et al. 2007), affinity for Ni(II) (Ringrose et al. 2008), polyelectrolyte HemogloBind reagent (Alvarez-Llamas et al. 2009), and using recombinant antibodies in combination with Protein A and strepatividin resin (Ettorre et al. 2006). In addition, protein equalizer technologies, namely ProteoMiner (combinatorial peptide ligand libraries) and HemoVoid (silica-based protein enrichment matrix) are used to enrich low abundant protein species by depleting and/or reducing the concentration of abundant hemoglobin (Roux-Dalvai et al. 2008; Simó et al. 2008; Boschetti and Righetti 2008; Righetti et al. 2006; Walpurgis et al. 2012).

Removal of Highly Abundant Proteins in Plant/Animal Tissues

Similar to the masking of low to medium molecular weight proteins on the 2-DE gel, the digestion of complex protein mixtures in the presence of high-abundance proteins is often caused by higher proteolytic peptides of the high abundant proteins and thus limits dynamic range in the mass spectrometer (Fonslow et al. 2013). Therefore, to analyze low to medium abundance of proteins, it is paramount to remove the high abundant proteins from the sample. Examples of methods to deplete these proteins include:

a. Precipitation/fractionation: Preferential extraction of low abundance proteins without depleting highlyabundant storage proteins has been reported in soybean seed using isopropanol (Natarajan et al. 2009). The authors evaluated different concentrations of isopropanol and concluded that 40% isopropanol enriches lower abundance proteins and is also suitable for 2-DE separation and further identification by MS. However, depletion of highly abundant soybean seed storage proteins by precipitating with 10mM calcium and abundant *RuBisCO* in soybean leaves by a combination of 10 mM Ca(2+) and 10 mM phytate has been reported (Krishnan and Natarajan 2009; Krishnan et al. 2009). The authors removed ~87% of highly abundant seed storage proteins and identified ~541 previously inconspicuous proteins by using a fractionation technique with 10mM calcium as well as the removal of ~85% of highly abundant soybean leaf proteins, *RuBisCO,* and identification of ~230 previously inconspicuous proteins through fractionation technique with 10 mM Ca(2+) and 10 mM phytate. In both cases, the authors used various concentrations of calcium and phytate to precipitate highly abundant proteins and enrich low abundance proteins. In some cases, the partitioning of *RuBisCO* with polyethylene glycol (PEG) fractionation improved detection of low-abundant proteins (Alam et al. 2013; Aryal et al. 2012). A selective depletion of highly abundant proteins, mainly mannose-binding lectin (agglutinin), from *Pinellia ternata* tubers using sequential extraction has been reported. In this investigation, the authors initially used acetic acid, followed by SDS-containing buffer. The SDS protein extract is mixed with phenol and further precipitated with methanol containing ammonium acetate. The protein extract was separated by 2-DE analysis (Wu et al. 2012).

b. Antibody based affinity methods: In mammalian proteomics, several strategies based on combinations of affinity depletion and enrichment of proteins with antibody arrays or ligand libraries followed by chromatographic prefractionation of proteins and peptides have been used to remove the highly abundant proteins (Liumbruno et al. 2010; Jmeian and El Rassi 2009). Although antibody based affinity methods of depleting abundant proteins are very common in mammalian systems (Walpurgis et al. 2012; Mortezai et al. 2010; Marco-Ramell and Bassols 2010; Desrosiers et al. 2007), the use of this method in plant systems is limited since it is difficult to acquire relatively pure antibodies for plant proteins. However, Agrisera (http://www.agrisera.com) has recently developed several antibodies for high abundance proteins which could effectively be used to test the feasibility of antibody-based depletion methods for plant proteins.

c. Filter aided separation methods: The development of filter aided depletion of highly abundant proteins in plant proteomics is proceedingat a slow pace, even though it has a positive impact on sample preparation. The use of this method for sample preparation in gel-free proteomics was reported in 2009 (Wiśniewski et al. 2009). Recently, a method for depleting high abundance proteins using Michaelis-Menten kinetics of trypsin digestion was also reported (Fonslow et al. 2013).

In this method, a molecular cut-off filter is used as a reactor-chamber between trypsin and high abundant proteins (Figure 3). However, a critical ratio between trypsin and the abundance of proteins is a pre-requisite for the success of this method in depleting the high abundance proteins.Taking advantages of both specific affinity (as in b) and filtration (as in c), another method has recently been used, namely "affinity depletion spin filter", for quantitative proteomic studies (Fratantoni et al. 2010).

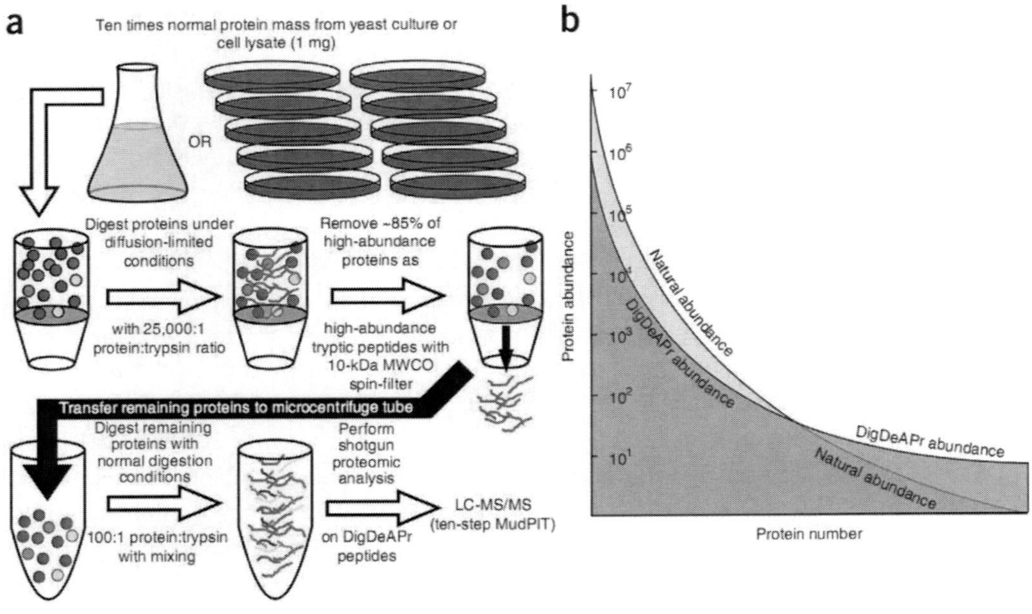

Figure 3. Schematic and description for removing abundant proteins (DigDeAPr method). (a) Ten times the normal mass of proteins are digested under trypsin-and diffusion-limited conditions for removal of abundant proteins as abundant peptides with a MWCO spin-filter. The remaining proteins are digested as normal for LC-MS/MS MudPITanalysis. (b) DigDeAPr changes the abundance profile of the proteome by starting with ten times more protein and digesting away ~85% of the most abundant proteins. The higher-abundance proteins are preferentially digested by trypsin and depleted as peptides (yellow region), which reduces their natural abundance to their DigDeAPr abundance (green region). By using ten times the protein mass to start, the DigDeAPr abundance of low-abundance proteins should be ten times higher than their natural abundance (blue region) (Source: Adapted from Fonslow et al. (2013)).

Conclusion

It is difficult to rely on a single method for extracting protein to obtain optimal resolution in the 2-DE or other separation methods due to a wide dynamic range of proteins and diversity of contaminants present in recalcitrant plant and mammalian tissues. Contaminants from cell wall associated compounds in plant samples, and biochemical properties of some mammalian tissues make the procedure further challenging as compared to microbial samples. Therefore, a combination of procedures outlined in this chapter for removing contaminants and depleting high abundant proteins may provide better resolved protein separation for a given proteome.

Table 1. Advantages and disadvantages of different methods

Methods	Extraction Methods	Advantages	Disadvantages
Removal of Contaminants	a. Trichloroacetic acid (TCA) / Acetone	Efficient for diverse samples [Islam et al. 2004; Florens and Washburn 2006]	Difficult to solubilize proteins [Clements 1970]
	b. Ammonium acetate in methanol followed by phenol extraction (NH4AC/methanol)	Easy to solubilize proteins [Islam et al. unpublished]	Less efficient [Islam et al. unpublished]
	c. Phenol extraction	Efficient for specific samples [Xu et al. 2003; Natarajan et al. 2005; Fic et al.2010]	Health hazards [Rodrigues et al. 2009]
Depletion of highly abundant protein	a. Precipitation/fractionati on	Efficient for specific proteins [Natarajan et al. 2009;Krishnan and Natarajan 2009; Alam et al. 2013]	Loss of other proteins [Rodrigues et al. 2009]
	b. Antibody based affinity methods	Highly sensitive with pure antibody [Liumbruno et al. 2010]	Loss of other proteins and limited to availability of pure antibody
	c. Filter aided separation method	Effective for LC based proteomics (Fonslow et al. 2013)	Not tested for gel-based proteomics

Acknowledgments

The authors thank Dr. Leslie Wanner for critical review of this manuscript. Funding for this research was provided by ARS project 1275-21000-224-00D.

References

Heazlewood JL, Howell KA, Millar AH (2003) Mitochondrial complex I from Arabidopsis and rice: orthologs of mammalian and fungal components coupled with plant-specific subunits. *Biochim. Biophys. Acta.,* 1604(3):159-169.

Hirano H, Islam N, Kawasaki H (2004) Technical aspects of functional proteomics in plants. *Phytochemistry,* 65(11):1487-1498.

Pandey A, Mann M (2000) Proteomics to study genes and genomes. *Nature,* 405(6788):837-846.

Millioni R, Tolin S, Puricelli L, Sbrignadello S, Fadini GP, Tessari P, Arrigoni G (2011) High abundance proteins depletion vs low abundance proteins enrichment: comparison of methods to reduce the plasma proteome complexity. *PLoS One,* 6(5):e19603.

Liumbruno G, D'Alessandro A, Grazzini G, Zolla L (2010) Blood-related proteomics. *J. Proteomics,* 73(3):483-507.

Iwahashi Y, Hosoda H (2000) Effect of heat stress on tomato fruit protein expression. *Electrophoresis,* 21(9):1766-1771.

Luo J, Ning T, Sun Y, Zhu J, Zhu Y, Lin Q, Yang D (2009) Proteomic analysis of rice endosperm cells in response to expression of hGM-CSF. J Proteome Res 8(2):829-837.

Salekdeh GH, Komatsu S (2007) Crop proteomics: aim at sustainable agriculture of tomorrow. *Proteomics,* 7(16):2976-2996.

Klubicová K, Danchenko M, Skultety L, Berezhna VV, Uvackova L, Rashydov NM, Hajduch M (2012) Soybeans grown in the Chernobyl area produce fertile seeds that have increased heavy metal resistance and modified carbon metabolism. *PLoS One,* 7(10):e48169.

Herman EM, Helm RM, Jung R, Kinney AJ (2003) Genetic modification removes an immunodominant allergen from soybean. *Plant Physiol.,* 132(1):36-43.

Danchenko M, Skultety L, Rashydov NM, Berezhna VV, Mátel L, Salaj T, Pret'ová A, Hajduch M (2009) Proteomic analysis of mature soybean seeds from the Chernobyl area suggests plant adaptation to the contaminated environment. *J. Proteome. Res.,* 8(6):2915-2922.

Ren N, Xie T, Xing D (2009) Composition of extracellular polymeric substances influences the autoaggregation capability of hydrogen-producing bacterium Ethanoligenens harbinense. *Bioresour. Technol.,* 100(21):5109-5113.

Agrawal GK, Hajduch M, Graham K, Thelen JJ (2008) In-depth investigation of the soybean seed-filling proteome and comparison with a parallel study of rapeseed. *Plant Physiol.,* 148(1):504-518.

Rabilloud T, Blisnick T, Heller M, Luche S, Aebersold R, Lunardi J, Braun-Breton C (1999) Analysis of membrane proteins by two-dimensional electrophoresis: comparison of the proteins extracted from normal or Plasmodium falciparum-infected erythrocyte ghosts. *Electrophoresis,* 20(18):3603-3610.

Xu Z, Xie Q, Zhou HM (2003) Trichloroacetic acid-induced molten globule state of aminoacylase from pig kidney. *J. Protein. Chem.,* 22(7-8):669-675.

Sivaraman T, Kumar TK, Jayaraman G, Yu C (1997) The mechanism of 2,2,2-trichloroacetic acid-induced protein precipitation. *J. Protein. Chem.,* 16(4):291-297.

Neilson KA, Gammulla CG, Mirzaei M, Imin N, Haynes PA (2010) Proteomic analysis of temperature stress in plants. *Proteomics,* 10(4):828-845.

Florens L, Washburn MP (2006) Proteomic analysis by multidimensional protein identification technology. *Methods Mol. Biol.,* 328159-175.

Isaacson T, Damasceno CM, Saravanan RS, He Y, Catalá C, Saladié M, Rose JK (2006) Sample extraction techniques for enhanced proteomic analysis of plant tissues. *Nat. Protoc.,* 1(2):769-774.

Thelen JJ, Peck SC (2007) Quantitative proteomics in plants: choices in abundance. *Plant Cell,* 19(11):3339-3346.

Clements RL (1970) Protein patterns in fruits. In. Academic Press, London; New York, NY, pp 159-177.

Islam N, Lonsdale M, Upadhyaya NM, Higgins TJ, Hirano H, Akhurst R (2004) Protein extraction from mature rice leaves for two-dimensional gel electrophoresis and its application in proteome analysis. *Proteomics,* 4(7):1903-1908.

Wilson R, Belluoccio D, Bateman JF (2008) Proteomic analysis of cartilage proteins. *Methods,* 45(1):22-31.

Desjardin C, Balliau T, Valot B, Zivy M, Wimel L, Guérin G, Cribiu E, Schibler L (2012) A method for proteomic analysis of equine subchondral bone and epiphyseal cartilage. *Proteomics,* 12(11):1870-1874.

Méchin V, Damerval C, Zivy M (2007) Total protein extraction with TCA-acetone. *Methods Mol. Biol.,* 3551-8.

Nandakumar MP, Shen J, Raman B, Marten MR (2003) Solubilization of trichloroacetic acid (TCA) precipitated microbial proteins via naOH for two-dimensional electrophoresis. *J. Proteome Res.,* 2(1):89-93.

Fic E, Kedracka-Krok S, Jankowska U, Pirog A, Dziedzicka-Wasylewska M (2010) Comparison of protein precipitation methods for various rat brain structures prior to proteomic analysis. *Electrophoresis,* 31(21):3573-3579.

Ashoub A, Berberich T, Beckhaus T, Brüggemann W (2011) A competent extraction method of plant proteins for 2-D gel electrophoresis. *Electrophoresis,* 32(21):2975-2978.

Natarajan S, Xu C, Caperna TJ, Garrett WM (2005) Comparison of protein solubilization methods suitable for proteomic analysis of soybean seed proteins. *Anal. Biochem.,* 342(2):214-220.

Natarajan SS, Xu C, Garrett WM, Lakshman DK, Bae H (2012) Assessment of natural variation of low abundant metabolic proteins in soybean seeds using proteomics. In. *Journal of Plant Biochemistry and Biotechnology,* pp 30-37.

Xu C, Caperna TJ, Garrett WM, Cregan PB, Bae H, Luthria DL, Natarajan SS (2007) Proteomic analysis of the distribution of the major seed allergens in sixteen soybean genotypes In. *Journal of the Science of Food and Agriculture,* pp 2511-2518.

Barbosa HS, de Souza DL, Koolen HHF, Gozzo FC, Arruda MAZ (2013) Sample preparation focusing on plant proteomics: extraction, evaluation and identification of proteins from sunflower seeds. In. *Anal Methods* pp 116-123.

Rodrigues SP, Ventura JA, Zingali RB, Fernandes PM (2009) Evaluation of sample preparation methods for the analysis of papaya leaf proteins through two-dimensional gel electrophoresis. *Phytochem. Anal.,* 20(6):456-464.

Rode C, Winkelmann T, Braun HP, Colditz F (2012) DIGE analysis of plant tissue proteomes using a phenolic protein extraction method. *Methods Mol. Biol.,* 854335-342.

Zhang LL, Feng RJ, Zhang YD (2012) Evaluation of different methods of protein extraction and identification of differentially expressed proteins upon ethylene-induced early-ripening in banana peels. *J. Sci. Food Agric.,* 92(10):2106-2115.

Boonmee A, Srisomsap C, Chokchaichamnankit D, Karnchanatat A, Sangvanich P (2011) A proteomic analysis of Curcuma comosa Roxb. rhizomes. *Proteome Sci.,* 943.

Keiblinger KM, Wilhartitz IC, Schneider T, Roschitzki B, Schmid E, Eberl L, Riedel K, Zechmeister-Boltenstern S (2012) Soil metaproteomics - Comparative evaluation of protein extraction protocols. *Soil. Biol. Biochem.,* 54(15-10):14-24.

Rodrigues EP, Torres AR, da Silva Batista JS, Huergo L, Hungria M (2012) A simple, economical and reproducible protein extraction protocol for proteomics studies of soybean roots. *Genet. Mol. Biol.,* 35(1 (suppl)):348-352.

Chatterjee M, Gupta S, Bhar A, Das S (2012) Optimization of an Efficient Protein Extraction Protocol Compatible with Two-Dimensional Electrophoresis and Mass Spectrometry from Recalcitrant Phenolic Rich Roots of Chickpea (Cicer arietinum L.). *Int. J. Proteomics, 2012536963.*

Tsugita A, Kamo M (1999) 2-D electrophoresis of plant proteins. *Methods Mol. Biol.,* 11295-97.

Wang W, Vignani R, Scali M, Cresti M (2006) A universal and rapid protocol for protein extraction from recalcitrant plant tissues for proteomic analysis. In. *Electrophoresis*, pp 2782-2786.

Wang X, Li X, Deng X, Han H, Shi W, Li Y (2007) A protein extraction method compatible with proteomic analysis for the euhalophyte Salicornia europaea. *Electrophoresis,* 28(21):3976-3987.

Bhattacharya D, Mukhopadhyay D, Chakrabarti A (2007) Hemoglobin depletion from red blood cell cytosol reveals new proteins in 2-D gel-based proteomics study. *Proteomics Clin. Appl.,* 1(6):561-564.

Kakhniashvili DG, Bulla LA, Goodman SR (2004) The human erythrocyte proteome: analysis by ion trap mass spectrometry. *Mol. Cell Proteomics.,* 3(5):501-509.

D'Amici GM, Rinalducci S, Zolla L (2012) Depletion of hemoglobin and carbonic anhydrase from erythrocyte cytosolic samples by preparative clear native electrophoresis. *Nat. Protoc.,* 7(1):36-44.

Prabakaran S, Wengenroth M, Lockstone HE, Lilley K, Leweke FM, Bahn S (2007) 2-D DIGE analysis of liver and red blood cells provides further evidence for oxidative stress in schizophrenia. *J. Proteome. Res.,* 6(1):141-149.

Ringrose JH, van Solinge WW, Mohammed S, O'Flaherty MC, van Wijk R, Heck AJ, Slijper M (2008) Highly efficient depletion strategy for the two most abundant erythrocyte soluble proteins improves proteome coverage dramatically. *J. Proteome. Res.,* 7(7):3060-3063.

Alvarez-Llamas G, de la Cuesta F, Barderas MG, Darde VM, Zubiri I, Caramelo C, Vivanco F (2009) A novel methodology for the analysis of membrane and cytosolic sub-proteomes of erythrocytes by 2-DE. *Electrophoresis,* 30(23):4095-4108.

Ettorre A, Rösli C, Silacci M, Brack S, McCombie G, Knochenmuss R, Elia G, Neri D (2006) Recombinant antibodies for the depletion of abundant proteins from human serum. *Proteomics,* 6(16):4496-4505.

Roux-Dalvai F, Gonzalez de Peredo A, Simó C, Guerrier L, Bouyssié D, Zanella A, Citterio A, Burlet-Schiltz O, Boschetti E, Righetti PG, Monsarrat B (2008) Extensive analysis of the cytoplasmic proteome of human erythrocytes using the peptide ligand library technology and advanced mass spectrometry. *Mol. Cell Proteomics,* 7(11):2254-2269.

Simó C, Bachi A, Cattaneo A, Guerrier L, Fortis F, Boschetti E, Podtelejnikov A, Righetti PG (2008) Performance of combinatorial peptide libraries in capturing the low-abundance proteome of red blood cells. 1. Behavior of mono- to hexapeptides. *Anal. Chem.,* 80(10):3547-3556.

Boschetti E, Giorgio Righetti P (2008) Hexapeptide combinatorial ligand libraries: the march for the detection of the low-abundance proteome continues. *Biotechniques,* 44(5):663-665.

Righetti PG, Boschetti E, Lomas L, Citterio A (2006) Protein Equalizer Technology : the quest for a "democratic proteome". *Proteomics,* 6(14):3980-3992.

Walpurgis K, Kohler M, Thomas A, Wenzel F, Geyer H, Schänzer W, Thevis M (2012) Validated hemoglobin-depletion approach for red blood cell lysate proteome analysis by means of 2D PAGE and Orbitrap MS. *Electrophoresis,* 33(16):2537-2545.

Fonslow BR, Stein BD, Webb KJ, Xu T, Choi J, Park SK, Yates JR (2013) Digestion and depletion of abundant proteins improves proteomic coverage. *Nat. Methods,* 10(1):54-56.

Natarajan SS, Krishnan HB, Lakshman S, Garrett WM (2009) An efficient extraction method to enhance analysis of low abundant proteins from soybean seed. *Anal. Biochem.,* 394(2):259-268.

Krishnan HB, Natarajan SS (2009) A rapid method for depletion of Rubisco from soybean (Glycine max) leaf for proteomic analysis of lower abundance proteins. *Phytochemistry,* 70(17-18):1958-1964.

Krishnan HB, Oehrle NW, Natarajan SS (2009) A rapid and simple procedure for the depletion of abundant storage proteins from legume seeds to advance proteome analysis: a case study using Glycine max. *Proteomics,* 9(11):3174-3188.

Alam I, Sharmin S, Kim KH, Kim YG, Lee J, Lee BH (2013) An improved plant leaf protein extraction method for high resolution two-dimensional polyacrylamide gel electrophoresis and comparative proteomics. *Biotech. Histochem.,* 88(2):61-75.

Aryal UK, Krochko JE, Ross AR (2012) Identification of phosphoproteins in Arabidopsis thaliana leaves using polyethylene glycol fractionation, immobilized metal-ion affinity chromatography, two-dimensional gel electrophoresis and mass spectrometry. *J. Proteome. Res.,* 11(1):425-437.

Wu X, Xiong E, An S, Gong F, Wang W (2012) Sequential extraction results in improved proteome profiling of medicinal plant Pinellia ternata tubers, which contain large amounts of high-abundance proteins. *PLoS One,* 7(11):e50497.

Jmeian Y, El Rassi Z (2009) Liquid-phase-based separation systems for depletion, prefractionation and enrichment of proteins in biological fluids for in-depth proteomics analysis. *Electrophoresis,* 30(1):249-261.

Jmeian Y, El Rassi Z (2009) Multicolumn separation platform for simultaneous depletion and prefractionation prior to 2-DE for facilitating in-depth serum proteomics profiling. *J. Proteome. Res.,* 8(10):4592-4603.

Mortezai N, Harder S, Schnabel C, Moors E, Gauly M, Schlüter H, Wagener C, Buck F (2010) Tandem affinity depletion: a combination of affinity fractionation and immunoaffinity depletion allows the detection of low-abundance components in the complex proteomes of body fluids. *J. Proteome. Res.,* 9(12):6126-6134.

Marco-Ramell A, Bassols A (2010) Enrichment of low-abundance proteins from bovine and porcine serum samples for proteomic studies. *Res. Vet. Sci.,* 89(3):340-343.

Desrosiers RR, Beaulieu E, Buchanan M, Béliveau R (2007) Proteomic analysis of human plasma proteins by two-dimensional gel electrophoresis and by antibody arrays following depletion of high-abundance proteins. *Cell Biochem. Biophys.,* 49(3):182-195.

Wiśniewski JR, Zougman A, Nagaraj N, Mann M (2009) Universal sample preparation method for proteome analysis. *Nat. Methods,* 6(5):359-362.

Fratantoni SA, Piersma SR, Jimenez CR (2010) Comparison of the performance of two affinity depletion spin filters for quantitative proteomics of CSF: Evaluation of sensitivity and reproducibility of CSF analysis using GeLC-MS/MS and spectral counting. *Proteomics Clin. Appl.,* 4(6-7):613-617.

Index

B

C

D

E

H

I

J

K

L

M

N

Q

R

S

V

W

X

Y

Z